Thomas N. Taylor   Edith L. Taylor
Editors

Antarctic Paleobiology

Its Role in the Reconstruction of Gondwana

With 74 Illustrations in 179 Parts

Springer-Verlag New York Berlin Heidelberg
London Paris Tokyo Hong Kong

Thomas N. Taylor
Edith L. Taylor
Byrd Polar Research Center and Department of Botany, The Ohio State University, Columbus, Ohio 43210, U.S.A.

Library of Congress Cataloging-in-Publication Data
Antarctic paleobiology: its role in the reconstruction of Gondwana/
Thomas N. Taylor and Edith L. Taylor
p. cm.
Bibliography: p.
Includes index.
1. Paleobotany—Antarctic regions—Congresses. 2. Palynology—
Antarctic regions—Congresses. 3. Gondwana (Geology)—Congresses.
I. Taylor, Thomas N. II. Taylor, Edith L.
QE950.A57 1989
560.9'998'9—dc20 89-11391

Softcover reprint of the hardcover 1st edition 1990

Typeset by TCSystems, Inc., Shippensburg, Pennsylvania.

9 8 7 6 5 4 3 2 1

ISBN-13: 978-1-4612-7929-7 e-ISBN-13: 978-1-4612-3238-4
DOI: 10.1007/978-1-4612-3238-4

Antarctic Paleobiology

Preface

Today, Antarctica is a vast wilderness of nearly 14,000,000 square kilometers, with nearly 95% of the continent covered by permanent ice and snow. The weather is extremely cold, with a temperature of −89.2°C being recorded from the interior in 1983; temperatures on the Antarctic Peninsula average about −10°C. The occurrence of high winds also serves to make the climate of Antarctica inhospitable to most forms of life. Because the continent receives so little precipitation, it is classified as a desert.

Life on Antarctica today is restricted to only a few invertebrate species that spend the entire year on the continent, and most of these are found on the peninsula and associated islands where the warming influence of the ocean is more moderating. With the exception of humans, no vertebrates are capable of living year round on the continent. In addition to various types of algae, the most common plants in Antarctica consist of bryophytes and lichens that occur around the edge of the continent and on the peninsula. Only two vascular plants are known to exist south of 65°S, a grass (*Deschampsia antarctica*) and a member of the pink family (*Colobananthus quitensis*).

In contrast to the environment in Antarctica today, the continent at one time possessed a very favorable climate that supported a highly diverse biota; some of these plants and animals can be found in rocks dating back to the Devonian. The fossil biota of Antarctica has only recently been appreciated as a major data resource that can greatly contribute to understanding the evolution and biogeography of organisms. In addition, it is becoming increasingly clear that the high-latitude fossil biota of Antarctica represents a unique assemblage of organisms that may be evaluated in the global context of both continental drift and plate tectonics. The near polar position of the Antarctic plate through much of geologic time and the changing biotas on the continent provide a unique opportunity to investigate an important set of physical parameters and their influence on the evolution of faunas and floras.

This volume of papers is the culmination of a three-day workshop held on the campus of The Ohio State University, June 13–15, 1988. The primary aim of the meeting was the discussion of the current status of paleobiology, principally paleobotany and palynology, in Antarctica, and the interrelationship of Antarctic floras to those of other Gondwana continents. In selecting the participants for this workshop, our aim was to provide a broad coverage of the major groups of plants (e.g., cycads, conifers, angiosperms) on the one hand, while on the other, seeking to evaluate the vegetational history and the physical and biological parameters that influence the distribution of floras through time and space. Finally, any discussion of biologic activity measured in terms of geologic time must be developed within a framework of the geologic history, including the tectonic and paleogeographic history of the region.

The ultimate goals of the workshop and this volume are to stimulate interdisciplinary activity on the fossil biotas of Antarctica and to develop a broad-based research initiative that is directed at the biologic, biostratigraphic, and biogeographic importance of Antarctic floras through time.

Also included in this volume is a bibliography of Gondwana paleobotany and palynology. We have tried to include all papers that describe paleobotanical finds in Antarctica or that review those finds. With each citation is a subject listing. Entries that include "Antarctica" either mention fossil plants found on the continent or include descriptions of floras, etc. We have also added some citations on southern hemisphere floras and reconstructions of Gondwana; however, no attempt has been made to completely cover the literature in these areas. Because literature on Antarctic research is scattered within many disciplines, there are no doubt references to fossil plants that we have missed. We would appreciate receiving notice of these so that this bibliography may be periodically updated.

We gratefully acknowledge the support and financial assistance of the Division of Polar Programs of the National Science Foundation and the Office of Research and Graduate Studies and the Byrd Polar Research Center of The Ohio State University. We would like to thank all of the participants of the workshop for making it a success and the authors of papers for their contributions. Finally, we are indebted to the staff of Springer-Verlag for their skill and special efforts in handling the manuscripts.

Thomas N. Taylor
Edith L. Taylor

Contents

List of Contributors

Sergio Archangelsky
Paleobotany Division, Natural Sciences Museum, Buenos Aires (1405), Argentina

M.N. Bose
Department of Geology, Universitetet I Oslo, N-316, Oslo-3, Norway

James W. Collinson
Department of Geology and Mineralogy and Byrd Polar Research Center, The Ohio State University, Columbus, Ohio 43210, U.S.A.

Peter R. Crane
Department of Geology, Field Museum of Natural History, Chicago, Illinois 60605, U.S.A.

Geoffrey T. Creber
Department of Biology, Royal Holloway and Bedford New College, Egham, Surrey TW20 OEX, United Kingdom

T. Delevoryas
Department of Botany, University of Texas, Austin, Texas 78713, U.S.A.

Andrew N. Drinnan
Department of Geology, Field Museum of Natural History, Chicago, Illinois 60605, U.S.A.

Dianne Edwards
Department of Geology, University of Wales, College of Cardiff, Cardiff, CF1 3YE, Wales, United Kingdom

William R. Hammer
Department of Geology, Augustana College, Rock Island, Illinois 61201, U.S.A.

Judith Totman Parrish
Department of Geosciences, The University of Arizona, Tucson, Arizona 85721, U.S.A.

Kathleen B. Pigg
Department of Botany and Microbiology, Arizona State University, Tempe, Arizona 85287, U.S.A.

Geoffrey Playford
Department of Geology and Mineralogy, University of Queensland, St. Lucia, Brisbane 4067, Australia

Robert A. Spicer
Department of Earth Sciences, Oxford University, Oxford, OX1 3PR, United Kingdom

Ruth A. Stockey
Department of Botany, University of Alberta, Edmonton, Alberta T6G 2E9, Canada

Edith L. Taylor
Byrd Polar Research Center and Department of Botany, The Ohio State University, Columbus, Ohio 43210, U.S.A.

Thomas N. Taylor
Department of Botany and Byrd Polar Research Center, The Ohio State University, Columbus, Ohio 43210, U.S.A.

Elizabeth M. Truswell
Bureau of Mineral Resources, Canberra ACT 2601, Australia

1—Depositional Setting of Late Carboniferous to Triassic Biota in the Transantarctic Basin

James W. Collinson

Introduction

Fossil plants are the primary means of correlating sedimentary rock sequences in Antarctica of Late Paleozoic to Early Mesozoic age. A high-resolution zonation will be difficult to attain because of the widespread heating effects of Jurassic diabase intrusions. However, careful stratigraphic collecting from selected sections could greatly improve correlations. Except for R.A. Askin's (i.e., Kyle 1977) study in southern Victoria Land in which she compared Antarctic palynomorph floras to those of eastern Australia, no systematic biostratigraphic studies have been attempted. Most plant collections have come from a few isolated localities with only a general knowledge of their stratigraphic position.

Despite an inadequate biostratigraphic framework, it is possible to make correlations throughout the Transantarctic and Ellsworth Mountains because of similarities in lithostratigraphy from one region to another. Plant fossils help to confirm these correlations. Regional similarities exist because deposition occurred in a large basin subject to the same external controls. Tectonics along the pacific margin of Gondwanaland, eustasy, and climate produced a broad similarity in the Gondwana sequence across Antarctica and the other southern continents. The Transantarctic basin was comparable to the Sydney–Bowen basin in Australia and the Karoo basin in South Africa. This chapter will discuss the evolution of the Transantarctic basin and the depositional settings of paleofloras.

Transantarctic Basin

The Transantarctic basin developed in the mid-Paleozoic on the margin of the East Antarctic craton (Fig. 1.1). Deposits in this basin are exposed throughout the Transantarctic and Ellsworth Mountains. Important structural elements affecting the basin were the East Antarctic craton, cratonic highs, a volcanic arc in West Antarctica, and a fold–thrust belt between the arc and the craton. The basin developed on the recently cratonized margin of East Antarctica. A former continental margin in the Late Proterozoic to Early Paleozoic, this belt was folded, metamorphosed, and extensively intruded by granites during the Late Cambrian–Early Ordovician Ross orogeny (summarized by Stump 1987). The Ross high, a cratonic uplift along the present Ross Sea coast of Victoria Land, was episodically active as early as the Devonian. It isolated the Victoria sub-basin during much of the Transantarctic basin's history. The Ross, Queen Maud, and other basement highs were sources of coarse feldspathic and quartzose detritus. The volcanic arc, which developed on a convergent paleo-Pacific margin, was active as early as the Devonian, supplying calc-alkaline volcanic detritus initially to the outer part of the basin (Ellsworth Mountains) and later to the entire basin. Rem-

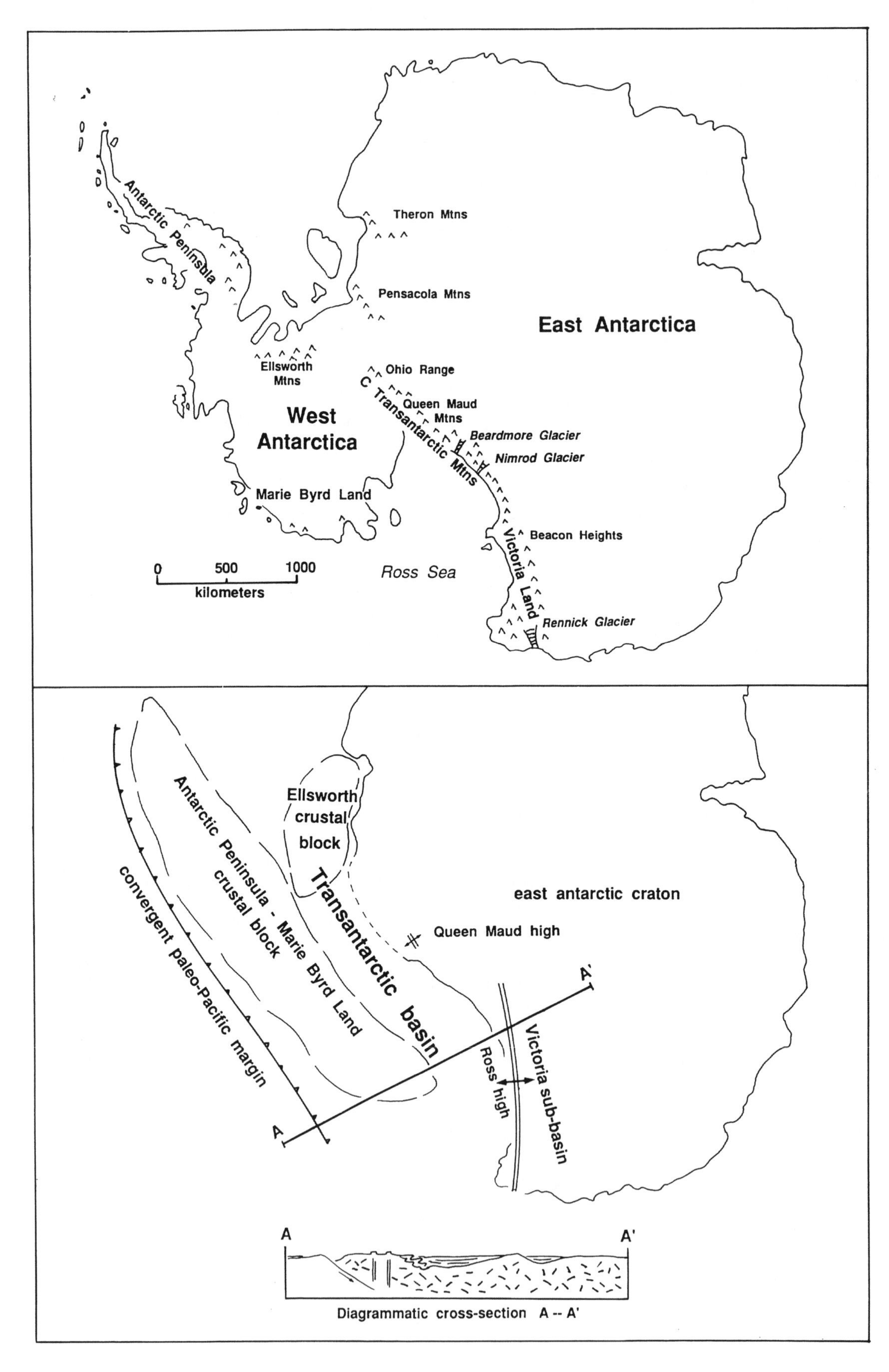

Antarctic Peninsula
Theron Mtns
Pensacola Mtns
East Antarctica
Ellsworth Mtns
Ohio Range
Queen Maud Mtns
West Antarctica
Transantarctic Mtns
Beardmore Glacier
Nimrod Glacier
Marie Byrd Land
Beacon Heights
Victoria Land
0 500 1000
kilometers
Ross Sea
Rennick Glacier
Ellsworth crustal block
Antarctic Peninsula - Marie Byrd Land crustal block
convergent paleo-Pacific margin
Transantarctic basin
east antarctic craton
Queen Maud high
Ross high
Victoria sub-basin
A
A'
Diagrammatic cross-section A -- A'

nants of the fold–thrust belt are folded Permian rocks in the Ellsworth and Pensacola Mountains along the margin of the East Antarctic craton. The oldest evidence of uplift in West Antarctica is a paleocurrent reversal and a change from cratonic to volcanic provenance in Upper Permian sandstones in the Beardmore Glacier area (Isbell in press). The fold–thrust belt apparently developed just inside the arc, uplifting and eroding the volcaniclastic apron of recently deposited volcanic sediments.

The Transantarctic basin evolved from a back-arc basin into a foreland basin with the development of a thrust and fold belt in West Antarctica during the mid-Permian (Collinson and Isbell 1987; Collinson in press). The Ross high blocked volcanic detritus from the Victoria sub-basin until the Middle Triassic when it was buried.

Stratigraphy

The overall similarity of sequences from the Central Transantarctic Mountains to the Pensacola and Ellsworth Mountains (Table 1.1) suggests deposition in a continuous basin. The sequence in the Victoria sub-basin (Table 1.2) is quite different below the Middle Triassic. Elliot (1975) published a detailed summary of the stratigraphy of Antarctica.

Central Transantarctic Mountains

Virtually the same lithologic units are traceable with minor facies changes from the Nimrod Glacier to the Ohio Range. The most complete section, 2800 m thick, is in the Beardmore Glacier region. This sequence consists of (1) Devonian quartzose sandstone (Alexandra Formation); (2) Permian glacial diamictite (Pagoda Formation), dark shale interbedded with fine-grained sandstone (Mackellar Formation), cross-bedded medium-grained sandstone (Fairchild Formation), and coal measures (Buckley Formation); and (3) Triassic sandstone and shale (Fremouw and Falla Formations). Devonian sandstones were removed by pre-Permian erosion over most of the Queen Maud Range. In the Ohio Range, a thin remnant of Lower Devonian sandstone (Horlick Formation) remains. Devonian and Lower Permian sandstones are feldspathic to quartzose, reflecting a basement source. Uppermost Permian coal measures in the Beardmore Glacier region are volcaniclastic, indicating a major change in provenance. Increasingly abundant volcanic tuffs upward in the Triassic sequence suggest that the source was in part contemporaneous volcanism. The Triassic and uppermost Permian were removed by post-Triassic erosion from the central Queen Maud Mountains to the Ohio Range.

◁

Figure 1.1. (a) Locality map. (b) Map at same scale showing major structural features and diagrammatic cross section. Crustal blocks in West Antarctica follow the reconstruction of Storey (in press), except that the Ellsworth crustal block has been restored to its hypothetical original position by a clockwise rotation of 90°, as suggested by the paleomagnetic data of Watts and Bramall (1981). For detailed discussions of crustal blocks in West Antarctica, see Dalziel and Elliot (1982).

Ellsworth and Pensacola Mountains

The 13,000-m-thick Paleozoic section in the Ellsworth Mountains is the thickest in Antarctica. The 3200-m-thick Crashsite Quartzite Group contains an Upper Cambrian shelly fauna near the base and a Devonian marine fauna near the top (Webers and Sporli 1983). The effects of the Early Ordovician Ross orogeny are not apparent in this section. A similar 3000-m-thick siliciclastic sequence occurs in the Pensacola Mountains, where the postorogenic unfossiliferous Neptune Group is overlain by the Dover Sandstone. Sandstone similar to the Dover in an isolated nunatak contains Middle Devonian plant fossils (Schopf 1968).

Upper Carboniferous(?)–Permian strata in the Ellsworth Mountains are 2000 to 2500 m thick, indicating thickening in the direction of the paleo-Pacific margin. Unlike the sequence in the Transantarctic Mountains, the basal glacial diamictites (Whiteout Conglomerate) are conformable with the underlying Crashsite Quartzite (Wyatt Earp Formation; Ojakangas

Table 1.1. Stratigraphy of the Transantarctic basin.[a]

	Ellsworth Mountains (Craddock et al. 1986)	Ohio Range (Long 1965, Bradshaw et al. 1984)	Beardmore Glacier (Barrett et al. 1986, Hammer et al. 1987)
Triassic Upper ? Middle ? Lower			Falla Formation: 160–530 m; volcaniclastic sandstone & tuff; *Dicroidium* flora; palynomorph subzones C–D; braided fluvial Fremouw Formation: 620–750 m; volcaniclastic and quartzose sandstone; *Dicroidium* flora; palynomorph subzones A–B; *Cynognathus* and *Lystrosaurus* zones; trace fossils; braided fluvial
Permian Upper ? Lower	Polarstar Formation: 1,000+ m Upper member: volcaniclastic sandstone, carbonaceous shale, and coal; *Glossopteris* flora; fluviodeltaic Middle member: dark shale and fine volcaniclastic sandstone; trace fossils; marine deltaic Lower member: dark shale; marine basinal	Mount Glossopteris Formation: 750 m; feldspathic sandstone, carbonaceous shale, and coal; *Glossopteris* flora; fluviodeltaic	Buckley Formation: 750 m Upper member: Volcaniclastic sandstone, carbonaceous shale, and coal; *Glossopteris* flora; braided fluvial Lower member: Feldspathic sandstone, carbonaceous shale, and coal; *Glossopteris* flora; braided fluvial. Fairchild Formation: 130–220 m; feldspathic sandstone; palynomorph stage 2–3(?); braided fluvial
	Whiteout Conglomerate: 1,000+ m; glaciomarine and glacial diamictite	Discovery Ridge Formation: 195 m; dark shale and fine feldspathic sandstone; marine-deltaic Buckeye Formation: 308m; glaciomarine and glacial diamictite	Mackellar Formation: 60–140 m; dark shale and fine feldspathic sandstone; palynomorph upper stage 2; trace fossils; marine-deltaic (fresh or brackish) Pagoda Formation: 126–200 m; palynomorph upper stage 2; glacial diamictite
Devonian and older(?)	Crashsite Group: 3200 m; quartzose and volcaniclastic sandstone; marine invertebrate fauna	Horlick Formation: 56 m; quartzose sandstone; marine invertebrate fauna	Alexandra Formation: 300 m; quartzose sandstone, marine(?)

[a] Palynomorph correlations are based on eastern Australian Permian stages and Antarctic Triassic subzones from Kyle and Schopf 1982.

and Matsch 1981). This suggests the possibility that Carboniferous sedimentary rocks, not identified elsewhere in Antarctica, may be present here. The 1000+-m-thick diamicite sequence grades abruptly upward into dark laminated shale interbedded with fine-grained volcaniclastic sandstone that is increasingly abundant upward. Coal measures cap the section. The Permian sequence in the Pensacola Mountains is less complete, but similar.

Table 1.2. Stratigraphy of the Victoria sub-basin, Transantarctic basin.[a]

	Beacon Heights (McElroy & Rose (1987)	Rennick Glacier (Collinson et al. 1986, Grindley & Oliver 1983)
Triassic Upper ? Middle ? Lower	Lashly Formation: 325+ m Upper member: volcaniclastic, quartzose sandstone, and coal; *Dicroidium* flora; palynomorph subzones C–D; meandering fluvial Lower member: volcaniclastic sandstone; palynomorph subzones A–B; braided fluvial	Section Peak Formation: 160 m; quartzose and volcaniclastic sandstone; *Dicroidium* flora; palynomorph subzones C–D; braided fluvial
	Feather Conglomerate: 215 m; quartzose sandstone; palynomorph subzone A; trace fossils; braided fluvial	Not exposed
Permian Upper ?	Weller Coal Measures: 222 m; feldspathic sandstone, carbonaceous shale, and coal; *Glossopteris* flora; palynomorph stage 4; meandering fluvial	Takrouna Formation: 0–280+ m; feldspathic sandstone, carbonacaeous shale, and coal; *Glossopteris* flora; trace fossils; braided to meandering fluvial
Lower	Metschel Tillite: 0–85 m; glacial diamictite	Unnamed unit: 0–350 m; glacial diamictite
Devonian and older?	Taylor Group: 1100 m; quartzose and feldspathic sandstone; marine and braided to meandering fluvial; trace fossils, fish fauna, palynomorphs, and *Haplostigma* stems.	Gallipoli Volcanics: 0–270 m; silicic volcanics and volcaniclastics; plant fragments

[a] Palynomorph correlations are based on eastern Australian Permian and Antarctic Triassic Subzones from Kyle & Schopf 1982.

Victoria Land

The most complete sedimentary section in Victoria Land is in the Beacon Heights area where a 1100-m-thick Devonian sequence dominated by quartzose sandstone is overlain by a 900-m-thick sequence of Permian and Triassic sandstone and shale (McElroy and Rose 1987). The Devonian is far more diverse in southern Victoria Land than in the adjacent Central Transantarctic Mountains. Rather than a single unit, the Taylor Group comprises five formations. Basal Permian tillite (Metschel) is discontinous, filling paleovalleys. Permian coal measures (Weller) are disconformably overlain by a Triassic sandstone sequence, the upper part of which contains coal measures. Permian sandstones are feldspathic; Lower Triassic sandstones (Feather Conglomerate) are quartzose. Middle to Upper Triassic sandstones (Lashly Formation) are volcaniclastic and are similar to their counterparts in the Central Transantarctic Mountains.

Lithostratigraphic units are difficult to correlate in northern Victoria Land because of the lack of a complete vertical section. The only Devonian sedimentary rocks occur with silicic volcanics at three widely scattered localities. Permian glacial deposits occur only in a small area on the east side of the Rennick Glacier, while the Permian fluvial deposits (Takrouna Formation) vary on opposite sides of the Rennick Glacier (Collinson et al. 1986). The east side is characterized by coarse feldspathic sandstones and the west side by coal measures similar to those in southern Victoria Land. Upper Triassic sandstone (Section Peak Formation) containing volcanic detritus overlies Ross granite in the escarpment along the polar plateau at the head of the Rennick Glacier.

Depositional History of the Transantarctic Basin

Late Ordovician—Carboniferous

The incompleteness of Ordovician to Devonian rocks in most of Antarctica makes paleogeographic reconstructions highly speculative. Post-Devonian erosion, possibly by

Gondwanan ice sheets in the Late Carboniferous, removed much of the record in the Central Transantarctic Mountains (See Fig. 1.1). Sequences in the Ellsworth and Pensacola Mountains, the thickest and most complete, are similar to the South African Table Mountain, Bokkeveld, and Witteberg Groups, which represent a range of alluvial, shallow marine, and coastal environments typical of nearshore areas of a subsiding continental shelf (Tankard et al. 1982). Related sequences in the rest of the Transantarctic Mountains are thinner and probably began in the Devonian. The thin remnant of Lower Devonian shallow marine beds resting on a granitic basement in the Ohio Range is the only clear evidence for the age and environment of deposition of such rocks in the Central Transantarctic Mountains. The Alexandra Formation in the Beardmore Glacier region, lacking fossils, gives little indication of age or environment of deposition. Mudcracks suggest subaerial exposure (Barrett et al. 1986), but these could have occurred in tidal flat, lacustrine, or alluvial settings.

The Devonian section in southern Victoria Land is better known than any other in Antarctica, yet the depositional setting is highly controversial. Trace fossil enthusiasts (Bradshaw 1981, Gevers and Twomey 1982) argue for marine deposition in the lower half of the sequence, whereas others (Barrett and Kohn 1975; Plume 1982) support an alluvial origin for most of the section based on sedimentary structures, such as desiccation features, paleosols, and uniformity of crossbed directions. Barrett (1981) suggested that the lower part of the sequence was probably alluvial, the middle part was possibly shallow marine, and the upper part was unquestionably alluvial.

Isolated plant-bearing sedimentary rocks associated with silicic to intermediate volcanic sequences in northern Victoria Land (Dow and Neall 1974, Grindley and Oliver 1983, Adams et al. 1986) suggest the existence of widespread nonmarine conditions. The significance of this volcanism is poorly understood, but similar plant-bearing beds in Marie Byrd Land associated with calcalkaline volcanics are related to the Late Devonian–Early Carboniferous(?) volcanic arc that bordered the paleo-Pacific margin (Grindley et al. 1980).

Late Carboniferous—Early Permian: Gondwanan Cycle Begins

The similar sequence of glacigenic deposits overlain by basinal and fluvial–deltaic deposits in the Central Transantarctic, Pensacola, and Ellsworth Mountains reflects the waning stages of Gondwanan glaciation. Similarities with the Karoo basin in South Africa and the Paraná basin in Brazil indicate that this was a widespread event. The dark shales recorded a marine transgression, probably related to the postglacial rise in sea level. A regressive sequence of deltaic to fluvial sandstones rapidly prograded across the basin.

A variety of facies are represented in glacigenic deposits (Lindsay 1970; Miller and Waugh 1986; in press; Miller, 1989). For paleogeographic reconstruction, the most important are those that indicate a terrestrial to marine transition. Such a transition has been documented in the Ellsworth Mountains (Ojakangas and Matsch 1981). Terrestrial facies, including lodgment tills similar to those that occur throughout the Central Transantarctic Mountains, occur at the southern end of the Ellsworth Mountains, but toward the north, these give way to a 1000+-m-thick sequence of glaciomarine conglomerates. This transition may represent the grounding line of ice on a marine shelf. A transition from glacial to glaciomarine deposits occurs vertically within the transgressive sequence in the Ohio Range (Aitchison et al. 1988). The same transition may occur elsewhere in the Central Transantarctic Mountains, where glacigenic deposits grade upward into the postglacial dark shale. Rare dropstones extend only a few meters up into the dark shales. Striations at the base of glacigenic deposits and paleocurrent indicators within glacigenic deposits trend parallel to the margin of East Antarctica and mostly point toward the present Weddell Sea. Ice streams were apparently diverted into an elongate basin parallel to the cratonic margin along a subsiding marine shelf. Glacial clasts were derived either locally or from the East Antarctic craton.

The abrupt change from conglomerate to dark shale at all localities from the Ellsworth Mountains to the Nimrod Glacier suggests a radical change in depositional conditions, probably the elimination of floating ice as the ice margin receded onto East Antarctica. A rising sea level flooded the entire basin. A variety of features in the dark shale and fine-grained sandstone sequence, including thin turbidite beds, a low-diversity ichnofauna, and low sulfur/carbon ratios suggest fresh-water to brackish conditions in the Beardmore Glacier region (Miller and Frisch 1987). Coalescing submarine fans formed by the underflow of cold, sediment-ladened stream water quickly filled the enclosed end of the basin. Salinities were apparently greater toward the Ohio Range, where sedimentologic evidence of marine conditions exist (Aitchison et al. 1988). A marine trace fossil fauna in the Ellsworth Mountains sequence suggests even more open marine conditions (Collinson et al. in press), but the lack of strong evidence, such as body fossils, makes these conclusions speculative. However, Visser (1988) described the same situation caused by postglacial flooding in the Karoo basin.

The Early Permian paleogeographic map (Fig. 1.2) shows fluvial sandstones prograding across the basin. Strata overlying the dark shale sequence are dominated by coarse feldspathic sandstone, probably derived from the reworking of glacial outwash. This contrasts with the Ellsworth Mountains sequence in which all postglacial sandstones are volcaniclastic. Volcaniclastic sandstones and shales containing *Glossopteris* also occur on the volcanic arc side of the basin near the south end of the Antarctic Peninsula (English Coast) (Laudon et al. 1987; Laudon 1987). Volcanic detritus was probably blocked from reaching the Central Transantarctic Mountains area by the seaway.

Lenticular quartz-pebble conglomerates mark the base of the coal measures throughout the Central Transantarctic Mountains. Orthoclase and microcline are also common in this unit. These conglomerates were derived from basement highs. Evidence for a Queen Maud high are the several hundred meters of conglomerate that interfinger with the dark shale sequence in the Queen Maud Mountains (Minshew 1967). Initial compression along the West Antarctic fold–thrust belt may have activated this and other cratonic highs.

The preservation of carbonaceous material and the deposition of coal was probably related to climatic changes, including increases in temperature and humidity (Krissek and Horner 1987; in press). Floras below the coal measures are poorly preserved or nonexistent over much of Antarctica. Microfloras in glacial deposits or dark shales are invariably correlative with Australian Stage 2, probably representing a sparse polar flora. The rarity of fossil wood in these deposits suggests a landscape with few trees. In contrast, the coal measures contain abundant fossil wood and leaves and more diverse microfloras representing eastern Australian Stages 3–5 (Kyle and Schopf 1982).

Both the coal measures and the underlying fluvial strata were deposited in a braided stream environment. The major difference is the preservation of carbonaceous material, although the shale/sandstone ratio increases upward in the sequence. This climate-related change probably affected the entire basin simultaneously, suggesting that fluvial sediments had prograded as far as the Ohio Range by late Early Permian. In the Ohio Range, the dark shale sequence (Discovery Ridge Formation) is directly overlain by coal measures (Mount Glossopteris Formation), which have been interpreted as being deposited in a marginal marine to deltaic environment (Bradshaw et al., 1984). Several deltaic cycles occur in the upper part of the black shale sequence (Polarstar Formation) in the Ellsworth Mountains, suggesting that fluvial conditions developed there even later. The marine aspect of the Lower Permian in the Ohio Range suggests that it was deposited where the basin was deepening. Deltaic–fluvial coal measures of quartz and feldspathic composition and containing *Glossopteris* floras occur in the Pensacola Mountains (Williams 1969) and in a 700-m-thick section in the Theron

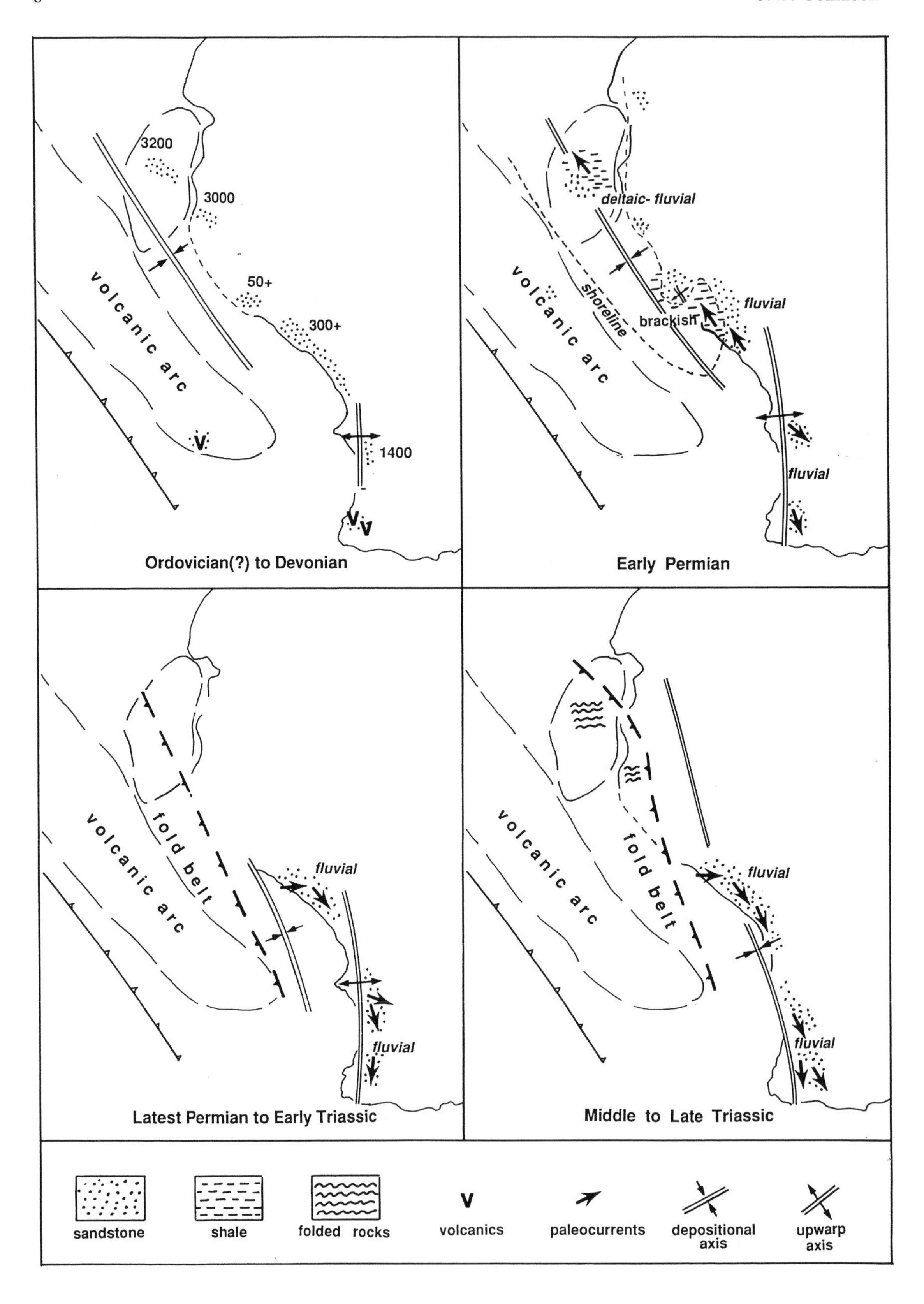

3200
3000
50+
300+
1400
volcanic arc
V
Ordovician(?) to Devonian
deltaic- fluvial
shoreline
volcanic arc
brackish
fluvial
fluvial
Early Permian
fold belt
volcanic arc
fluvial
fluvial
Latest Permian to Early Triassic
volcanic arc
fold belt
fluvial
fluvial
Middle to Late Triassic
sandstone
shale
folded rocks
V
volcanics
paleocurrents
depositional axis
upwarp axis

Mountains (Brook 1972). These may help define the mid-Permian shoreline (Fig. 1.2).

Late Permian—Early Triassic: The Foreland Basin

Sandstones in the uppermost Permian in the Queen Maud Mountains and Beardmore Glacier region are dominated by volcanic detritus. This change in provenance from cratonic to volcanic is most striking in the Beardmore Glacier region midway through the Buckley Formation. The abrupt change in composition is accompanied by a 180° reversal in paleocurrent directions (Collinson and Isbell 1986; Isbell in press). Barrett et al. (1986) had previously recognized this paleocurrent reversal, but had associated it with uplift associated with the Permian-Triassic disconformity. The sequence through which these changes occurred is conformable. Events outside the immediate basin produced the changes. Alluvial fans generated by folding and thrusting in the West Antarctic orogenic belt may have flooded the region with volcanic detritus and reversed the drainage direction (Collinson and Isbell 1987). According to Isbell (in press), deposition of the Upper Permian (upper Buckley member) is attributed to unconfined braided streams, such as in the Himalayan foreland (Brahmaputra River, Coleman 1969; Kosi River, Wells and Dorr 1987). Because the drainage system was choked by sediment, broad sandy stream channels migrated by avulsion rather than meandering. During major floods, the entire surface was inundated. Thick flood basin deposits were preserved by the rapid subsidence characteristic of a foreland basin.

Compressional plant fossils of the *Glossopteris* flora are abundant in all the carbonaceous shale units. Silicic permineralization in flood-basin waters charged with silica from dissolution of silicic volcanic detritus was probably responsible for the excellent preservation of plant material at localities such as Skaar Ridge (Taylor EL et al. 1986). Groups of upright tree stumps occur along some fine-grained bedding surfaces.

The Ross high formed a drainage divide between the Central Transantarctic Mountains and Victoria Land (Barrett and Kohn 1975). Permian coal measures in Victoria Land do not contain volcanic detritus. The Weller Coal Measures in the Beacon Heights area is a coarse-grained meandering stream deposit (Isbell, personal communication). At localities away from the high, such as the Allan Hills, fine-grained flood-plain deposits assume a greater proportion of the sequence (Collinson et al. 1983). Compressional examples of the *Glossopteris* flora are abundant in carbonaceous siltstones and shales. Permineralized logs are common in channel sandstones. Kyle and Schopf (1982) reported a Stage 4 microflora.

In northern Victoria Land, on the east side of the Rennick Glacier, adjacent to the Ross high, the Takrouna Formation was deposited by sandy braided streams (Collinson et al. 1986). Carbonaceous fine-grained sediments occur only as rare channel-fill deposits. On the west side of the glacier, the formation is characterized by finer grained, possibly meandering stream deposits, with preservation of abundant carbonaceous material. Compressional fossils of the *Glossopteris* flora are abundant in these rocks (Hammer 1986), but the microflora was mostly destroyed by the intensity of heating by Jurassic intrusions.

Lower Triassic rocks contain very little carbonaceous material. Like the Permian, they were deposited by sandy braided streams. Paleocurrent directions are much the same as in Upper Permian rocks. The major change was probably climatic, toward conditions warm enough to support a diverse reptilian fauna, but dry enough that organic carbon was oxidized.

In the Queen Maud Mountains and Beardmore Glacier area, Lower Triassic deposits contain thick flood-plain sequences composed mainly of greenish-gray, rarely

◁

Figure 1.2. Paleogeographic maps at four stages in the evolution of the Transantarctic basin. Numbers in the Ordovician(?) to Devonian map show thicknesses in meters. Distribution of known exposures is shown by lithologic symbols.

reddish-gray, fine-grained beds (Collinson and Elliot 1984). Abundant root structures in fine-grained deposits and silicified logs in sandstone channels indicate that vegetation was present. The reptiles and amphibians preserved in these rocks (see Hammer, this volume) probably required abundant vegetation for their sustenance. However, to date, no palynomorphs have been recovered. Some of the dark greenish-gray mudstones look promising. Leaf compressions are also lacking.

Lower Triassic rocks in the Victoria sub-basin (Feather Conglomerate), dated on their microflora (Kyle and Schopf 1982), also represent sandy braided stream deposits. Fine-grained flood-plain deposits are rare in the lower part, but the upper part (Fleming Member) contains greenish-gray fine-grained beds containing root structures. The Feather is conglomeratic only in the Beacon Heights area near the Ross high. Away from this area, the formation becomes sandstone and locally even siltstone (Barrett and Kohn 1975). Further evidence that the high divides drainage is the complete lack of volcanic detritus here that is so common in Lower Triassic sandstones of the Central Transantarctic Mountains.

The Permo-Triassic boundary was previously placed between the lower and upper (Fleming) members of the Feather Conglomerate (Barrett and Kohn 1975). This assignment was not based on fossils, but on a 90° change in paleocurrent directions that was correlated with the paleocurrent reversal in the Beardmore Glacier area. Because the Beardmore evidence for the reversal occurs well within the Permian section, these criteria are unsuited to locate the Permian–Triassic boundary. The change from coal measures to noncarbonaceous beds from Permian to Triassic, reflecting climatic change, seems to be a more likely indicator of time. The Feather Conglomerate contains an Early Triassic microflora (Klye and Schopf 1982).

Middle—Late Triassic: End of the Gondwanan Cycle

A single drainage system probably extended from the central Queen Maud Mountains to Victoria Land by Middle Triassic time (Elliot 1975; Collinson et al. 1987). Similar sandy braided stream deposits dominated by volcanic detritus occur in both the Central Transantarctic Mountains (uppermost Fremouw Formation) and the Victoria sub-basin (lower Lashly Formation), suggesting that the Ross high was buried. Preservation of fine-grained carbonaceous deposits containing the *Dicroidium* flora suggests a return to more humid conditions. The exquisitely preserved silicified plant remains near the top of the Fremouw Formation in the Beardmore Glacier area occur in peat rafts buried within a sandy braided stream deposit (Taylor TN et al. 1986). Large (20 m) silicified logs are scattered along the same horizon. Possibly a forested, swampy island was undercut and destroyed during a major flood. Peat rafts became grounded and buried by sand as floodwaters waned. Plant material was quickly permineralized by groundwaters rich in silica from dissolution of the abundant silicic volcanic detritus.

Upper Triassic rocks in the Central Transantarctic Mountains (Falla Formation) are similar to those in the Lower Triassic in that braided stream channel sandstones are preserved with a high proportion of fine-grained flood-plain deposits. The major difference is that Upper Triassic rocks are carbonaceous, containing thin coals. In addition to the *Dicroidium* flora, these rocks contain a rich microflora (Kyle and Fasola 1978; Farabee et al. 1989).

In southern Victoria Land, coal measure sandstones (upper Lashly Formation) contain point-bar surfaces characteristic of meandering stream deposits. Large fossil logs occur in channel sandstones. The *Dicroidium* flora and palynomorphs occur in fine-grained carbonaceous beds (Kyle and Schopf 1982). In northern Victoria Land, similar sandstones (Section Peak Formation), which rest directly on a granite basement (Collinson et al. 1986), contain a diverse Late Triassic microflora (Norris 1965).

The Gondwanan cycle in the Transantarctic basin ended with increasing volcanism. Basaltic lavas and pyroclastic breccias overlie an unconformity with 500 m of local relief in southern Victoria Land, suggesting that regional uplift accompanied volcanism (Collinson et al. 1983). The latest Triassic–Early Jurassic sequence in the Central Transantarctic Mountains is dominated by silicic tuffs (upper Falla Formation) unconformably overlain by lahar debris, pyroclastic breccia, tuffs, and volcaniclastic sandstones (Prebble Formation; Barrett et al. 1986). A unit of volcanic breccia, tuffs, and tuffaceous sandstone mostly of basaltic composition (Exposure Hill Formation) rests unconformably on Upper Triassic sandstones in northern Victoria Land (Elliot et al. 1986). This sequence of pyroclastic rocks apparently represents an early phase of the tholeiitic magmatism of the overlying flood basalts of early Middle Jurassic age associated with the breakup of Gondwanaland.

Toward Establishing a Better Biostratigraphy in Antarctica

Compiling a detailed plant fossil biostratigraphy will require careful cooperation between sedimentologists and paleobotanists during the section measuring process. Measuring, describing, and sampling of accurate stratigraphic sections under Antarctic conditions is extremely arduous. It is not surprising that sampling for plant fossils has been inadequate. In sequences in which carbonaceous beds are present, plant fossils are abundant in the upper part of virtually every sandstone–shale cycle.

Coal rank and vitrinite reflectance values of rocks throughout the Transantarctic basin indicate extensive heating and alteration of sedimentary rocks by the widespread intrusion of Jurassic diabase. The only area not intruded by diabase is in the Ellsworth Mountains, but these rocks suffered an equivalent degree of alteration during deformation and burial. In the Transantarctic Mountains, diabase supplied heat to drive chemical reactions in fluids within the rock, so the degree of alteration was apparently as dependent on permeability as distance from a dike or sill (Vavra 1984). This helps to explain the difficulty in predicting the suitability of samples based on their distance from an intrusion. Some good samples have been found fairly close to diabase. If fluids were the more important factor, localities where interbedded sandstones have suffered little chemical (zeolite) alteration should be more favorable. Most collections have come from immediately below sandstones, because shales are exposed there. The shales are relatively impermeable, so finding less altered samples may require collecting well below channel-form sandstones. Such samples may require trenching, but care must be taken to avoid areas where solifluction has occurred.

More data on coal rank and vitrinite reflectance may reveal that some areas are less altered than others. Sections in these areas should be better for biostratigraphic sampling. For example, Krissek and Horner (1987) have identified least-altered samples based on several criteria, including kaolinite content, low illite crystallinity, vitrinite reflectance, and high organic carbon content. Some sections have suffered relatively little alteration, and these are the ones that should be extensively sampled for palynomorphs.

Summary

The Transantarctic basin includes undeformed sedimentary rocks of Paleozoic and early Mesozoic age in the Transantarctic and Ellsworth Mountains. The basin had its beginnings as early as Late Ordovician after the Ross orogeny. By Devonian time at the latest, a volcanic arc developed along the paleo-Pacific margin. Widespread but incomplete sequences of Devonian sandstone occur throughout the basin. Devonian volcanics occur in Marie Byrd Land and northern Victoria Land. Carboniferous sedimentary rocks are unknown from Antarctica, but may be present in the Ellsworth Mountains where the Devonian to Permian sequence may be conformable. Otherwise, Late Carboniferous(?)–Early Permian glacigenic deposits disconformably overlie the Devonian or Ross basement. A marine grounding line for glacigenic deposits apparently passed through the Ellsworth Mountains and the Ohio Range. As the margin of the East Antarctic ice sheet retreated onto the continent, dark marine shales transgressed as far as the Nimrod Glacier area. A variety of evidence suggests an inland sea that ranged from open marine in the Ellsworth Mountains to brackish or fresh water at the enclosed end in the Beardmore Glacier area. The shallow, closed end of the seaway quickly filled by progradation of turbidites followed by braided fluvial deposits. Coal measures appeared in the late Early Permian with a warming of climatic conditions. In the middle to Late Permian, deformation and uplift in a fold–thrust belt in West Antarctica between the volcanic arc and the craton caused a reversal of slope in the Beardmore Glacier area. Abundant volcanic detritus was transported to the East Antarctic side of the basin, except to the Victoria subbasin, which was isolated by the Ross high. Early Triassic depositional patterns followed those of the Late Permian, except carbonaceous material was not preserved, possibly because of less humid climatic conditions. By the Middle Triassic, volcaniclastic sandstones from West Antarctica buried the Ross high and drainage extended from the Central Transantarctic Mountains to Victoria Land. Deposition in the Transantarctic basin ended with uplift and magmatic activity throughout the Transantarctic Mountains.

Most stratigraphic correlations in Antarctica are based on fossil plants. Establishing a more detailed biostratigraphy will require systematic collecting of selected stratigraphic sections that were least affected by secondary alteration following burial and intrusion by Jurassic diabase.

Acknowledgments. I am grateful to David H. Elliot and John I. Isbell for reviewing the manuscript. Research was funded under National Science Foundation grant DPP-8716414.

References

Adams CJ, Whitla PF, Findlay RH, and Field BF (1986) Age of the Black Prince Volcanics in the Central Admiralty Mountains and possibly related hypabyssal rocks in the Millen Range: *in* Stump E, Geological Investigations in Northern Victoria Land, Antarctic Research Series, Vol 46, American Geophysical Union, Washington, DC, pp 203–210

Aitchison JC, Brandshaw MA, and Newman J (1988) Lithofacies and origin of the Buckeye Formation: Late Paleozoic glacial and glaciomarine sediments, Ohio Range, Transantarctic Mountains, Antarctica: Palaeogeography, Palaeoclimatology, Palaeoecology 64:93–104

Barrett PJ (1981) History of the Ross Sea region during the deposition of the Beacon Supergroup 400–180 million years ago: Journal of the Royal Society of New Zealand 11(4):447–568

Barrett PJ and Kohn BP (1975) Changing sediment transport directions from Devonian to Triassic in the Beacon Supergroup of south Victoria Land, Antarctica: *in* Campbell KSW (ed), Gondwana Geology: Australian National University Press, Canberra, pp 15–35

Barrett PJ, Elliot DH, and Lindsay JF (1986) The Beacon Supergroup and Ferrar Group in the Beardmore area: *in* Turner MD and Splettstoesser JF (eds), Geology of the Central Transantarctic Mountains: Antarctic Research Series, Vol 36, American Geophysical Union, Washington DC, pp 339–429

Bradshaw MA (1981) Paleoenvironment interpretations and systematics of Devonian trace fossils from the Taylor Group (lower Beacon Supergroup). Antarctica: New Zealand Journal of Geology and Geophysics 24:615–652

Bradshaw MA, Newman J, and Aitchison JC (1984) Preliminary geological results of the 1983–84 Ohio Range Expedition: New Zealand Antarctic Record 5(3):1–17

Brook D (1972) Stratigraphy of the Theron Mountains, British Antarctic Survey Bulletin 29:67–89

Coleman JM (1969) Brahmaputra River: Channel processes and sedimentation: Sedimentary Geology 3:129–239

Collinson JW (in press) The palaeo-Pacific margin as seen from East Antarctica: *in*Thomson MRA, Crame JA, and Thomson JW (eds), Geological Evolution of Antarctica: Cambridge University Press, Cambridge

Collinson JW and Elliot DH (1984) Triassic stratigraphy of the Shackleton Glacier area: *in* Turner MD and Splettstoesser JP (eds), Geology of the Central Transantarctic Mountains: Antarctic Research Series, Vol 36, American Geophysical Union, Washington DC, pp 103–117

Collinson JW and Isbell JL (1986) Permian-Triassic sedimentology of the Beardmore Glacier region: Antarctic Journal of the US 21(5):29–30

Collinson JW and Isbell JL (1987) Evidence from the Beardmore Glacier region for a late Paleozoic/early Mesozoic foreland basin: Antarctic Journal of the US 22(5):17–19

Collinson JW, Pennington DC, and Kemp NR (1983) Sedimentary petrology of Permian–

Triassic rocks in Allan Hills, central Victoria Land: Antarctic Journal of the US 18(5):20–22
Collinson JW, Pennington DC, and Kemp NR (1986) Stratigraphy and petrology of Permian and Triassic fluvial deposits in northern Victoria Land, Antarctica: *in* Stump E (ed), Geological Investigations in Northern Victoria Land: Antarctic Research Series, Vol 46, American Geophysical Union, Washington DC, pp 211–242
Collinson JW, Kemp NR, and Eggert JT (1987) Comparison of the Triassic Gondwana sequences in the Transantarctic Mountains and Tasmania: *in* McKenzie GD (ed), Gondwana Six: Stratigraphy, Sedimentology and Paleontology, Geophysical Monograph Series, Vol 41, American Geophysical Union, Washington DC, pp 51–61
Collinson JW, Vavra CL, and Zawiskie JM (in press) Sedimentology of the Polarstar Formation, Permian, Ellsworth Mountains, Antarctica: *in* Webers GF, Craddock C, and Splettstoesser JF (eds), Geology and Paleontology of the Ellsworth Mountains, Antarctica: Geological Society of America Memoir 170
Craddock C, Webers GF, Rutford RH, Spörli KB, and Anderson JJ (1986) Geologic map of the Ellsworth Mountains, Antarctica: Geological Society of America, Boulder, Colorado
Dalziel IWD and Elliot DH (1982) West Antarctica: problem child of Gondwanaland: Tectonics 1(1):3–19
Dow JAS and Neall VE (1974) Geology of the lower Rennick Glacier, northern Victoria Land, Antarctica: New Zealand Journal of Geology and Geophysics 17:659–714
Elliot DH (1975) Gondwana basins in Antarctica: *in* Campbell KSW (ed), Gondwana Geology: Australian National University Press, Canberra, pp 493–536
Elliot DH, Haban MA, and Siders MA (1986) The Exposure Hill Formation, Mesa Range: *in* Stump E (ed), Geological Investigations in Northern Victoria Land: Antarctic Research Series, Vol 46, American Geophysical Union, Washington DC, pp 267–278
Farabee, MJ, Taylor TN, and Taylor EL (1989) Pollen and spore assemblages from the Falla Formation (Upper Triassic), Central Transantarctic Mountains, Antarctica. Rev. Palaeobot. Palynol. 61 (in press)
Gevers TW and Twomey A (1982) Trace fossils and their environment in Devonian (Silurian?) lower Beacon strata in the Asgard Range, Victoria Land, Antarctica: *in* Craddock C (ed), Antarctic Geoscience: University of Wisconsin Press, Madison, pp 639–647
Grindley GW and Oliver PJ (1983) Post-Ross orogeny cratonisation of northern Victoria Land: *in* Oliver RL, James PR, and Jago JB (eds), Antarctic Earth Science: Australian Academy of Science, Canberra, pp 133–139
Grindley GW, Mildenhall DC, and Schopf JM (1980) A mid-Late Devonian flora from the Ruppert Coast, Marie Byrd Land, West Antarctica: Journal of the Royal Society of New Zealand 10:271–285
Hammer WR (1986) Takrouna Formation fossils of northern Victoria Land: *in* Stump E (ed), Geological Investigations in Northern Victoria Land: Antarctic Research Series, Vol 46, American Geophysical Union, Washington DC, pp 243–247
Hammer WR, Ryan WJ, and DeFauw SL (1987) Comments on the vertebrate fauna from the Fremouw Formation (Triassic), Beardmore Glacier region Antarctica: Antarctic Journal of the US 22(5):32–33
Isbell JI (in press) Evidence for a low-gradient alluvial fan from the palaeo-Pacific margin in the Upper Permian Buckley Formation, Beardmore Glacier region, Antarctica: *in* Thomson MRA, Crame JA, and Thomson JW (eds), Geological Evolution of Antarctica: Cambridge University Press, Cambridge
Krissek LA and Horner TC (1987) Provenance evolution recorded by fine-grained Permian clastics, central Transantarctic Mountains: Antarctic Journal of the US 22(5):26–28
Krissek LA and Horner TC (in press) Clay mineralogy and provenance of fine-grained Permian clastics, central Transantarctic Mountains: *in* Thomson MRA, Crame JA, and Thomson JW (eds), Geological Evolution of Antarctica: Cambridge University Press, Cambridge
Kyle RA (1977) Palynostratigraphy of the Victoria Group of south Victoria Land, Antarctica: New Zealand Journal of Geology and Geophysics 20(6):1081–1102
Kyle RA and Fasola A (1978) Triassic palynology of the Beardmore Glacier area of Antarctica: Palinologia 1:313–319
Kyle RA and Schopf JM (1982) Permian and Triassic palynostratigraphy of the Victoria Group, Transantarctic Mountains, *in* Craddock C (ed), Antarctic Geoscience: University of Wisconsin Press, Madison, pp 649–659
Laudon TS (1987) Petrology of sedimentary rocks from the English Coast, eastern Ellsworth Land: *in* Abstracts: 5th International Symposium on Antarctic Earth Sciences, Cambridge (23–28 August 1987), p 89
Laudon TS, Lidke DJ, Delevoryas T, and Gee CT (1987) Sedimentary rocks of the English Coast, eastern Ellsworth Land, Antarctica: *in* McKenzie GD (ed), Gondwana Six: Structure, Tectonics and Geophysics: Geophysical Monograph Series, Vol 40, American Geophysical Union, Washington DC, pp 183–189
Lindsay JF (1970) Depositional environment of Paleozoic glacial rocks in the Central Transantarctic Mountains: Geological Society of America Bulletin 81:1403–1410

Long WE (1965) Stratigraphy of the Ohio Range, Antarctica: *in* Hadley JB (ed), Geology and Paleontology of the Antarctic, Antarctic Research Series, Vol 6, American Geophysical Union, Washington DC, pp 71–116

McElroy CT and Rose G (1987) Geology of the Beacon Heights area, southern Victoria Land, Antarctica. 1:50,000: New Zealand Geological Survey miscellaneous series map 15(1 sheet) and notes, 47 pp

Miller JMG (1989) Glacial advance and retreat sequences in a Permo-Carboniferous section, Central Transantarctic Mountains: Sedimentology 36 (in press)

Miller JMG and Waugh BJ (1986) Sedimentology of the Pagoda Formation (Permian), Beardmore Glacier area: Antarctic Journal of the US 21(5):45–46

Miller MF and Frisch RS (1987) Early Permian paleogeography and tectonics of the central Transantarctic Mountains: Inferences from the Mackellar Formation: Antarctic Journal of the US 22(5):24–25

Miller JMG and Waugh BJ (in press) Permo-Carboniferous glacial sedimentation in the central Transantarctic Mountains and its palaeotectonic implications: *in* Thomson MRA, Crame JA, and Thomson JW (eds), Geological Evolution of Antarctica: Cambridge University Press, Cambridge

Minshew VH (1967) Geology of the Scott Glacier and Wisconsin Range areas, Central Transantarctic Mountains, Antarctica: Unpubl. PhD dissertation, The Ohio State University, Columbus, 268 pp

Norris G (1965) Triassic and Jurassic miospores and acritarchs from the Beacon and Ferrar groups, Victoria Land, Antarctica: New Zealand Journal of Geology and Geophysics 8:236–277

Ojakangas RW and Matsch CL (1981) The Late Palaeozoic Whiteout Conglomerate: A glacial and glaciomarine sequence in the Ellsworth Mountains, West Antarctica: *in* Hambrey JJ and Harland WB (eds), Earth's Pre-Pleistocene Glacial Record: Cambridge University Press, Cambridge, pp 241–244

Plume RW (1982) Sedimentology and palaeocurrent analysis of the basal part of the Beacon Supergroup (Devonian [and older?] to Triassic) in south Victoria Land, Antarctica: *in* Craddock C (ed), Antarctic Geoscience: University of Wisconsin Press, Madison, pp 571–580

Schopf JM (1968) Studies in Antarctic paleobotany, Antarctic Journal of the US 3(5):176–177

Storey, BC (in press) The crustal blocks of West Antarctica within Gondwanaland; Reconstruction and break-up model: *in* Thomson MRA, Crame JA, and Thomson JW (eds), Geological Evolution of Antarctica: Cambridge University Press, Cambridge

Stump, E (1987) Construction of the Pacific margin of Gondwana during the Pannotios cycle: *in* McKenzie GD (ed), Gondwana Six: Structure, Tectonics and Geophysics: Geophysical Monograph Series, Vol 40, American Geophysical Union, Washington DC, pp 77–87

Tankard AJ, Jackson MPA, Eriksson KA, Hobday DK, Hunter DR, and Minter WEL (1982) Crustal Evolution of Southern Africa, Springer-Verlag, New York, Heidelberg, Berlin, 523 pp

Taylor EL, Taylor TN, Collinson JW, and Elliot DH (1986) Structurally preserved Permian plants from Skaar Ridge, Beardmore Glacier region: Antarctic Journal of the US 21(5):27–28

Taylor TN, Taylor EL, and Collinson JW (1986) Paleoenvironment of Lower Triassic plants from the Fremouw Formation: Antarctic Journal of the US 21(5):26–27

Vavra CL (1984) Triassic Fremouw and Falla Formations, central Transantarctic Mountains, Antarctica: Institute of Polar Studies Report 87, Ohio State University, Columbus, 98 pp

Visser JNJ (1988) The palaeogeography of part of southwestern Gondwana during the Permo-Carboniferous glaciation: Palaeogeography, Palaeoclimatology, Palaeoecology 61:205–219

Watts DR and Bramall AM (1981) Palaeomagnetic evidence for a displaced terrane in western Antarctica. Nature 293:638–641

Webers GF and Sporli KB (1983) Palaeontological and stratigraphic investigations in the Ellsworth Mountains, West Antarctica: *in* Oliver RL, James PR, and Jago JB (eds), Antarctic Earth Science: Australian Academy of Science, Canberra, pp 261–264

Wells NA and Dorr JA Jr (1987) Shifting of the Kosi River, northern India: Geology 15:204–207

Williams PL (1969) Petrology of the upper Precambrian and Paleozoic sandstones in the Pensacola Mountains, Antarctica: Journal of Sedimentary Petrology 39:1455–1465

2—Gondwanan Paleogeography and Paleoclimatology

Judith Totman Parrish

Introduction

The supercontinent Gondwana would, by its sheer size, be expected to have had a profound influence on climate. In general, the continental interior would be expected to have been dry and temperatures strongly seasonal, especially at times of low sea level. Changes in the position of the continent relative to the South Pole controlled the timing of Southern Hemisphere continental glaciation and, consequently, global sea level.

New continental reconstructions of the position of Gondwana in the Paleozoic (Scotese 1986; Van der Voo 1988) require revision of previous predictions of the paleoclimatic history of Gondwana and have raised important new controversies. The purpose of this chapter is not to review the vast literature on climatically significant data from the Gondwanan rock record but to reassess previous climate predictions (Ziegler et al. 1981a; Parrish 1982; Rowley et al. 1985; Raymond et al. 1985), particularly for the Paleozoic, in light of the new reconstructions. These predictions will provide a framework for future interpretation of Gondwanan biogeographic and sedimentological data.

Paleogeography of Gondwana

Before Gondwana began breaking up in the late Paleozoic, the continent comprised continental crust that now is divided into numerous distinct plates (Figure 2.1). The major plates are Australia, Africa (including the Arabian plate), Antarctica, India, and South America. In addition, Gondwana included the following plates, all of which were distinct at some time in their history (hyphenated names indicate regions that constituted single, unnamed plates; Scotese and Denham 1988): England, Avalon (Nova Scotia and southeastern Newfoundland), Iran–Turkey, Apulia (Italy, Yugoslavia, and Greece), Iberia, western Europe (France, Belgium, Netherlands, Germany), Tibet, South China, and various crustal fragments that now constitute West Antarctica. Malaysia, Indochina, North China, and Japan also might have formed a continuous crust with Gondwana (Scotese 1986).

Gondwanan Climate—General Considerations

Although the details of Gondwanan climatic patterns are dependent on its geography and position at particular times, certain generalities may be stated about paleoclimate on the continent. The reader is referred to Parrish (1982), Parrish et al. (1982), Rowley et al. (1985), and Parrish and Peterson (1988) for more detailed discussions of the climate processes described in the following sections.

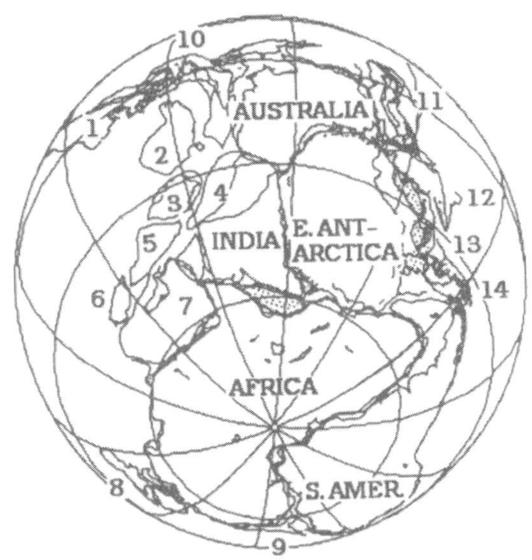

Figure 2.1. Index map of the plates of Gondwana. 1, North China; 2, South China; 3, Indochina; 4, Tibet; 5, Iran; 6, Turkey (Iran and Turkey behaved as one plate in the Late Paleozoic and Early Mesozoic); 7, Arabia; 8, Central Europe, Apulia, and Iberia (separate plates); 9, Florida; 10, Burma/Malaya and Japan (two plates); 11, North and South New Zealand (separate plates); 12, Chatham Rise; 13 (shaded area), Marie Byrd Land, Thurston and Berkner Islands, Whitmore and Ellsworth Mountains (all separate plates); and 14, Antarctic Peninsula and South Shetland Islands (separate plates). The shaded areas in the center of the continent are Madagascar (between India and Africa) and Sri Lanka (between India and East Antarctica).

Size

Gondwana was a large continent. The area of Gondwanan continental crust was greater than 100×10^6 km^2. The area of exposed land during the Late Cambrian, a time of relatively low sea level, was about 65×10^6 km^2 (J.T. Parrish, unpublished data from Franconian paleogeographic reconstruction by Scotese et al. 1979). By contrast, the exposed land area of Eurasia is about 58×10^6 km^2 (Espenshade and Morrison 1978). The size of the continent would have had two important effects, disruption of zonal atmospheric circulation and establishment of monsoonal circulation.

Disruption of Zonal Circulation

Positioned with its coastline on the pole, Gondwana would have stretched nearly to the equator. This position maximizes the disruption of the zonal component of circulation, which is best expressed over the oceans, for example, the Southern Ocean at present. The zonal circulation is controlled by the equator-to-poles thermal gradient and by the rotation of the Earth (e.g., Petterssen 1969). The heat exchange between equator and poles is effected in a series of cells—the tropical Hadley cells, the midlatitude Ferrel cells, and the polar cells. These result in three surface wind belts in each hemisphere, the equatorial easterlies, midlatitude westerlies, and polar easterlies. Stretching from pole to equator, Gondwana would have interrupted the paths of all three wind systems, creating oceanic subtropical high-pressure and high midlatitude low-pressure cells and increasing poleward heat and moisture transport in the Southern Hemisphere (Figure 2.2).

In general, rainfall is controlled by the zonal pattern and is heaviest on the windward sides

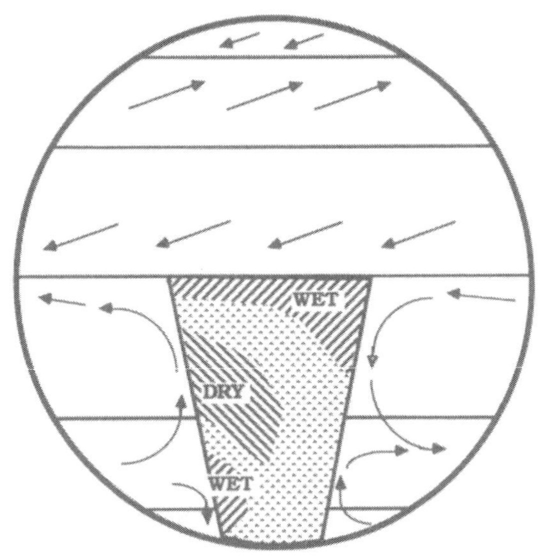

Figure 2.2. Schematic diagram of zonal circulation (top) and the effect of the continent on zonal circulation and the resulting precipitation patterns (bottom). Arrows are surface winds.

of mountains and where airflow crosses the coastline toward the continent. In the absence of other climatic perturbations, such as monsoonal circulation, discussed in the next Section, the wettest regions would have been on the east side of the continent between 30° north and south and above 70° latitude and on the west side of the continent between 40° and 60° latitude.

The Gondwanan Monsoon

Through much of its history, Gondwana probably was large enough to create the strongly seasonal circulation pattern known as a monsoon (e.g., Singh 1987). First, seasonality would have been strong because of the isolation of the continental interior from the ameliorating influences of the surrounding ocean. The low thermal inertia of land ensures a maximal response to seasonal fluctuations in insolation. Seasonality would have been particularly strong when the continent was centered in midlatitudes, rather than directly over the pole, because summer heating is greater at lower latitudes and winter cooling greater at higher latitudes (Young 1987). At midlatitudes, the continent would have covered regions of maximum cooling and heating.

Both temperature and, especially, precipitation would have fluctuated strongly with the seasons. Climatic data for Calcutta, India, are illustrative of the strong seasonality of an intense monsoonal circulation. In Calcutta, the cold-month mean temperature is 19°C and the warm-month mean temperature 30°C. The same region has very high annual rainfall, about 160 cm/yr, but more than 90% of that falls between May 1 and October 31 (Rumney 1970).

Aridity

The large size of the continent also would have meant that the interior of the continent was arid. Even in the absence of mountains, moisture-laden (i.e., warm) winds flowing landward lose their moisture over land because heating drives the air masses upward (in winter, the moisture in landward-flowing air does not condense unless the air is forced upward by mountains or by a warm air mass from another source). Thus, in low latitudes, rainfall not only would have been seasonal, but also confined to coastal regions. This process is illustrated by numerical experiments for idealized continents by Barron (1986) and for Pangaea by Kutzbach and Gallimore (1989). The experiments by Barron (1986; Figure 2.3) illustrate the importance of moisture source. The principal source is at the equator, where the air is warmest and thus contains the most moisture. Overall precipitation is low in the equatorial-continent experiment because this source of moisture is lacking. Overall precipitation is higher both at the equator and in the coastal regions of the land masses in the polar-continents experiment.

Although the interior of a large continent would receive little precipitation at any latitude, climate would not be uniformly arid over the entire continent. At high altitudes and latitudes, where temperatures are cooler and evaporation rates correspondingly lower, the climate may appear to have been humid (on the evidence of plants, for example), even though precipitation was relatively low. Conversely, at low altitudes and latitudes, aridity will be more severe with low rainfall because temperatures and evaporation rates are lower.

Polar Position

For much of its history, Gondwana lay over or near the South Pole, and the great Paleozoic glaciations have been attributed to this position (Crowell and Frakes 1970). Maximum glaciation occurred when the South Pole was near the coast, rather than in the center of the continent (Crowell and Frakes 1970; Crowell 1982). These observations support recent results of numerical model simulations of Gondwanan climate, which predict that glaciation is most intense when the pole is near the coast (Crowley et al. 1987; Figure 2.4). Glaciation depends on a supply of moisture as well as on cold temperatures (e.g., Crowell and Frakes 1970). Although temperatures would have been very low in the Gondwanan interior

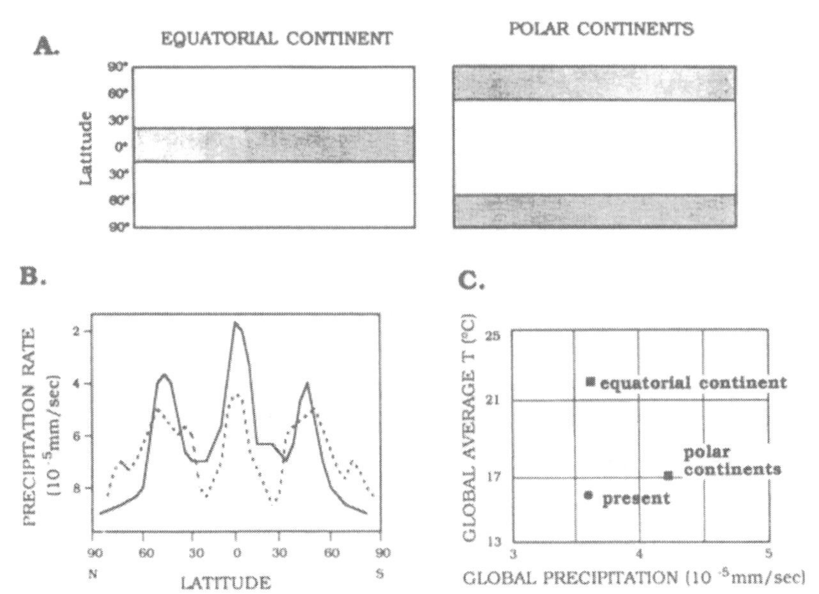

Figure 2.3. Results of numerical experiments on idealized continent–ocean distribution, adapted from Barron (1986). (a) Idealized continents. (b) Latitudinal distribution of precipitation. Dotted line, equatorial continent; solid line, polar continents. (c) Global temperature and precipitation for the present and the two idealized cases. Note that total global precipitation is high in the polar continents case, although temperature is low.

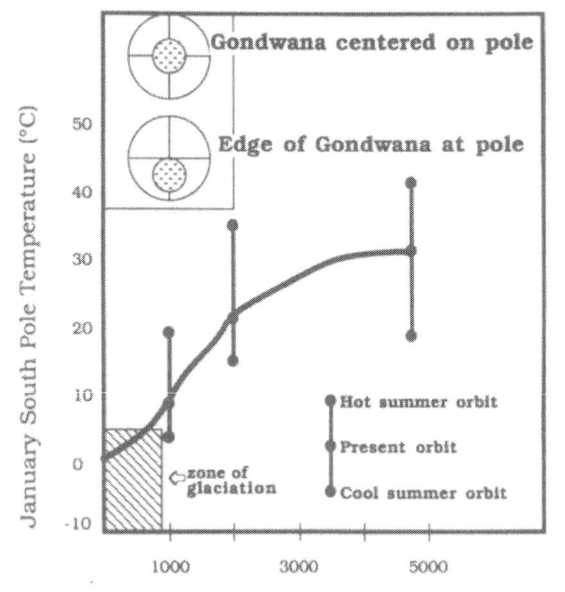

Figure 2.4. Results of numerical experiments on idealized Gondwana (Crowley et al. 1987). "Hot" and "cold" summer orbits refer to the effects of changes in the eccentricity of the Earth's orbit.

when the continent was centered on the pole, the region of lowest temperatures also would have been farthest from the moisture supply. In contrast to Antarctica, which lies mostly south of 70°S, Gondwana was large enough to completely encompass and extend beyond the region bounded by the 60°S parallel (Figure 2.5). Thus, the coldest regions of the continent would have been isolated from the coastal sources of moisture. In addition, the presence of land-locked marginal seas would enhance glaciation, first by acting as a source of moisture and later by accumulating sea ice and increasing albedo (Crowell and Frakes 1970).

Predicted Paleoclimatic History of Gondwana

In the reconstructions for Gondwana presented in this section (Figure 2.6 through 2.12), the plate positions are taken directly or modified from Terra Mobilis 2.0™ (Scotese

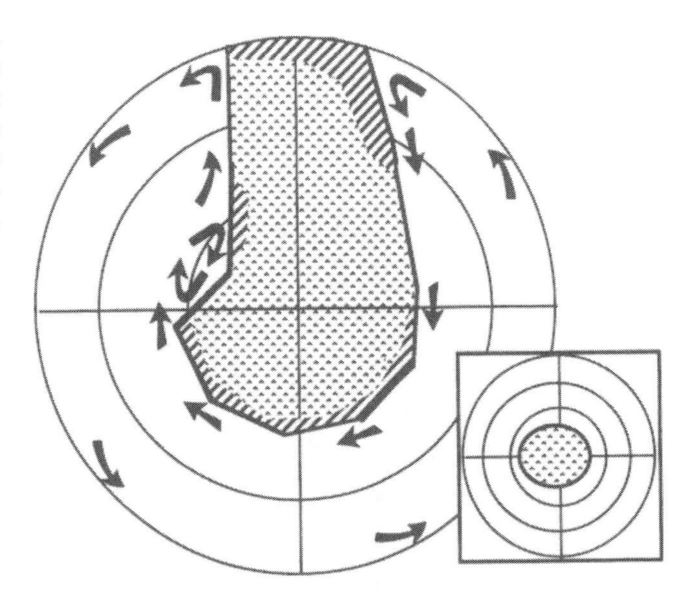

Figure 2.5. Schematic diagram of climate in Gondwana when the pole is centered over the continent, as it was in the Silurian (Figure 2.7). Shading as in Figure 2.2. Compare the isolation of the pole in Gondwana and Antarctica (inset). Latitudinal lines in both figures are at 0°, 30°, and 60°.

and Denham 1988), and the paleogeography is from maps presented in Scotese et al. (1979), with modifications suggested by Scotese (1986). In the following discussions, present coordinates are used when referring to modern geography (e.g., eastern Australia) and paleocoordinates when referring to ancient geography (e.g., western Gondwana).

Late Ordovician

In the late Ordovician (Figure 2.6), the South Pole lay in northern Africa, just inland from the coastline, and Australia straddled the equator. Thus, during this interval, Gondwana

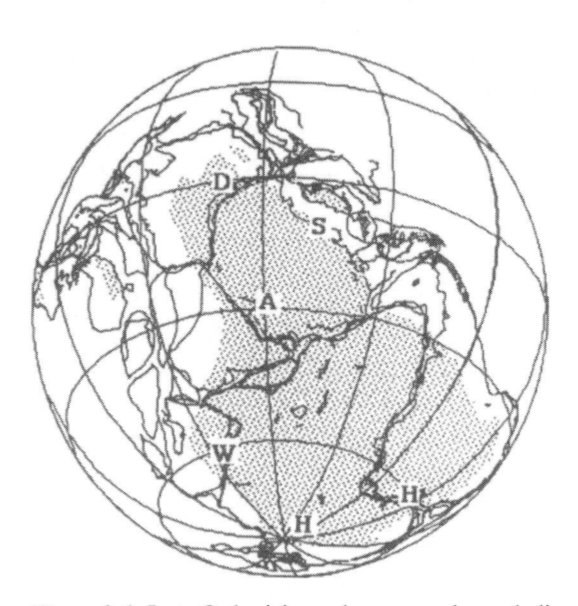

Figure 2.6. Late Ordovician paleogeography and climate of Gondwana. Plate positions from Scotese (1986) and Scotese and Denham (1988); shading is paleogeography from Scotese et al. (1979). Precipitation (qualitative): A, arid; D, dry; H, humid; W, wet; S, strongly seasonal; see text for further explanation. Latitudinal lines are spaced 30° apart.

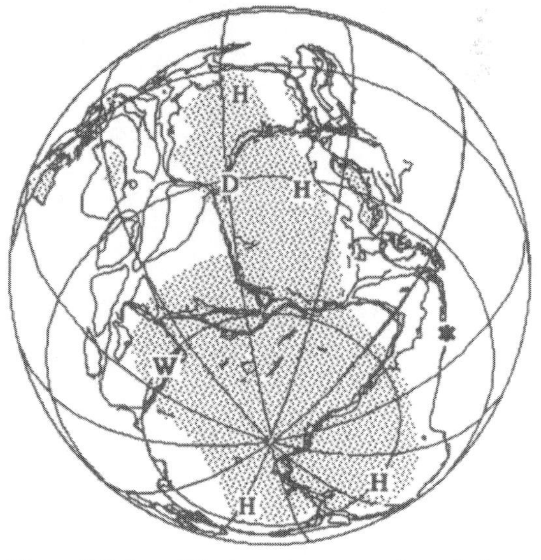

Figure 2.7. Late Silurian paleogeography and climate of Gondwana. Plate positions, paleogeography, and symbols as in Figure 2.6. Star is position of South Pole from Figure 2.8.

Figure 2.8. "Middle" (probably early Late) Silurian paleogeography and climate of Gondwana. Plate positions from Van der Voo (1988); parts of Gondwana were not reconstructed by him. Shading is paleogeography from Late Silurian of Scotese et al. (1979). Star is position of South Pole from Figure 2.7.

Figure 2.10. Early Carboniferous paleogeography and climate of Gondwana. Plate positions, paleogeography, and symbols as in Figure 2.6.

would have disrupted zonal circulation more than at any subsequent time until the late Paleozoic. The size of the continent and its distribution across high- and low-latitudinal zones would have maximized summer heating and winter cooling, creating a strong monsoonal circulation. The summer monsoon would have been most intense, owing partly to the cross-equatorial pressure contrast with the northern subtropical high-pressure belt (Parrish and Peterson 1988), but cross-

Figure 2.9. Early Devonian paleogeography and climate of Gondwana. Plate positions from Scotese and Denham (1988) modifed according to Scotese (1988). Paleogeography and symbols as in Figure 2.6.

Figure 2.11. Late Carboniferous paleogeography and climate of Gondwana. Plate positions, paleogeography, and symbols as in Figure 2.6.

Figure 2.12. Earliest Triassic paleogeography and climate of Gondwana. Plate positions, paleogeography, and symbols as in Figure 2.6.

equatorial heat exchange also would have occurred in the winter. The shape of the northeastern coastline and the termination of the continent just north of the equator would have shunted the equatorial current into the northern hemisphere.

The entire southern coastline of Gondwana, from the region of Saudi Arabia to northern South America, would be expected to have been relatively wet, although the highest precipitation would have been in the west. India and easternmost Antarctica would have been relatively arid. Rainfall in western East Antarctica and southern Africa would have been highly seasonal, with a summer monsoon maximum. Rainfall in southern Australia might have been bimodal, with maximum rainfall in the spring and autumn, during the transition from summer to winter monsoon, when equatorial circulation is least disrupted.

Late Silurian

By the Late Silurian (Figure 2.7), the South Pole was in western equatorial Africa, about as far from the coastal regions as it would be at that time. Monsoonal circulation would have been weakened by the rise in sea level in northern India and East Antarctica. The continent occupied most of the area south of 60°, limiting the amount of moisture that could penetrate to the cold pole. According to Crowley et al.'s (1987) models, glaciation should have been at a minimum at this time.

Precipitation would have been relatively high year-round along the entire coastline south of about 50° and in the summer in eastern and northern Australia. Moderate rainfall, especially in the summer, would be expected in western East Antarctica, whereas eastern East Antarctica and western Australia were probably dry.

An alternative "middle" Silurian reconstruction for parts of Gondwana has been published by Van der Voo (1988), based on data by Hargraves et al. (1987; Figure 2.8). This reconstruction places the pole in southernmost Chile, several thousand kilometers from the Scotese and Denham (1988) pole position. (The two reconstructions probably represent the same interval; the Late Silurian reconstruction of Scotese et al. (1979) and Scotese and Denham (1988) is for the Wenlock Stage, which is early Late, informally middle, Silurian.)

The Van der Voo (1988) reconstruction has important implications for Gondwanan climate. Disruption of zonal circulation and monsoonal circulation would have been as strong as in the Ordovician because the main exposed land area would have been in midlatitudes. In addition, Crowley et al.'s (1987) model would predict glaciation. Although some evidence for glaciation in South America exists (Crowell et al. 1980), the event was apparently minor compared with the earlier one. A number of other inconsistencies are raised by the Van der Voo (1988) reconstruction. Although it is beyond the scope of this chapter to review all the paleoclimatic evidence, the following points need to be addressed.

1. Localities for the *Clarkeia* brachiopod fauna (Cocks 1972; =Malvinokaffric Realm of Boucot and Johnson 1973) lie within a relatively narrow latitudinal band, between 40°S and 60°S, in the Scotese reconstruction (Ziegler et al. 1977, 1981a; Scotese et al. 1979; Scotese 1986; Scotese and Den-

ham 1988), whereas the Van der Voo (1988) reconstruction requires distribution of the fauna from 80°S to the equator. This biogeographic pattern is rare in the fossil record. The only comparable distribution is that of *Monotis* in the Triassic (Westermann 1973), and although the Silurian and Triassic share the characteristic of having extremely cosmopolitan faunas, to suggest that *Clarkeia* was a similarly widespread form is tenuous, particularly with the paucity of corroborating evidence.

2. Ziegler et al. (1977, 1979) showed that carbonates, mixed carbonates–clastics, and clastics form coherent belts parallel to latitude in the Scotese reconstructions and evaporites in Australia occur where expected in the subtropics in the western, low-latitude portion of Gondwana. Moreover, reef carbonates, which are abundant everywhere in the Silurian tropics, are absent from northern Africa. As with the *Clarkeia* fauna, the Van der Voo (1988) reconstruction would orient these belts longitudinally. Moreover, it would place coarse clastics on the western side of the continent, where the climate would have been driest and less erosion would have taken place to produce those clastics (Ziegler et al. 1981b). The Australian evaporites would be at 50°–60°S, on the outer fringes of the dry belt, in the Van der Voo (1988) reconstruction. Although Van der Voo (1988, p. 318) stated that Australia's position is essentially insensitive to the new poles, in fact the rotation of Gondwana would eliminate the paleogeographic conditions required to maintain aridity in that part of Gondwana.
3. As Van der Voo (1988) pointed out, the new pole position requires a hairpin movement of the continent at "rapid" drift rates; these rates are conservatively estimated to be at least 61 cm/yr (a minimum of 8000 km from the Middle–Upper Ordovician pole to the "middle" Silurian pole, over a minimum of 13 my, that is, from the end of the Ordovician to the beginning of the "middle" Silurian), an order of magnitude greater than any present drift rate.

Early Devonian

More confusion has arisen over Devonian plate positions than for any other time (Scotese et al. 1979; Heckel and Witzke 1979; Ziegler et al. 1981a; Scotese 1984; Barrett 1985) and, although much of the controversy has centered around the position of North America, the position of Gondwana also has been modified repeatedly. Although Scotese and Denham (1988) placed the South Pole roughly in Angola, a position in east-central Argentina is now favored (Scotese 1988), similar to the position illustrated by Van der Voo (1988; Figure 2.9). The principal effect is to place northern Africa closer to the equator, although the new position also would favor glaciation, according to Crowley et al.'s (1987) model.

Sea level was high, particularly in South America, during the Early Devonian (Scotese et al. 1979). The westerlies and polar easterlies would have occurred over land and sea, supplying abundant moisture to the continent. If land-locked sea is also favorable to glaciation, as suggested by Crowell and Frakes (1970), the Early Devonian should have been an ideal time for a major glacial event. However, although glacigenic deposits (Caputo and Crowell 1985) and cold-water faunas (Copper 1977) have been reported from South America, evidence for extensive glaciation is lacking, and the paleogeography suggests that any glaciers were montane.

Early Carboniferous

In the Early Carboniferous (Figure 2.10), the pole was roughly in the midpoint of the longitudinal coastline between New Guinea and northern South America. The effect of this position was to confine the exposed area of Gondwana to a region poleward of about 25°S, where it might have been out of the influence of equatorial circulation, depending on the continent's relation to North America. Various reconstructions (Scotese et al. 1979; Raymond et al. 1985; Scotese 1986) are ambiguous about an emergent connection between the two plates (Figure 2.10).

Glaciation is favored by the coastal position of the pole (Crowley et al. 1987) and by the presence of a large, land-locked sea in its vicinity (Crowell and Frakes 1970). Although the continent did not extend to very low latitudes, it might have been large enough to induce monsoonal circulation, in which case northern Africa would have had highly seasonal rainfall.

Late Carboniferous

By the Late Carboniferous (Figure 2.11), Gondwana had collided with Laurussia to form Pangaea. Glaciation was at a maximum at this time (Frakes and Crowell 1969, 1970; Crowell and Frakes 1971ab, 1972, and related papers). In addition to the collision with Laurussia, the drop in sea level and northward movement of the continent created conditions favorable to monsoonal circulation, although the summer heating would have been moderated by the ice cap, and the monsoonal deflection of the equatorial low-pressure system might have been countered by the equatorial highlands that resulted from the collision (Rowley et al. 1985).

By the end of the Carboniferous, the interior of northern Pangaea was relatively dry (e.g., Phillips et al. 1985; Cecil et al. 1985), suggesting, along with data on eolian sandstones (Parrish and Peterson 1988), that full monsoonal conditions characteristic of Pangaea (Robinson 1973; Parrish et al. 1982, 1986) had been established. From this time until the Late Jurassic, Gondwana did not exist as a separate paleogeographic or paleoclimatic entity.

Late Permian through Middle Jurassic

Apart from relatively minor changes of sea level and position relative to the South Pole (Scotese et al. 1979; Ziegler et al. 1983), the paleogeography of Gondwana was similar from the Late Permian through the Middle Jurassic (Figure 2.12). The Pangaean monsoon was predicted by Parrish et al. (1986) to have been at a maximum in the Triassic, but detailed corroborating evidence from the Northern Hemisphere (e.g., Hallam 1984; Rowley et al. 1985; Parrish and Peterson 1988) has not been matched by data from the Southern Hemisphere. For example, the Triassic humidity maximum recognized on the Colorado Plateau in deposition of the fluvial Chinle Formation (e.g., Dubiel 1987) might be expected to have a counterpart in northern South America. Nevertheless, redbeds, which are characteristic of global sedimentation at this time (e.g., Robinson 1973; Turner 1980; Van Houten 1982), also are abundant in the Southern Hemisphere (Robinson 1973), especially in postglacial rocks (e.g., Rocha-Campos 1971).

Late Jurassic

Gondwana reassumed an independent identity in the late Jurassic with the opening of western Tethys to the paleo-Pacific Ocean (Ziegler et al. 1983; Figure 2.13). Although this break might have weakened the monsoonal system in the Southern Hemisphere as it destroyed the Pangaean monsoon (Parrish et al. 1982; Parrish and Doyle 1984), Gondwana was probably large enough to sustain a monsoonal climate on its own (Parrish et al. 1982). The position of Gondwana in the Late Jurassic was similar to that in the Ordovician, with the South Pole near the coast and Africa extending north of the equator. However, Southern Hemisphere glacial deposits are unknown (Hambrey and Harland 1981). Although "glacial erratics" have been reported from much lower latitudes in the Northern Hemisphere

Figure 2.13. Late Jurassic paleogeography and climate of Gondwana. Plate positions, paleogeography, and symbols as in Figure 2.6.

(Frakes and Francis 1988), the glacial origin of these deposits is highly controversial (Chumakov 1981; Kemper 1987), not least because of the lack of glacigenic deposits in the south, where a paleogeographic setting conducive to glaciation in previous times failed to produce glaciation in the Jurassic.

Summary and Conclusions

Predictions for continental glaciation and monsoonal circulation are summarized in Table 2.1; Devonian predictions depend on the paleogeographic reconstructions used for the paleoclimatic predictions. For the Scotese (1986) and Scotese and Denham (1988) reconstructions, continental glaciation is both predicted and observed on Gondwana in the Ordovician and Carboniferous, and predicted, but not observed, in the Jurassic. Monsoonal circulation is predicted for the Ordovician and Permian through Early Jurassic, peaking in the Triassic (Parrish et al. 1986). Although the Pangaean monsoon postdated for the most part the Permo-Carboniferous glaciation, continental glaciation and monsoonal circulation are not mutually exclusive. For example, in numerical experiments with a global circulation model, Kutzbach and Guetter (1986) found that, although the North American ice sheet affected circulation strongly throughout the last glacial, the strength of the Asian monsoon responded much more strongly to orbitally induced changes in insolation. Overall, however, the monsoonal circulation may be weaker during continental glaciation, and low-latitude continental interiors may be drier (Manabe and Hahn 1977).

Table 2.1. Predictions for continental glaciation and monsoonal circulation on Gondwana.

	Predictions	
Time	Continental glaciation	Monsoonal circulation
Late Ordovician	Yes	Yes
Late Silurian	No	No
Early Devonian	Yes	No
Early Carboniferous	Yes	Weak
Late Carboniferous	Yes	Yes
Permian–Early Jurassic	No	Strong
Late Jurassic	Yes	Yes

References

Barrett SF (1985) Early Devonian continental positions and climate: A framework for paleophytogeography. In Tiffney BR (ed) Geological Factors and the Evolution of Plants, Yale University Press, New Haven, pp 93–127

Barron EJ (1986) Mathematical climate models: Insights into the relationship between climate and economic sedimentary deposits. In Parrish JT, Barron EJ Paleoclimates and Economic Geology, Society of Economic Paleontologists and Mineralogists, Short Course, Vol 18, pp 31–83

Boucot AJ, Johnson JG (1973) Silurian brachiopods. In Hallam A (ed) Atlas of Palaeobiogeography, Elsevier, Amsterdam, pp 59–65

Caputo MV, Crowell JC (1985) Migration of glacial centers across Gondwana during Paleozoic Era. Geological Society of America Bulletin 96:1020–1036

Cecil CB, Stanton RW, Neuzil SG, Dulong FT, Ruppert LF, Pierce BS (1985) Paleoclimate controls on Late Paleozoic sedimentation and peat formation in the central Appalachian basin (U.S.A.). International Journal of Coal Geology 5:195–230

Chumakov NM (1981) Scattered stones in Mesozoic deposits of North Siberia, U.S.S.R. in Hambrey MJ, Harland WB (eds) Earth's Pre-Pleistocene Glacial Record, Cambridge University Press, Cambridge, pp 264.

Cocks LRM (1972) The origin of the Silurian *Clarkeia* shelly fauna of South America, and its extension to West Africa. Palaeontology 15:623–630

Copper P (1977) Paleolatitudes in the Devonian of Brazil and the Frasnian-Famennian mass extinction. Palaeogeography, Palaeoclimatology, Palaeoecology 21:165–207

Crowell JC (1982) Continental glaciation through geologic time. In Berger WH, Crowell JC (eds) Climate in Earth History, National Academy Press, Washington, DC, pp 77–82

Crowell JC, Frakes LA (1970) Phanerozoic glaciation and the causes of ice ages. American Journal of Science 268:193–224

Crowell JC, Frakes LA (1971a) Late Palaeozoic glaciation of Australia. Journal of the Geological Society of Australia 17:115–155

Crowell JC, Frakes LA (1971b) Late Paleozoic glaciation: Part IV, Australia, Geological Society of America Bulletin 82:2515–2540

Crowell JC, Frakes LA (1972) Late Paleozoic glaciation: Part V, Karroo Basin, South Africa. Geological Society of America Bulletin 83:2887–2912

Crowell JC, Rocha-Camos AC, Suarez-Soruco R

(1980) Silurian glaciation in central South America. In Cresswell MS, Vella P (eds) Gondwana Five, Proceedings of the Fifth International Gondwana Symposium, Wellington, New Zealand, February, 1980, pp 105–110
Crowley TJ, Mengel JG, Short DA (1987) Gondwanaland's seasonal cycle. Nature 329:803–807
Dubiel RF (1987) Sedimentology of the Upper Triassic Chinle Formation, southeastern Utah. Unpubl. PhD Dissertation, University of Colorado, Boulder.
Espenshade EEB, Morrison JL (eds) (1978) Goode's World Atlas, 15th edition. Rand McNally and Company, Chicago.
Frakes LA, Crowell JC (1969) Late Paleozoic glaciation, I. South America. Geological Society of America Bulletin 80:1007–1042
Frakes LA, Crowell JC (1970) Late Paleozoic glaciation, II. Africa exclusive of the Karroo basin. Geological Society of America Bulletin 81:2261–2284
Frakes LA, Francis JE (1988) A guide to Phanerozoic cold polar climates from high-latitude ice-rafting in the Cretaceous. Nature 333:547–549
Hallam A (1984) Continental humid and arid zones during the Jurassic and Cretaceous. Palaeogeography, Palaeoclimatology, Palaeoecology 47: 195–223
Hambrey MJ, Harland WB (eds) (1981) Earth's Pre-Pleistocene Glacial Record, Cambridge University Press, Cambridge.
Hargraves RB, Dawson EM, Van Houten FB (1987) Paleomagnetism and age of mid-Paleozoic ring complexes in Niger, West Africa, and tectonic implications. Royal Astronomical Society Geophysical Journal 90:705–729
Heckel PH, Witzke BJ (1979) Devonian world palaeogeography determined from distribution of carbonates and related lithic palaeoclimate indicators. Special Papers in Palaeontology 23:99–123
Kemper E (1987) Das Klima der Kreide-Zeit. Geologisches Jahrbuch A96:5–185
Kutzbach JE, Guetter PJ (1986) The influence of changing orbital parameters and surface boundary conditions on climate simulations for the past 18,000 years. Journal of the Atmospheric Sciences 43:1726–1759
Kutzbach JE, Gallimore RG (1989) Pangaean climates: Megamonsoons of the megacontinent. Journal of Geophysical Research 94:3341–3358.
Manabe S, Hahn DG (1977) Simulation of the tropical climate of an ice age. Journal of Geophysical Research 82:3889–3911
Parrish JT (1982) Upwelling and petroleum source beds, with reference to the Paleozoic. American Association of Petroleum Geologists Bulletin 66:750–774
Parrish JT, Doyle JA (1984) Predicted evolution of global climate in Late Jurassic–Cretaceous time. International Organization of Paleobotany Conference, Edmonton, August, 1984, Abstracts
Parrish JT, Peterson F (1988) Wind directions predicted from global circulation models and wind directions determined from eolian sandstones of the western United States—A comparison. Sedimentary Geology 56:261–282
Parrish JT, Ziegler AM, Scotese CR (1982) Rainfall patterns and the distribution of coals and evaporites in the Mesozoic and Cenozoic. Palaeogeography. Palaeoclimatology, Palaeoecology 40: 67–101
Parrish JM, Parrish JT, Ziegler AM (1986) Permian–Triassic paleogeography and paleoclimatology and implications for therapsid distributions. In Hotton NH III, MacLean PD, Roth JJ, Roth EC (eds) The Ecology and Biology of Mammal-like Reptiles, Smithsonian Press, Washington, DC, pp 109–132
Petterssen S (1969) Introduction to Meteorology, McGraw-Hill, New York.
Phillips TL, Peppers RA, DiMichele WA (1985) Stratigraphic and interregional changes in Pennsylvanian coal-swamp vegetation: Environmental inferences. International Journal of Coal Geology 5:43–109
Raymond A, Parker WC, Parrish JT (1985) Phytogeography and paleoclimate of the Early Carboniferous. In Tiffney BR (ed) Geological Factors and Evolution of Plants, Yale University Press, New Haven, pp 169–222
Robinson PL (1973) Palaeoclimatology and continental drift. In Tarling DH, Runcorn SK (eds) Implications of Continental Drift to the Earth Sciences, Vol I, Academic Press, London, pp 449–476
Rocha-Campos AC (1971) Upper Paleozoic and Lower Mesozoic paleogeography, and paleoclimatological and tectonic events in South America. In Logan A, Hills LV (eds) The Permian and Triassic Systems and their Mutual Boundary, Canadian Society of Petroleum Geologists Memoir, pp 398–424
Rowley DB, Raymond A, Parrish JT, Lottes AL, Scotese CR, Ziegler AM (1985) Carboniferous paleogeographic, phytogeographic, and paleoclimatic reconstructions. International Journal of Coal Geology 5:7–42
Rumney GR (1970) Climatology and the World's Climates, Macmillan, New York.
Scotese CR (1984) An introduction to this volume: Paleozoic paleomagnetism and the assembly of Pangaea. In Van der Voo R, Scotese CR, Bonhommet N (eds) Plate Reconstruction from Paleozoic Paleomagnetism, Geodynamics Series, Vol 12, American Geophysical Union, Washington, DC, pp 1–10
Scotese CR (1986) Phanerozoic reconstructions: A new look at the assembly of Asia. University of Texas Institute for Geophysics Technical Report 66.

Scotese CR, Denham CR (1988) Terra Mobilis™: Plate tectonics for the Macintosh®.
Scotese CR, Bambach RK, Barton C, Van der Voo R, Ziegler AM (1979) Paleozoic base maps. Journal of Geology 87:217–277
Singh K (1987) The Indian monsoon in literature. In Fein JS, Stephens PL (eds) Monsoons, John Wiley and Sons, New York, p 35–49
Turner P (1980) Continental Red Beds. Developments in Sedimentology 29, Elsevier, Amsterdam.
Van der Voo R (1988) Paleozoic paleogeography of North America, Gondwana, and intervening displaced terranes: Comparison of paleomagnetism with paleoclimatology and biogeographical patterns. Geological Society of America Bulletin 100:311–324
Van Houten FB (1982) Redbeds. McGraw-Hill Encyclopedia of Science and Technology 5/e:441–442
Westermann GEG (1973) The late Triassic bivalve *Monotis*. In Hallam A (ed) Atlas of Palaeobiogeography, Elsevier, Amsterdam, pp 251–258
Young JA (1987) Physics of monsoons: The current view. In Fein JS, Stephens PL (eds) Monsoons, John Wiley and Sons, New York, pp 211–243
Ziegler AM, Hansen KS, Johnson ME, Kelly MA, Scotese CR, Van der Voo R (1977) Silurian continental distributions, paleogeography, climatology, and biogeography. Tectonophysics 40:13–51
Ziegler AM, Scotese CR, McKerrow WS, Johnson ME, Bambach RK (1979) Paleozoic paleogeography. Annual Review of Earth and Planetary Sciences 7:473–502
Ziegler AM, Bambach RK, Parrish JT, Barrett SF, Gierlowski EH, Parker WC, Raymond A, Sepkoski JJ, Jr (1981a) Paleozoic biogeography and climatology. In Niklas KJ (ed) Paleobotany, Paleoecology, and Evolution, Vol 2, Praeger Publishers, New York, pp 231–266
Ziegler AM, Barrett SF, Scotese CR (1981b) Palaeoclimate, sedimentation and continental accretion. Philosophical Transactions of the Royal Society of London A301:253–264
Ziegler AM, Scotese CR, Barrett SF (1983) Mesozoic and Cenozoic paleogeographic maps. In Brosche P, Sündermann J (eds) Tidal Friction and the Earth's Rotation, Vol II, Springer-Verlag, Berlin, pp 240–252

3—Reconstructing High-Latitude Cretaceous Vegetation and Climate: Arctic and Antarctic Compared

Robert A. Spicer

Introduction

Polar climate exerts a strong influence on global atmospheric and oceanic conditions, and therefore a sound knowledge of the thermal regime, precipitation/evaporation, and the presence or amount of ice at high latitudes is essential for understanding a wide range of global systems. Probably the most sensitive indicator of atmospheric conditions, and one that bequeaths an extensive and abundant fossil record, is land vegetation. Accurate reconstructions of polar vegetation through time, and appropriate interpretations of the climatic signal contained in the resultant plant fossil record, is therefore critical to our understanding of global paleoclimate, the evolution and distribution of global biotas, and successful retrodiction of sources of fossil fuels and climate-related mineral deposits.

The present "icehouse" world furnishes us with living vegetation that can be regarded only as a partial model for ancient plant ecosystems. Critically, modern plants growing at today's cold high latitudes provide a poor analog of polar vegetation at times of global warmth. In order to understand these ancient systems, we have to turn to the fossil record and take care in extrapolations from living plants. We have to utilize fully the more time-stable aspects of morphologies and anatomies related to plant physiology and the physical processes of sedimentation, and make minimal use of taxon-specific tolerances.

Because terrestrial vegetation is directly exposed to the atmosphere, the structure (physiognomy) of the vegetation and characteristics of the individual plants (leaf architecture—margin organization and venation patterns; tree rings—thickness, variation, proportion of early wood to late wood, etc.) can provide an accurate climate signal with respect to temperature (mean annual and mean annual range), precipitation/evaporation, seasonality, and even light distribution throughout the year (Wolfe 1979, 1985; Creber and Chaloner 1985; Parrish and Spicer 1988a,b). The reliability of this signal is dependent, however, on the vegetation being in equilibrium with the environment. This condition is only achieved in climax vegetation: precursor and pioneer communities provide a less reliable climate signal. For paleoclimatological purposes, communities in a state of flux have to be identified by means of detailed facies analysis and taphonomic considerations (Spicer 1989).

Regional vegetation is composed of a mosaic of communities. In particular, riparian vegetation has a distinct floristic composition and physiognomy that is rarely typical of the vegetation as a whole. Riparian vegetation, however, is sampled preferentially by fluvial systems biasing fossil samples of the regional flora. An expanding body of data (reviewed in Spicer 1989) now exists that allows some of these biases to be compensated for, and therefore provides a means of improving the accuracy of paleoclimatic interpretations. In the absence of precise taphonomic constraints (not always possible with restricted outcrop extent and logistic limitations), a useful alternative is to sample as many different depositional environments (and therefore facies) as possible, so that facies biases can be accomodated in reconstructions (e.g., Spicer and Parrish 1986).

Incremental changes of environment with latitude are pronounced at high latitudes. This is expressed strongly in the variations in the daily ratio of light to darkness. For this reason, any study of high-latitude environments has to take place in a framework that is well constrained for paleolatitude, particularly with respect to the rotational rather than the magnetic pole.

Research to integrate megafossil assemblages with both marine and nonmarine sedimentary and palynofacies throughout the Late Cretaceous of the North Slope is still in progress, but results to date provide critical insights into plant community structure and climate of Cretaceous Arctic forests that are consistent with, and help explain, more limited observations made at high paleolatitudes in the Southern Hemisphere. The Alaskan work also aids in the formulation of research approaches in the more hostile environment of Antarctica.

The Northern Alaska Cretaceous Record

Paleolatitude

In order to make vegetation reconstruction and paleoclimate interpretations as accurate as possible, detailed facies analysis, taphonomic data, and floristic analysis are essential integrated elements in the Alaskan work.

The lack of appreciable relative movement between northern Alaska and cratonic North America since the Early Cretaceous (Jones 1983; Sweeny 1983) allows abundant palaeomagnetic data from North America and "best fit" global plate reconstructions to be used to define the Cretaceous paleolatitude of Northern Alaska with some precision. These data, and those from the North Slope itself (Witte et al. 1987) position the North Slope of Alaska between 75°N and 85°N (Smith et al. 1981; Ziegler et al. 1983) during the Late Cretaceous, with a possible northward drift from Cenomanian to Maastrichtian times. Furthermore, best-fit studies of the distribution of climate-sensitive deposits, such as coals and evaporites, show that for the Maastrichtian the rotational and magnetic poles were coincidental to within 4° (Lottes 1987).

Stratigraphy

Primary uplift of the east–west trending Brooks Range took place between Berriasian and Aptian times and initiated sediment shedding into an asymmetric foredeep to the north known as the Colville Basin (Mull 1985). By Albian times, two large river-dominated deltaic complexes had developed, the Corwin Delta to the west and the Umiat Delta to the east (Huffman et al. 1985). Progradation of the Corwin Delta in a northeasterly direction, and the Umiat Delta northward, infilled the basin throughout the Late Cretaceous and Early Tertiary. The resultant package of highly fossiliferous intertonguing marine and nonmarine sediments, the Nanushuk and Colville Groups, provides a detailed sequential record of the Arctic terrestrial and marine biota and environments throughout the Late Cretaceous.

Although plant collections were made in the 1960s (Smiley 1966, 1967, 1969a,b), insufficient data were obtained to enable the plants to be placed in a facies framework, and taxonomic descriptions have never been published. Since 1976, I have been engaged in collecting plant megafossils from southwestern, central, and northern Alaska. Originally, this work was for biostratigraphic purposes, and it became apparent that the evolutionary radiation of angiosperms and their geographic spread into Alaska during the early part of the Late Cretaceous provided an extremely powerful biostratigraphic tool for continental deposits (Spicer 1983), Minimal post-Cretaceous tectonic deformation, lithostratigraphic correlation, and the marine paleontological record (Ahlbrandt 1979; Huffman 1985; Detterman et al. 1975; Brosgé and Whittington 1966; Cobban and Gryc 1961; Sliter

1979) together with some palynological and radiometric constraints (Frederiksen 1986; Lanphere and Tailleur 1983) have provided a framework for calibrating North Slope megafossil biostratigraphy and tracking temporal changes in Late Cretaceous vegetation and climate.

In contrast, conventional palynostratigraphic techniques have enjoyed only limited success on the North Slope, and the currently published resolution is rather coarse, being little better than stage level (Witmer et al. 1981). Facies control on megafossil assemblages is strong (Spicer 1987), indicating a heterogeneous vegetation. Thus, a strong ecological signal, coupled with extensive reworking of palynomorphs (e.g., Detterman and Spicer 1981) during delta progradation, is probably responsible for the relatively poor stratigraphic resolution.

Cretaceous Arctic Vegetation

Published interpretations of Cretaceous North Slope plant communities and climate based on palynomorphs (May and Shane 1985) are of limited value because all the samples contain marine elements indicating either local reworking or sorting of the style documented by Muller (1959). In Alaska, offshore differential pollen sorting was biased against the angiosperms. This in turn resulted in a highly distorted vegetational interpretation with poor community resolution, which bears very little relation to the observed megafossil suites (Spicer and Parrish 1986; Parrish and Spicer 1988a).

Albian–Cenomanian Assemblages

Although palynodebris from nonmarine facies is under study (Spicer, et al. 1987; Parrish, et al. 1987), interpretations here are based on megafossils. Prior to the angiosperm influx in latest Albian times, the lower floodplain environment was dominated by *Podozamites* with ferns (*Onychiopsis, Sphenopteris,* and *Birisia*) and *Equisetites* forming most of the ground cover in preserved overbank and river margin environments. *Equisetites* rhizomes are a ubiquitous feature below coal seams and paleosols, suggesting that the plant was a primary colonizer. Subordinate but widespread coniferous elements included *Athrotaxopsis* and *Elatocladus*. Fluvial sandstones yield *Ginkgo* leaves, but other ginkgophytes (*Sphenobaiera*) are largely confined to overbank deposits. Cycads, such as *Nilssonia decursiva* and *N. alaskana,* have restricted distributions; *N. decursiva* is consistently found in clay and siltstones associated with nonmarine/marine transitions, whereas *N. alaskana* was found in baked shales associated with a burned coal.

In the central North Slope, angiosperms first appear in the uppermost part of the Lower Killik Tongue of the Chandler Formation (latest Albian), and in the Upper Killik and Niakogon Tongues (middle to late Cenomanian), they dominate riparian assemblages in the form of "platanoids," such as members of the *Protophyllum, Pseudoprotophyllum,* and *Pseudoaspidiophyllum* complex. In spite of a relatively coarse matrix (fine sand) and evidence of moderately high-energy deposition (ripples, poorly developed grain-size sorting and bedding, leaves "rolled" within the sediments), very few leaves showed signs of biological or mechanical degradation (*sensu* Spicer 1981). This lack of preburial degradation is all the more remarkable in view of the fact that many leaves measured 10–25 cm in width.

Overbank and lacustrine units of the Niakogon Tongue (late Cenomanian) yield a variety of entire-, toothed-, and lobed-margined angiosperm leaves. Both simple and compound leaves occur. Leaf size range is large, and leaves show little evidence of preburial degradation, even though many forms appear to have been thin textured in life.

Complete conifer leafy shoots of taxodiaceous form are common in lacustrine units together with some isolated leaves of *Podozamites,* the delicate *Ginkgo concinna,* and a variety of ferns. Erect stems of *Equisetites* occur frequently.

The nonmarine rocks of the Late Cretaceous Alaskan North Slope include an estimated 2.70 trillion tons of coal reserves (Sable and Stricker 1985); seams several meters thick are common. Very few megafossils are preserved in coal seams. The only identifiable remains seem to be those of *Podozamites* leaves, together with compressed and structurally preserved logs measuring up to 55 cm in diameter. The thick (up to 5 m), bituminous, low-ash, low-sulphur coals of the Nanushuk Group exhibit complex microlithologies that have yielded at least six different palynofacies (Grant et al. 1988). Some undoubtedly represent, at least in part, detrital peats and sapropels, whereas others are autochthonous accumulations of woody, resinous, elements. Fusain is notably absent in all facies.

Logs, many with roots attached, are a common component of fluvial, overbank, and mire environments. Where preservation allows, they exhibit well-defined growth rings characterized by a paucity of late wood in relation to early wood (Spicer 1987; Parrish and Spicer 1988b). In more than 50 specimens representing different logs, 3 vesselless taxa have been recognized, all of which display the same ring characteristics of extensive early wood, wide rings, moderate mean sensitivies, and few false rings (Parrish and Spicer 1988b).

Coniacian Assemblages

During Turonian times, much of the North Slope deltas suffered a marine transgression; nonmarine sediments are rare. Coniacian plant assemblages from riparian and lacustrine environments of the Tuluvak Tongue of the Prince Creek Formation are a rich mixture of conifers, *Ginkgo*, angiosperms, ferns, and the ubiquitous *Equisetites*. Cycads were apparently absent. Entire-margined angiosperm leaf forms appear to be slightly more diverse than in the Cenomanian, but sampling is limited.

Campanian–Maastrichtian Assemblages

The upper part of the nonmarine Kogosukruk Tongue of the Prince Creek Formation is well exposed along 80 km of the Colville River between Uluksrak Bluffs and Ocean Point, northern Alaska. These coal-bearing rocks are considered to be of late Campanian to Maastrichtian age based on the palynoflora (Frederiksen et al. 1986) and the underlying marine Rogers Creek member, which was dated using invertebrates (Brosgé and Whittington 1966). They were deposited at a paleolatitude of 80 to 85°N (Smith et al. 1981; Ziegler et al. 1983).

A variety of sedimentary facies are present within the unit. Fluvial (including channel-fill and overbank) deposits comprise the bulk of the sediments (Phillips 1987), but lacustrine and mire environments are also represented. Both autochthonous and allochthonous coals occur throughout the unit. Those nearer the base of the unit tend to be bituminous, resin-rich, and thicker (typically 0.5 m but rarely more than 1.0 m thick) than those nearer the top, which also are of lower rank. Paleosols and bentonites occur throughout the unit.

In comparison to over 80 plant megafossil forms recovered from the Albian–Cenomanian Nanushuk Group (Spicer and Parrish 1986), only 8 forms have been recovered from 62 localities in the Kogosukruk Tongue. These consist of two conifer foliage forms (*Parataxodium wigginsii* Arnold and Lowther and a cupressaceous shoot), one terrestrial angiosperm leaf (*Hollickia quercifolia* (Hollick) Krassilov), one aquatic ?angiosperm (*Quereuxia angulata* (Lesquereux) Kryshtofovich, also referred to as "*Trapa*"), one ?angiosperm fruit, two ferns, and *Equisetites*. This marked drop in diversity cannot be attributed to ecological sampling bias because both sedimentary and diagenetic facies are comparable between the Nanushuk Group and the Kogosukruk Tongue (Parrish and Spicer 1988a).

Based on the megafossils, no major vegetational change is evident during the period of deposition of the Kogusukruk Tongue. However, the later coals appear to contain less woody material than do the earlier ones, which may reflect a diminution of the climax forest ecosystem and more open vegetation. Regionally, the vegetation was dominated by the deciduous conifer *Parataxodium wigginsii*, which, because of the size and morphology of associated leafy shoots, branches, and logs, is interpreted to have reached small tree stature. Structurally preserved (calcified and silicified to minimally altered) coniferous logs rarely exceed 20 cm in diameter and the maximum diameter observed was 50 cm.

No angiosperm wood has been identified, and it appears that the angiosperms formed only a minor understory or lake margin component. Channel fill sediments are devoid of platanoid leaves suggesting that the platanoid riparian communities of the Cenomanian (Spicer and Parrish 1986) had ceased to exist in this area.

Angiosperm pollen diversity increases in latest Maastrichtian times (T. Ager, written communication 1986), but the absence of complementary leaf fossils suggests this increase may reflect diversification among herbaceous elements. Fusain is common throughout the Kogosukruk Tongue.

Vegetation Summary

Remains of conifers occur in all facies, suggesting they dominated the regional vegetation throughout the Late Cretaceous although diversity gradually declined. Other ubiquitous elements, such as ferns and *Equisetites*, undoubtedly formed much of the gound cover. Angiosperms had a more heterogeneous distribution both in time and space. Initially occupying riparian sites, platanoids had migrated into lacustrine margin situations by the Coniacian. By the end of the Cretaceous, they were all but eliminated from the North Slope. Entire-margined forms, mostly restricted to overbank and understorey environments, may have exhibited greatest diversity in the Coniacian but disappeared by the Campanian. Other elements to disappear were first the cycads by the Coniacian and then the ginkgophytes by the Campanian.

The depauperate Maastrichtian vegetation was probably more open than that of earlier times, with trees of smaller stature and herbaceous angiosperms. Fires were common and may have contributed to the openness.

Climate

Extensive coals, the abundance of large structurally preserved logs, and the diversity of the fossil flora point to a thriving Late Cretaceous forest ecosystem very close to the North Pole. On the basis of studies of plant fossils, the climate of northern Alaska has been interpreted as warm temperate to subtropical (Smiley 1967; May and Shane 1985), although no quantitative estimates of temperatures have been given. Furthermore, some elements of high-latitude floras have been interpreted as being evergreen. The survival of evergreen plants through a warm, protracted polar winter dark period has been doubted because of the likelihood that stored metabolites from summer photosynthesis would be rapidly exhausted (Wolfe 1980; Spicer 1987). The presence of evergreen plants might suggest a more even light profile throughout the year, and therefore, that in the past Earth's obliquity may have been less than at present (Wolfe 1980; Douglas and Williams 1982).

Users of numerical global climate models have experienced difficulty in simulating warm poles (e.g., Barron 1983; Barron and Washington 1982), casting doubt on the predictive value of the models in a geologic context. Reduced obliquity compounds this problem because it would lead to a cooling of the poles. However, the required boundary conditions of temperature are poorly known, having been extrapolated from marine isotopic temperatures from much lower latitudes. In addition, numerical climate models are known to be poor at predicting critical climatic elements, such as precipitation at high latitudes (e.g., Crowley et al. 1986). To date, nonvegetational, high-latitude palaeoclimate indicators have provided only spurious data.

Deciduousness and the Polar Light Regime

Herbaceous elements, such as ferns and sphenophytes, could survive long, dark winters by dying back to underground organs. Overwintering is a problem confined to plants with extensive secondary tissues. None of the North Slope Cretaceous angiosperm leaves can be demonstrated to have been evergreen. The antiquity of the forms prevents taxonomic affinity being used as a reliable guide, and no forms appear to have been coriaceous. The strongest candidates for an evergreen habit are among the conifers and cycads. However, Kimura and Sekido (1975) have shown the *Nilssonia* plant to have been vinelike and deciduous, quite unlike frost-sensitive evergreen relict modern cycads, and most conifers are represented by abscissed shoots typical of modern deciduous conifers such as *Metasequoia* and *Taxodium*. The most notable exception is a small cupressaceous-like leafy shoot that, because of its xeromorphy, may represent a plant that could have entered dormancy without shedding leaves. Furthermore, the lack of preburial degradation in all taxa suggests a synchronous entry into the sedimentary record and therefore deciduousness. Evidence suggests all taxa were either deciduous or died back to perrenating organs each winter (Spicer and Parrish 1986; Spicer 1987; Parrish and Spicer 1988b). This obviates the need to invoke reduced obliquity to explain polar forests.

Tree-ring data suggest a highly seasonal climate and, in particular, that the change from summer to winter was rapid. This is consistent with the change in day length at high latitudes today and suggests a similar pattern of light distribution and therefore obliquity. It follows that the North Slope Cretaceous plants experienced a three- to four-month-long dark winter with perhaps a bounding period of twilight of one to two months.

Precipitation/Evaporation

Wide rings with few false rings in Cenomanian wood suggest benign summer growth conditions. Large thin leaves, thick coals, the lack of fusain and a water-rich sedimentary system all suggest abundant, and probably uniform, precipitation. Rare false rings have been attributed to root waterlogging (Parrish and Spicer 1988).

Kogosukruk woods exhibit proportionately less early wood than those of the Cenomanian and false rings are common. This suggests less benign, more variable summer conditions. Abundant fusain throughout the unit suggests that the vegetation repeatedly dried enough to burn, but evidence for drought (e.g., mudcracks) is lacking.

Temperatures

Quantitative air temperature estimates for Cretaceous high latitudes based on plants necessarily involve a degree of uncertainty, but are undoubtedly more reliable than those currently gained by any other method. Angiosperm nearest living relative (NLR) methods are inappropriate because of the evolutionary novelty of the group, but among the more conservative gymnosperms, coherent time trends can be demonstrated across unrelated taxa, such as conifers, cycadophytes, and ginkgophytes, and more prolonged survivorship at lower latitudes suggests progressive cooling (Spicer 1987).

The invasion of the angiosperms produced abnormally high diversities during Cenomanian times, and difficulty in delimiting biological species renders temperature estimates based on diversity (e.g., Barron 1983) highly unreliable (Spicer 1987). Angiosperm leaf physiognomy, in particular the ratio of entire-margined leaves to toothed leaves, is related to mean annual temperature in climax vegetation that is not limited by water (Bailey and Sinnott 1916; Wolfe 1979). Cretaceous calibration of this relationship is difficult for the newly evolved angiosperms, but early evolution of major tooth types and a positive correlation of paleolatitude with the proportion of toothed leaves suggest a relationship was established as early as the Cenomanian in spite of a low equator-to-pole temperature gradient (Wolfe and Upchurch 1987).

Because most evergreens today are entire-margined, polar light regimes and deciduousness may bias leaf margin analyses and lower the apparent temperature. However, this effect is counteracted partially by the enhanced diversity of angiosperm leaf types produced by relatively lax inter-plant competition enjoyed by the novel angiosperms. Although more data are needed, the apparent straight-line nature of the relationship of leaf margin ratios to paleolatitude also suggests that light-induced deciduousness may not unduly bias leaf margin temperature estimates.

Vegetational physiognomy is independent of leaf margin analysis, and yet for the Late Cretaceous of the North Slope, comparable results are obtained. The rationale for the temporal stability of vegetational physiognomic techniques is outlined in Wolfe (1979), and using Wolfe's definitions of forest types, the Cenomanian North Slope forests would fall into the category of low montane mixed coniferous forest. This translates into a likely mean annual temperature (MAT) of 10°C, which is the same as that derived from leaf margin analysis (Spicer and Parrish 1986).

The Kogosukruk floras contain too few angiosperm leaf forms to make any direct quantitative estimates of mean annual paleotemperature, and the flora is markedy depauperate compared to that of the Cenomanian or Coniacian—major groups such as the cycadophytes and ginkgophytes are absent, leaf size is small, and entire-margined angiosperms are also absent. This strongly suggests a lower MAT. Based on the available evidence, the Maastrichtian MAT is estimated to have been between 2 and 8°C (Parrish and Spicer 1988a) and most likely 5°C (Spicer 1987).

Winter Temperatures and Mean Annual Range of Temperature (MAR)

Estimating winter temperatures from plants when growth ceases is difficult. Forests with large trees in the Cenomanian argue against permafrost, a conclusion supported by the lack of sedimentological evidence for extensive sea or river ice throughout the Alaskan Late Cretaceous. However, huge coal reserves in the face of a limited growth season and minimal postabscission degradation of megafossils suggests that postmortem decay was minimal and therefore winter temperatures were probably below 4°C (Spicer 1987). Diurnally induced freezing conditions at middle latitudes today are caused by a lack of insolation for nights less than 18 hr in duration. Winter darkness of 3–4 months duration is therefore likely to have given rise to frosts even with the 10°C MAT of the Cenomanian. Frosts were even more likely during the Maastrichtian (Brouwers et al. 1987).

The Antarctic Cretaceous Plant Record

By comparison with Alaska, the Cretaceous plant record from Antarctica is poor. Several accounts document the occurrence of leaves and wood from the Antarctic Peninsula region (Birkenmajer and Zastawniak 1986; Hernandez and Azcarate, 1971; Jefferson 1982; Jefferson and Macdonald 1981; Francis 1986), but taxonomic or paleoclimatic interpretations are presently inadequate. Additionally, problems exist with stratigraphic resolution and paleolatitude. Recent paleogeographic reconstructions (Lawver et al. 1985) place the Antarctic Peninsula at about 60°S, which is outside the region where light constraints are most strongly felt.

Climate and Vegetation

Most information comes from woods. Jefferson (1982) reported that in situ trees from the Lower Cretaceous of Alexander Island have uneven growth rings with widths up to 7.78 mm. He attributed the uneven growth pattern to reaction wood due to gravitational stimulus, but also admitted that it might have been caused by mutual shading, implying a partially closed canopy forest. While clearly some of Jefferson's trees were displaced from the vertical, many were not. By far, the most common cause of growth-ring asymmetry in modern forests is asymmetric shading, and any assumption that Cretaceous trees differed from the modern in this respect is unwarranted.

The wide growth rings indicate that the trees were not growing under significant stress. Their biological productivity was moderately high. Creber and Chaloner (1985) suggest that the apparent wide spacing of Jefferson's trees was in response to low angles of illumination. This implies a degree of stress on the trees that is not substantiated by the ring widths. Biological organisms usually pack to the maximum density concomitant with their ability to survive. Wide tree spacing with high productivity therefore suggests that the community was in a state of development and had not yet attained the densest packing tolerable with the prevailing light conditions. Jefferson's trees were in a volcanic terrain and preserved as the result of periodic inundations of volcaniclastic material. In some cases, disturbance was by fluvial processes, in others apparently by direct and local volcanic activity (charring in association with a volcanic ash layer). Whatever the cause, the vegetation was subject to periodic catastrophic disturbance. If disturbance was frequent, successive cohorts of seedlings would have been eliminated, so that wide tree spacing was maintained. Jefferson (1981) also reports evidence of buried ground cover. This would be expected in recovery vegetation, but not in a system that was light limited and in which the trees had to be widely spaced in order to survive because shading at ground level would have been more acute than at crown level. In view of these critical questions, a strong case can be made for reexamining the fossil forests of Alexander Island in some detail to determine the composition of the ground cover, frequency of disturbance, and taphonomic factors related to deciduousness.

The climate signal exhibited by collections of isolated woods from both marine and nonmarine environments for the Cretaceous and Tertiary has been synthesized by Francis (1986). She identified podocarp and araucarian-like woods and concluded that the Cretaceous Antarctic Peninsula experienced a cool temperate climate that, although qualitative, is consistent with predictions from the Alaskan studies and isotopic data from the marine waters around the Antarctic peninsula (Pirrie and Marshall 1988). Parrish and Spicer (1988b) suggested that maritime influences would have resulted in relative climatic constancy throughout the year compared to the continental interior. Although the data are sparse, the Cretaceous vegetation of the Antarctic Peninsula appears to have been conifer dominated with abundant ferns, and therefore broadly similar to the Alaskan vegetation.

The present inadequacy of Antarctic paleobotanical studies demands that southern high-latitude Cretaceous climates may have to be interpreted in large part from predictions based on Alaskan work. It is probably reliable to assume symmetry of light regimes (insolation) at both poles, but temperatures and precipitation might have been influenced by the strong asymmetry in the polar land/sea distribution, for example, a closed Arctic Ocean versus continental Antarctica. Both Alaskan North Slope and Antarctic Peninsula and continental margin climates were ameliorated by maritime influences, but more extreme conditions were undoubtedly experienced inland. At present, paleobotanical data for such environments are lacking and may well remain so.

The Alaskan North Slope MAT estimates were for conditions at sea level. If a modern day saturated adiabatic lapse rate (SALR) of 6°C per 1000 m altitude is applied, then MATs of 0°C or below might have been experienced above 1700 m during the Cenomanian and above 1000 m during the Maastrichtian at latitudes greater than 75°. This suggests that montane glaciation was a feature of Antarctica throughout the Late Cretaceous. However, coastal Antarctica might still have supported cool temperate coniferous forests mostly enjoying a moderate to high water supply during the growing season, if not year round. Like high northern latitudes, high summer productivity would have been linked with a minimal winter biomass loss leading to a high net organic accumulation, either in the form of potential terrestrial coals or onshore-derived but shelf-deposited organics. To date, no substantial Cretaceous coal deposits have been found in Antarctica, but this may be because of the lack of suitable basins to accumulate and preserve the organics. These predictions require verification and the implications clearly demand continued, indeed enhanced, paleobotanical activity in Antarctica.

References

Ahlbrandt TS (ed) (1979) Preliminary geologic, petrologic, and paleontologic results of the study of Nunushuk Group rocks, North Slope, Alaska. Circ. US Geol. Surv. 794

Bailey IW, Sinnott EW (1916) The climatic distribution of certain types of angiosperm leaves. Am. J. Bot. 3:24–39

Barron EJ (1983) A warm, equable Cretaceous: the nature of the problem. Earth Science Reviews 19:305–338

Barron EJ, Washington WM (1982) Atmospheric circulation during warm geologic periods: Is the equator-to-pole surface-temperature gradient the controlling factor? Geology 10:633–636

Birkenmajer K, Zastawniak E (1986) Plant remains of the Dufayel Island Group (early Tertiary?), King George Island, South Shetland Islands (West Antarctica). Acta Palaeobotanica 26:33–54

Brosgé WP, Whittington CL (1966) Geology of the Umiat-Maybe Creek Region, Alaska. Prof. Pap. US Geol. Surv. 303-H:501–638

Brouwers EM, Clemens WA, Spicer RA, Ager TA, Carter LD, Sliter WV (1987) Dinosaurs on the North Slope Alaska: Reconstructions of high-latitude, latest Cretaceous environments: Science 237:1608–1610

Cobban WA, Gryc G (1961) Ammonites from the Seabee Formation (Cretaceous) of northern Alaska. Journ. Paleontol. 35:176–190

Creber GT, Chaloner WG (1985) Tree growth in the Mesozoic and early Tertiary and the reconstructions of palaeoclimates. Palaeogeogr. Palaeoclimatol. Palaeoecol. 52:35–60

Crowley TJ, Short, DA, Mengel JG, North GR (1986) Role of seasonality in the evolution of climate during the last 100 million years. Science 231:579–584.

Detterman RL, Spicer, RA (1981) New stratigraphic assignment for rocks along Igilatvik (Sabbath) Creek, William O. Douglas Arctic Wildlife Range, Alaska. Circ. US Geol. Surv. 823-B: 11–12

Detterman RL, Reiser HN, Brosgé WP, Dutro JT Jr (1975) Post-Carboniferous stratigraphy, northeastern Alaska. Prof. Pap. US Geol. Surv. 886: 1–46

Douglas JG, Williams GE (1982) Southern polar forests: the early Cretaceous floras of Victoria and their palaeoclimatic significance. Palaeogeogr. Palaeoclimatol. Palaeoecol. 39:171–185

Francis JE (1986) Growth rings in Cretaceous and Tertiary wood from Antarctica and their paleoclimatic implications. Palaeontology 29:665–684

Frederiksen NO, Ager TA, Edwards LE (1986) Early Tertiary marine fossils from northern Alaska: implications for Arctic Ocean paleogeography and faunal evolution (Comment). Geology 14:802–803

Grant PR, Spicer RA, Parrish JT (1988) Palynofacies of Northern Alaskan Cretaceous coals. 7 Int. Palynol. Congr., Brisbane, Abstr.: 60

Hernandez P, Azcarate V (1971) Estudio paleobotánico preliminar sobre restos de una tafoflora de la Peninsula Byers (Cerro Negro), Isla Livingston, Islas Shetland del Sur, Antártida. Instituto Antártido Chileno Ser Científica 2:15–20

Huffmann AC Jr (ed) (1985) Geology of the Nanushuk Group and related rocks, North Slope, Alaska. Bull. US Geol Surv. 1614:1–129

Huffmann AC Jr, Ahlbrandt TS, Pasternack I, Stricker GD, Fox JE (1985) Depositional and sedimentologic factors affecting the reservoir potential of the Cretaceous Nanushuk Group, central North Slope, Alaska. Bull. US Geol. Surv. 1614:61–74

Jefferson TH (1981) Palaeobotanical contributions to the geology of Alexander Island, Antarctica. Unpubl. PhD dissertation, Cambridge University, England

Jefferson TH (1982) Fossil forests from the Lower Cretaceous of Alexander Island, Antarctica. Palaeontology 25:681–708

Jefferson TH, MacDonald DIM (1981) Fossil wood from South Georgia. Bull. British Antarctic Survey 54:57–64

Jones PB (1983) The cordilleran connection—a link between Arctic and Pacific sea-floor spreading. Jour. Alaska Geol. Soc. 2:41–55

Kimura T, Sekido S (1975) *Nilssoniocladus* n. gen. (Nilssoniaceae n. fam.) newly found from the early Lower Cretaceous of Japan. Palaeontographica 153 Abt. B:111–118

Lanphere MA, Tailleur IL (1983) K-Ar ages of bentonites in the Seabee Formation, Northern Alaska: A Late Cretaceous (Turonian) Time Scale Point. Cretaceous Research 4:361–370

Lawver LA, Sclater JG, Meinke L (1985) Mesozoic and Cenozoic reconstructions of the South Atlantic. Tectonophysics 114:233–254

Lottes AL (1987) Paleolatitude determinations: comparison of paleoclimatic and paleomagnetic methods. GSA Abstract with Program 19:749

May FE, Shane JD (1985) An analysis of the Umiat delta using palynologic and other data, North Slope, Alaska. Bull. US Geol. Surv. 1614:97–120

Mull CG (1985) Cretaceous tectonics, depositional cycles, and the Nanushuk Group, Brooks Range and Arctic Slope, Alaska. Bull. US Geol. Surv. 1614:7–36

Muller J (1959) Palynology of Recent Orinocco Delta and shelf sediments. Micropalaeontology 5:1–32

Parrish JT, Spicer RA (1988a) North Polar Late

Cretaceous temperature curve: evidence from plant fossils. Geology 16:22–25
Parrish JT, Spicer RA (1988b), Fossil woods of the Nanushuk Group (Albian-Cenomanian), Northern Alaska. Palaeontology 31:19–34
Parrish JT, Spicer RA, Grant P (1987) Near-polar climate in the latest Cretaceous. Geol Soc. Amer. Abstracts with Program 19:800
Phillips RL (1987) Late Cretaceous to Early Tertiary deltaic to marine sedimentation, North Slope, Alaska, Amer. Assoc. of Petrol. Geol. Bull. 71:601–602
Pirrie D, Marshall JD (1988) Late Cretaceous fossil preservation and oxygen isotope paleotemperatures: preliminary results from James Ross Island, Antarctica. Origin and Evolution of the Antarctic Biota (meeting held in London and Cambridge, 24–26 May 1988) Abstracts Volume: 34
Sable EG, Stricker GD (1985) Coal in the National Petroleum Reserve in Alaska (NPRA): framework, geology, and resources. AAPG Bull. 69:677
Sliter WV (1979) Cretaceous foraminifera from the North Slope of Alaska, Circ. US Geol. Surv. 794:89–112
Smiley CJ (1966) Cretaceous floras from the Kuk River area, Alaska, stratigraphic and climatic interpretations. Bull. Geol. Soc. Amer. 77:1–14
Smiley CJ (1967) Paleoclimatic interpretations of some Mesozoic floral sequences. Bull. Amer. Assoc. Petrol. Geol. 51:849–863
Smiley CJ (1969a) Cretaceous floras of the Chandler–Colville region, Alaska—stratigraphy and preliminary floristics. Bull. Amer. Assoc. Petrol. Geol. 53:482–502
Smiley CJ (1969b) Floral zones and correlations of Cretaceous Kukpowruk and Corwin Formations, northwestern Alaska. Bull. Amer. Assoc. Petrol. Geol. 53:2079–2093
Smith AC, Hurley AM, Briden JC (1981) Phanerozoic paleocontinental world maps. Cambridge University Press, Cambridge.
Spicer RA (1981) The sorting and deposition of allochthonous plant material in a modern environment at Silwood Lake, Silwood Park, Berkshire, England. US Geol. Surv. Prof. Pap. 1143:1–69
Spicer RA (1983) Plant megafossils from Albian to Paleocene rocks in Alaska. Contract report to the Office of National Petroleum Reserves in Alaska. US Geol. Surv.: 1–521
Spicer RA (1987) The significance of the Cretaceous flora of northern Alaska for the reconstruction of the climate of the Cretaceous. Geol. Jb. A96:265–291
Spicer RA (1989) The formation and interpretation of plant fossil assemblages. Advances in Botanical Research, 16:95–191 Academic Press, London
Spicer RA, Parrish JT (1986) Paleobotanical evidence for cool north polar climates in the mid-Cretaceous (Albian-Cenomanian). Geology 14:703–706
Spicer RA, Grant PR, Parrish JT (1987) The Late Cretaceous polar coals: a sedimentological, vegetational, and climatological study of the Umiat Delta, Alaska. Deltas Sites and Traps for Fossil Fuels, Geological Society of London Abstracts:39
Sweeney JF (1983) Evidence for the origin of the Canada Basin margin by rifting in Early Cretaceous time. Jour. Alaska. Geol. Soc. 2:17–23
Witmer RJ, Mickey MB, Harga H (1981) Biostratigraphic correlation of selected test wells of NPRA. USGS open-file Report 81–1165:1–85
Witte WK, Stone DB, Mull CG (1987) Paleomagnetism, paleobotany, and paleogeography of the Cretaceous North Slope, Alaska. *In* Tailleur I, Weimer P (eds) Alaskan North Slope Geology, Vol. 2, SEPM and Alaska Geological Society, Bakersfield, California, pp 571–579
Wolfe JA (1979) Temperature parameters of humid to mesic forests of eastern Asia and relation to forests of other regions of the northern hemisphere and Australasia. US Geol. Surv. Prof. Pap.. 1106:1–37
Wolfe JA (1980) Tertiary climates and floristic relationships at high latitudes in the northern hemisphere. Paleogeogr. Palaeoclimatol. Palaeoecol. 30:313–323
Wolfe JA (1985) Distribution of major vegetational types during the Tertiary. Am. Geophys. Union. Geophys. Mon. 32:357–375
Wolfe JA, Upchurch GR (1987) North American nonmarine climates and vegetation during the Late Cretaceous. Palaeogeogr. Palaeoclimatol. Palaeoecol. 61:33–77
Ziegler AM, Scotese CR, Barrett SF (1983) Mesozoic and Cenozoic paleogeographic maps. *In* Broshe P, Sundermann J (eds) Tidal Friction and the Earth's Rotation. Springer Verlag, Berlin, pp 240–252

4—The South Polar Forest Ecosystem

Geoffrey T. Creber

The solar energy input into the very high southern latitudes determines the maximum productivity level that could have been achieved by Antarctic ecosystems in the geological past when there was not a major glaciation. The input of energy must supply all that is needed by the primary producers (the green plants) to carry out photosynthesis. In turn, the herbivorous animals make use of the high-energy materials (e.g., carbohydrates, proteins, etc.) synthesized by the plants. Finally, the carnivorous animals feed either on the herbivorous ones or on other carnivorous ones. Parasites feed either on living plants or animals and scavengers feed on the latter's dead remains. Every organism in the ecosystem is thus dependent on the energy input either directly or indirectly. The solar input of about 3500 megajoules/m^2/yr (Farman and Hamilton 1978; LaGrange 1963) for high latitudes in Antarctica will determine for the continent a temperate ecosystem. Inputs such as 6300 MJ/m^2/yr for 36°S and over 7000 MJ/m^2/yr for equatorial latitudes will clearly determine the more luxuriant productivities that are found in those lower latitudes.

It would seem likely therefore that the animals would be smaller than those in past geological lower latitude ecosystems. Alternatively, if larger fossils are found, it may be that they were fewer in number so that their biomass was limited to a certain upper threshold appropriate to the total energy input. Similarly, the volume of wood production by forests would be directly related to the solar input since the wood is a fuel and may be burned to release what was originally solar energy.

Estimation of the Productivity of Fossil Forests

If an area of forest floor becomes fossilized with the stumps in situ, it is possible to count the number of trees in a given area and to measure their diameters and ring-width sequences. From these data, as in modern forestry practice, the volume of wood produced annually per unit area of forest floor can be calculated (Creber and Chaloner, 1984a, 1984b, 1985; Creber 1986; Creber and Francis 1987; Hamilton and Christie 1971).

The mass of wood produced may be calculated by multiplying the volume by the density (0.4 g/cm^3). The total energy content is found by multiplying by the calorific value of conifer wood (21,000 J/g). Since this figure is usually only about 0.4% of the total solar input, the result of the calculation sets a lower limit to the annual solar energy input. The calculations enable estimates to be made of paleoclimatic parameters operating at the time of the growth of the original forest.

Since detailed measurements of growth rings under Antarctic field conditions are rather impractical as is the collection of large stump specimens, it is suggested that stump cross sections may be photographed with a scale, and the enlarged prints used to measure the ring-width series back in the laboratory. Small reference specimens from the stumps must also be collected in order to make corrections if microscopic examination of them shows that shrinkage of the wood may have occurred during fossilization. In general, stumps are somewhat resistant to shrinkage because the rings resist collapse in a periclinal (tangential) direction. Collapsed wood is much more common in horizontal trunks or isolated fragments that have been subjected to vertical pressure.

In cases where there are no fossilized stumps in situ, data from individual pieces of fossil wood may be used. The method is from Assmann (1970), who has used it on extant trees. He calculates the cross-sectional area added by each successive growth ring to the existing cross-sectional area of the trunk before the ring was formed. This additional area he calls the area increment, and it is, of course, directly proportional to the amount of wood produced by the whole trunk during the year. The advantage of the method is that the area increments do not depend on the ring widths alone but are very much influenced by the diameters of the cross sections of the trunks on which they accrue. The larger the initial trunk diameter, the larger will be the area increment for a given ring width.

The area increment added by a ring of width x mm to the existing area trunk cross section can be found by using the following formula:

$$\pi(r+x)^2 - \pi r^2$$

where r is the radius of the trunk before the new ring is added. If it is supposed that a tree has a sequence of ring widths of 1 mm per year, it is possible to calculate the area increments created by the addition of 1 mm to the existing radius annually. A graph of these increments plotted against years is a straight line of a certain slope. The same exercise can be carried out with ring widths of 3 and 5 mm to produce lines of greater slopes. By reference to Hamilton and Christie (1971), these lines can be seen to represent tree growths of high (5 mm), medium (3 mm), and low (1 mm) annual productivity. Fossil ring-width series can be plotted on the same graph as the sequences for the 1-, 3-, and 5-mm ring widths, and an assessment can then be made of the productivity represented by the fossil series.

Frequently, specimens of fossil wood may lack the central pith region. In these cases, the original trunk diameter can be estimated from the radius of curvature of the rings. If the original trunk had been very large, the outer rings may show virtually no curvature. However, it is still possible to estimate the productivity by plotting the sequence of area increments and comparing the slope with the 1-, 3-, and 5-mm ones as described previously. Clearly, it is better if these calculations are carried out on a large number of wood specimens from the same site, avoiding branch wood as far as possible, so that the average productivity can be calculated.

Estimation of Land Productivity

A knowledge of the trunk wood (bole wood) productivity enables an estimate to be formed of the total productivity of the entire forest. A number of authors (Forrest and Ovington 1970; Jordan 1971; Kanninen et al. 1982; Ovington 1961; Whittaker 1966) have studied extant forests composed of a wide variety of tree species and in many parts of the world. From their published data, it is evident that trunk wood productivity is about 40–50% of total forest productivity. Bazilevich et al. (1971) have produced world maps of forest productivity and interesting comparisons may be made between fossil forest productivity at the palaeolatitudes of growth and that of extant forests in their present day latitudes.

Permian

There are many records of Permian fossil wood from the Transantarctic Mountains in Antarctica at a palaeolatitude of about 70°S (Creber and Chaloner 1987; Doumani and Long 1962; Maheshwari 1972). An examination of a number of specimens of this wood collected by the late James Schopf (presently in the collections of Ohio State University,

Columbus, Ohio), reveals that maximum ring widths of about 1 cm were formed by the trees and the average ring width was about 2.2 mm. The latter would indicate a productivity approaching the medium level (3 mm) indicated previously. Clearly, the palaeoclimate relatively close to the South Pole at the time of growth was very different from that of the present day.

Mesozoic and Early Tertiary

It is in these periods that the largest amount of fossil wood is found in polar latitudes. Some is located in the same regions of Antarctica as those in which the Permian specimens are seen. Examination of Triassic material (also at Ohio State University) has yielded an average ring width of 2.3 mm with a maximum of 10 mm.

However, the discovery of the remains of the in situ stumps of a fossil forest in the Lower Cretaceous of Alexander Island, Antarctic Peninsula (Jefferson 1982) at a palaeolatitude of about 70°S has enabled a detailed estimate to be made of its productivity (Creber and Chaloner 1984a, 1984b, 1985). This forest appears to have been producing about 5 tons/ha (0.5kg/m^2) of trunk wood with a mean ring width of 2.5 mm. Using the principle explained previously, this would have been equivalent to a total forest productivity of about 10 tons/ha. Such a forest productivity at the present day occurs only in much lower latitudes and in regions where rainfall and ambient temperatures do not restrict tree growth (Creber and Francis 1987). As in the Triassic, the mean ring width also indicates a medium productivity close to the 3-mm level. A large number of other specimens of fossil wood of Cretaceous and Early Tertiary age have been described from the Antarctic Peninsula region (Francis 1986). These also showed a medium level of productivity with a mean ring width of about 2.3 mm, an indication of a productivity very similar to the forest on Alexander Island.

The annual solar input at the present day (LaGrange 1963; Farman and Hamilton 1978) in the region of Alexander Island is about 3500 MJ/m^2/yr. Since the energy content of the 0.5 kg/m^2 of wood produced by the forest would have represented about 0.3% of the total solar energy input, there would appear to be no problem about an adequate light supply. The rather wide spacing of the fossil trees (563 trees/ha) would have tended to minimize mutual shading. In an attempt to indicate the likely incidence of shadows in the forest, a reconstruction was made (Creber and Chaloner 1984b) that showed the shadows cast by a single "model" coniferous tree at 3 times during a midsummer day at 70°S. It was clear that no trees would have been severely adversely affected. It must also be borne in mind that in a polar summer the sun makes a complete circuit of the sky so that one tree will not shade another for much of the day as in temperate latitudes. Even on a cloudless day when hard shadows are cast, there may still be as much as 25% of the incident radiation in the form of diffuse light (Monteith 1973) that could be intercepted by the tree crown from all directions.

It has been demonstrated by Vaartaja (1959, 1962) that certain modern tree species, growing near the northern tree line, represent photoperiodic ecotypes. These ecotypes are found to concentrate their annual growth into the short polar summer and to grow very poorly if transplanted to lower latitudes. Similarly, Gregory and Wilson (1968) showed that the production of xylem cells by the cambium of *Picea glauca* (white spruce) trees growing near College, Alaska (65°N), was at a much higher rate than in trees of the same species growing at Petersham, Massachusetts (42°N). This facility enables the high-latitude members of this species to complete a growth ring in a much shorter time during the short polar summer. Another characteristic of trees near the northern timber line is their tall conical shape. Jahnke and Lawrence (1965) have shown that this form of crown is ideally suited to intercepting the maximum amount of light energy from sunlight at low angles to the horizon. Were the high latitude trees of past Antarctic floras to have adopted this tall habit, it would have been a further adaptation enabling them to grow successfully in forests.

Conclusion

It is seen that the only necessary factor for trees to grow at very high latitudes at the present day is an increase in the average ambient temperature. All of the other factors required for the evolution of polar forest trees are within the scope of adaptations that are known to be available to modern trees, such as photoperiodic ecotypes, higher rates of cambial activity during the growing season, and an appropriate crown architecture. Additionally, the trees could have availed themselves of the facility to adopt the deciduous habit as an adaptation to the long winter darkness, although there is evidence that the evergreen habit might have been sustainable in a long, dark winter. Larson (1964) showed that conifer branches may be covered by black bags for up to 140 days and still retain the capacity to regenerate their pigments and recommence photosynthesis when the bags were removed. Furthermore, if the long winters had been relatively mild, the trees might have used the adaptation shown by the Bristlecone pines studied by Mooney *et al.* (1966). They brought seedlings of the pines down from the White Mountains of California to sea level and found that the trees drastically reduced their respiration rate in the higher ambient temperature.

References

Assmann E (1970) The Principles of Forest Yield Study. Pergamon Press, Oxford.

Bazilevich NI, Drozdov AV, and Rodin LE (1971) World forest productivity. *In* Duvigneaud P (ed) Productivity of Forest Ecosystems. UNESCO, Paris: 345–353

Creber GT (1986) Tree growth at very high latitudes in the Permian and Mesozoic. Colloque International sur l'Arbre. Naturalia Monspeliensa: 487–493

Creber GT and Chaloner WG (1984a) Climatic indications from growth rings in fossil woods. *In* Brenchley PJ (ed) Fossils and Climate. Wiley, Chichester: 49–74

Creber GT and Chaloner WG (1984b) Influence of environmental factors on the wood structure of living and fossil trees. Botanical Review 50:357–448

Creber GT and Chaloner WG (1985) Tree growth in the Mesozoic and early Tertiary and the reconstruction of palaeoclimates. Palaeogeography, Palaeoclimatology, Palaeoecology 52:35–60

Creber GT and Chaloner WG (1987) The contribution of growth rings to the reconstruction of past climates. *In* Ward RG (ed) Application of Tree-Ring Studies: Current Research in Dendrochronology and Related Areas. British Archaeological Reports, International Series 333:37–67

Creber GT and Francis JE (1987) Productivity in fossil forests. *In* Jacoby GC (ed) Proceedings of the International Symposium on Ecological Aspects of Tree-Ring Analysis. Department of Energy, Carbon Dioxide Research Division, Washington, D.C.:319–326

Doumani GA and Long WE (1962) The ancient life of the Antarctic. Scientific American 207(3):168–184

Farman JC and Hamilton RA (1978) Measurements of radiation at the Argentine Islands and Halley Bay, 1963–1972. British Antarctic Survey Scientific Report No. 99

Forrest WG and Ovington JD (1970) Organic matter changes in an age series *Pinus radiata* plantation. Journal of Applied Ecology 7:177–186

Francis JE (1986) Growth rings in Cretaceous and Tertiary wood from Antarctica and their palaeoclimatic implications. Palaeontology 29:665–684

Gregory RA and Wilson BF (1968) A comparison of cambial activity of white spruce in Alaska and New England. Canadian Journal of Botany 58:687–692

Hamilton GJ and Christie JM (1971) Forest management tables (metric). Forestry Commission, Her Majesty's Stationery Office, London

Jahnke LS and Lawrence DB (1965) Influence of photosynthetic crown structure on potential productivity of vegetation, based primarily on mathematical models. Ecology 46:319–326

Jefferson TH (1982) Fossil forests from the Lower Cretaceous of Alexander Island, Antarctica. Palaeontology 25:681–708

Jordan CF (1971) Productivity of a tropical forest and its relation to a world pattern of energy storage. Journal of Ecology 59:127–143

Kanninen M, Hari P, and Kellomaki S (1982) A dynamic model for above-ground growth and dry matter production in a forest community. Journal of Applied Ecology 19:465–476

LaGrange JJ (1963) Trans-Antarctic Expedition 1955–58 Scientific Report No. 13. Meteorology I. Shackleton, South ice and the Journey across Antarctica. Transantarctic Expedition Committee

Larson PR (1964) Contribution of different aged needles to growth and wood formation of young red pines. Forest Science 10:224–238.

Maheshwari HK (1972) Permian wood from Antarctica and revision of some Lower Gondwana woods. Palaeontographica B, 138:1–43
Monteith JL (1973) Principles of Environmental Physics. Edward Arnold, London
Mooney HA, Billings WD and Brayton R (1966) Field measurements of the metabolic responses of bristle cone pine and big sagebrush in the White Mountains. Botanical Gazette 127:105–113.
Ovington JD (1961) Some aspects of energy flow in plantations of *Pinus sylvestris*. Annals of Botany 25:12–20
Vaartaja O (1959) Evidence of photoperiodic ecotypes in trees. Ecological Monographs 29:91–111
Vaartaja O (1962) Ecotypic variation in photoperiodism of trees with special reference to *Pinus resinosa* and *Thuja occidentalis*. Canadian Journal of Botany 40:849–856
Whittaker RH (1966) Forest dimensions and productivity in the Great Smokey Mountains. Ecology 47:102–121

5—Triassic Terrestrial Vertebrate Faunas of Antarctica

William R. Hammer

Since the first discovery of a fragmentary amphibian jaw at Graphite Peak in 1968 (Barrett et al. 1968), five expeditions have searched for terrestrial vertebrate fossils in the Transantarctic Mountains (Elliot et al. 1970, Colbert 1982, Cosgriff et al. 1978, Hammer et al 1986, 1987). During this interval, 11 productive sites (Fig. 5.1) in the central portion of the range, near the Beardmore and Shackleton Glaciers, have been identified. Other areas explored that have not yielded vertebrates include the Allan Hills near the Dry Valleys of Southern Victoria Land and sections of Northern Victoria Land. (Hammer and Zawiskie 1982, Hammer 1987).

The vertebrate collecting groups have concentrated their efforts on the Permo-Triassic portion of the Beacon Supergroup, which includes extensives sequences of mainly fluvial terrestrial sediments (Barrett et al. 1986, Elliot et al. 1972). Of the various stratigraphic units searched within this sequence, only the Triassic Fremouw Formation has yielded fossil bone (Cosgriff et al. 1978, Hammer and Zawiskie 1982, Hammer et al. 1986).

The Fremouw Formation was named by Barrett (1969) for a 614-m-thick sequence at Fremouw Peak near the Beardmore Glacier (Figs. 5.1 and 5.2). This formation can be traced from the Beardmore Glacier area to the Shackleton Glacier (Collinson and Elliot 1984), and Barrett has suggested that the maximum thickness may approach 800 m in the area of Graphite Peak (see Fig. 5.1) (Barrett 1969, Collinson and Elliot 1984). The formation has informally been divided into lower, middle, and upper members and mainly consists of alternating sandstone and mudstone (Fig. 5.2, Barrett 1969, Collinson and Elliot 1984). Since this is the only unit in which terrestrial vertebrates have been found and it is best exposed in the central Transantarctic Mountains, it is not surprizing that all of the collections come from this region.

The Fremouw was deposited by a major fluvial system that drained the Gondwanian orogen and Antarctic craton during the Triassic (Collinson et al. 1981). Sediment accumulated in a foreland basin wedged between the craton and the orogenic belt. Paleocurrent analyses indicate a principal flow direction to the northwest, toward North Victoria Land and Tasmania. Several flood plain and channel facies have been recognized throughout the formation (Collinson et al. 1981), with the upper Fremouw showing a relative paucity of flood plain deposits compared to the lower two members.

At 10 of the 11 vertebrate localities, all except Gordon Valley (Fig. 5.1), the bones occur in the first 100 m of the lower member of the Fremouw (Fig. 5.2). Of these 10 sites, Shenk Peak, *Thrinaxodon* Col, Collinson Ridge, and Kitching Ridge, near the Shackleton Glacier, and Graphite Peak and Coalsack Bluff, near the Beardmore Glacier (Fig. 5.1), have been the most productive. Bones are much less abundant at the other four localities; Lamping Peak, for example, has produced only one fragment, and the three pieces

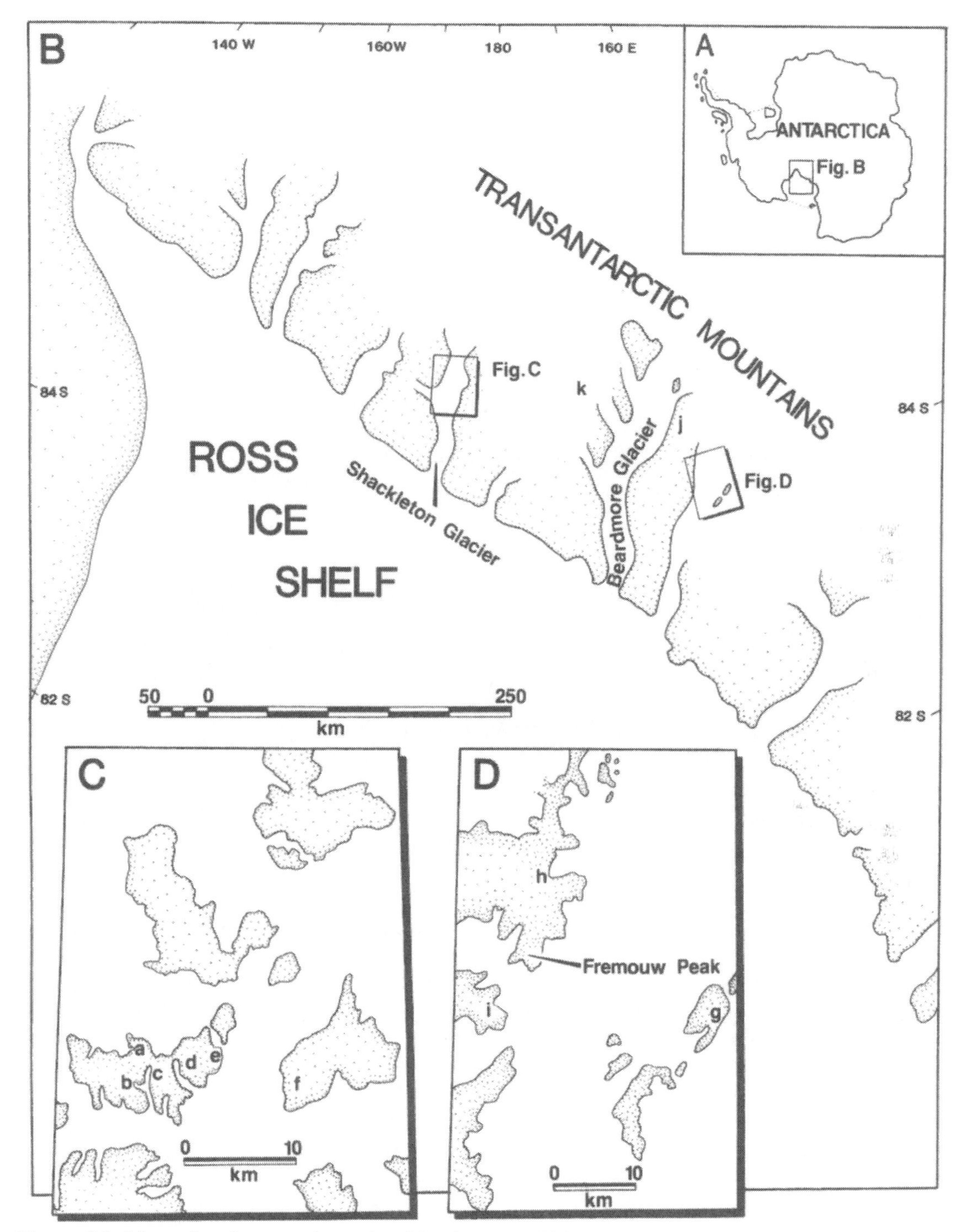

Figure 5.1. Maps showing vertebrate localties in the Central Transantarctic Mountains: a, Mt. Kenyon; b, *Thrinaxodon* Col; c, Shenk Peak; d, Collinson Ridge; e, Halfmoon Bluff; f, Kitching Ridge; g, Coalsack Bluff; h, Gordon Valley; i, Lamping Peak; j, Willey Point; and, k, Graphite Peak.

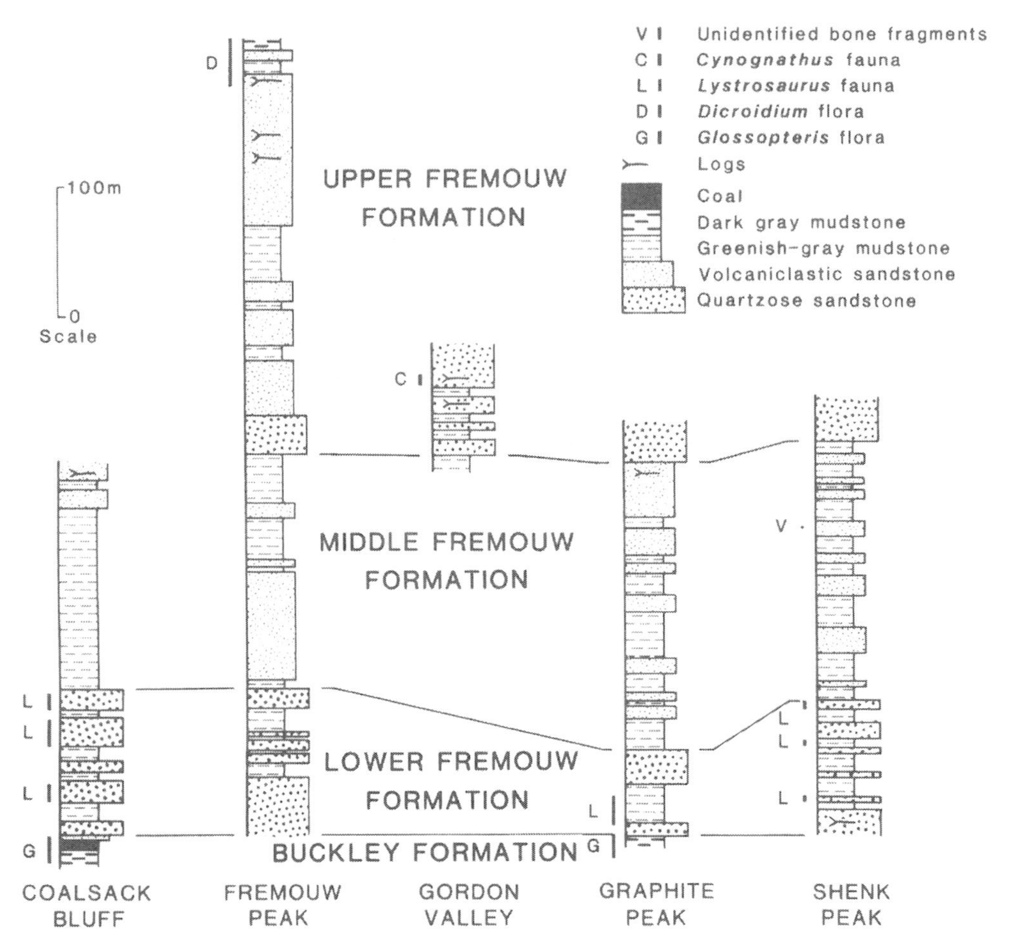

Figure 5.2. Stratigraphic sections correlating exposures of the Fremouw Formation at several key vertebrate localities to the type section at Fremouw Peak.

collected at Willey Point came from what could be an erratic block. Both Halfmoon Bluff and Mt. Kenyon have also produced smaller amounts of fragmentary remains; however, these sites are very close to the bone-rich beds at Collinson Ridge and *Thrinaxodon* Col, respectively.

The fossil bones from the lower Fremouw occur in both channel sandstone and floodplain mudstone facies (Fig. 5.2). The composite fauna from the various localities is an exceptionally large and varied one that closely corresponds with the fauna of the *Lystrosaurus* Zone (Early Scythian or earliest Triassic) of South Africa (Colbert 1972, 1982; Kitching et al. 1972; Hammer and Cosgriff 1981, Cosgriff et al. 1982). The fauna of the *Lystrosaurus* Zone is more diverse (Kitching 1977), but this is probably because of the extensive collecting within this unit over more than the last century. Synapsid (mammal-like) reptiles common to both units include *Myosaurus gracilis, Lystrosaurus murrayi, L. curvatus, L. maccaigi, Thrinaxodon liorhinus,* and *Ericiolacerta parva*. Other conspecific reptiles include the procolophonid *Procolophon trigoniceps* and the eosuchian *Prolacerta broomi* (Colbert 1982). Shared forms on the family level include galesaurid reptiles and lydekkerinid, rhytidosteid, brachyopid and capitosaurid amphibians (Cosgriff and Hammer 1983, 1984). Thecodonts, an order ancestral to the dinosaurs, are also represented in both faunas; however, the Antarctic thecodont is known only from very fragmentary postcranial material (Cosgriff 1983). A single partial tooth plate of the lungfish *Ceratodus* from the Shackleton Glacier area (Dziewa, 1980) represents the only group from the lower Fremouw not found in the *Lystrosaurus* Zone. Illustrations of most lower Fremouw taxa are provided in Colbert's (1982) excellent review of this fauna.

At the Gordon Valley locality (Figs. 5.1, 5.3), discovered during the austral summer of 1985–86, fossil vertebrates are found much higher stratigraphically than at any of the other sites. The bones occur as large clasts in a poorly sorted 8-m-thick quartzitic channel sand near the base of the upper member of the Fremouw Formation (Fig. 5.2). Log impressions and fragments of fossil wood are found at the same level, a feature not typical of the lower Fremouw; although wood does occur near the bone levels at some places in the lower part of the section. Since the Gordon Valley locality was found near the end of the 1985–86 season, less than two weeks of excavation was accomplished. During that time, nearly 50 specimens, mainly of large disarticulated cranial fragments, were recovered. Approximately half of these are identifiable to at least the family level, and in many cases, genus and species placements have also been possible.

As expected from its stratigraphic position, the collected assemblage from the Gordon Valley contains both therapsid reptiles and temnospondyl amphibians somewhat younger in age than the *Lystrosaurus* fauna of the lower Fremouw. The large, carnivorous cynodont reptile *Cynognathus* (Fig. 5.4) is represented by a partial mandible (American Museum of Natural History, AMNH # 24422) that shows the extensive development of the adductor fossa characteristic of this genus. In addition, the general size, shape, and nature of the teeth correspond to *Cynognathus*. *Cynognathus* is known from South African and South American deposits that have been assigned a Late Scythian (late Early Triassic) age. In fact, in South Africa, the next terrestrial fauna zone above the *Lystrosaurus* Zone has been traditionally called the *Cynognathus* Zone; however, Keyser and Smith (1978) have suggested the name *Kannemeyeria* Zone for this level, since this large anomodont is more common in the Late Scythian African deposits

Figure 5.3. Vertebrate locality in the Gordon Valley.

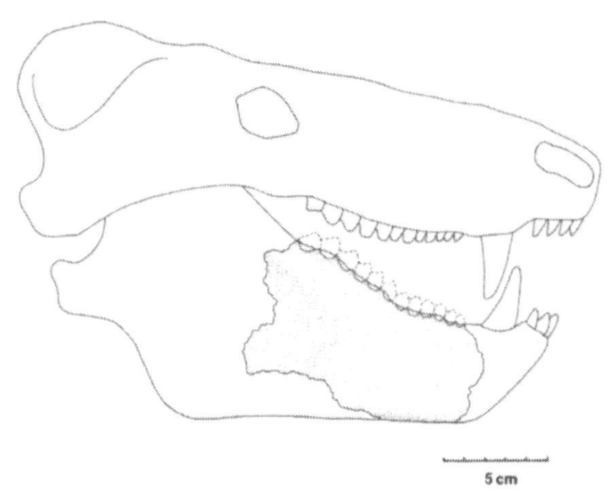

Figure 5.4. Skull drawing of *Cynognathus* sp., stipled section of the mandible represents the Antarctic specimen.

than the carnivorous *Cynognathus*. The Gordon Valley assemblage also includes a maxillary piece (AMNH # 24403) with a large tusk that is referrable to the anomodont family Kannemeyeriidae (Fig. 5.5); however, the specimen is too fragmentary to determine whether or not the Antarctic animal is congeneric with *Kannemeyeria*.

The occurrence of both *Cynognathus* and a large kannemeyeriid indicate that the new fauna from the upper Fremouw Formation may be approximately age equivalent with the *Cynognathus* (*Kannemeyeria*) Zone of South Africa. However, two new therapsids from this fauna suggest that this may not be the case. The mandible (AMNH # 24407) of a new carnivorous cynodont has an even

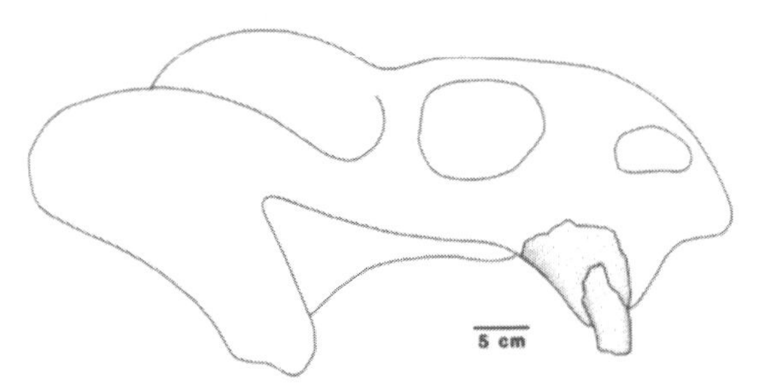

Figure 5.5. Skull drawing of *Kannemeyeria* from South Africa, stipled portion represents Antarctic specimen referred to the same family.

deeper, larger adductor fossa than the well-developed fossa typical of *Cynognathus*. Thus, this animal, which is obviously derived from a *Cynognathus* grade cynodont, shows a progressive feature that is not found in late Scythian forms from South Africa or elsewhere.

The right maxilla (AMNH # 24421) of a second new genus has features that relate it to the gomphodont cynodonts. The size of this animal is equivalent to the largest of the diademodontid gomphodonts typical of the *Cynognathus* Zone of South Africa. Also, the postcanine tooth row is offset medially from the canine, the postcanine teeth are transversely widened, and the snout is narrow relative to the orbital region of the skull, all general features typical of the herbivorous diademodontids. However, unlike members of this group, this animal has a sizable diastema behind the canine, giving it fewer postcanine teeth. In addition, all of the postcanine teeth are of the gomphodont type (i.e., transversely widened), the simple anterior type and the sectorial posterior type of postcanines found in diademodontids are missing. Because of these features, the general nature of the dentition resembles that of typical Middle Triassic traversodontids (Crompton 1972, Kemp 1982), rather than the diademodontids. This animal also appears to lack the large rostral maxillary salt gland depression (Grine et al. 1979) found in *Diademodon*.

Thus, while two of the four upper Fremouw therapsids correspond to taxa typical of the *Cynognathus* Zone, the other two appear to represent more derived forms. Perhaps Late Scythian Antarctic reptiles were simply more progressive than contemporary African animals. Some of the Middle Triassic therapsids typical of other southern continents may even have evolved in Antarctica earlier in the period and then later migrated to other areas of Gondwana. It is also possible that the upper Fremouw fauna could be slightly younger than the *Cynognathus* Zone, hence it contains holdovers from the late Scythian mixed with younger, more derived therapsids. Further collection of this assemblage should lead to an expanded faunal list that in turn will make the biostratigraphic position of the upper Fremouw much clearer.

Two of the three temnospondyl amphibians currently under description from the Gordon Valley locality belong to the family Capitosauridae; the third appears to be a member of the Benthosuchidae. The capitosaurs are both very large, one nearly complete skull (AMNH # 24411) is 61 cm in length (Figs. 5.6 and 5.7). This specimen shows an open otic notch typical of more primitive members of the family, such as *"Parotosuchus" (Parotosaurus)* (Morales, 1987a), yet it is generally larger and somewhat broader relative to its length than most parotosuchians. Overall, captiosaur taxonomy is in a state of flux, and good diagnostic

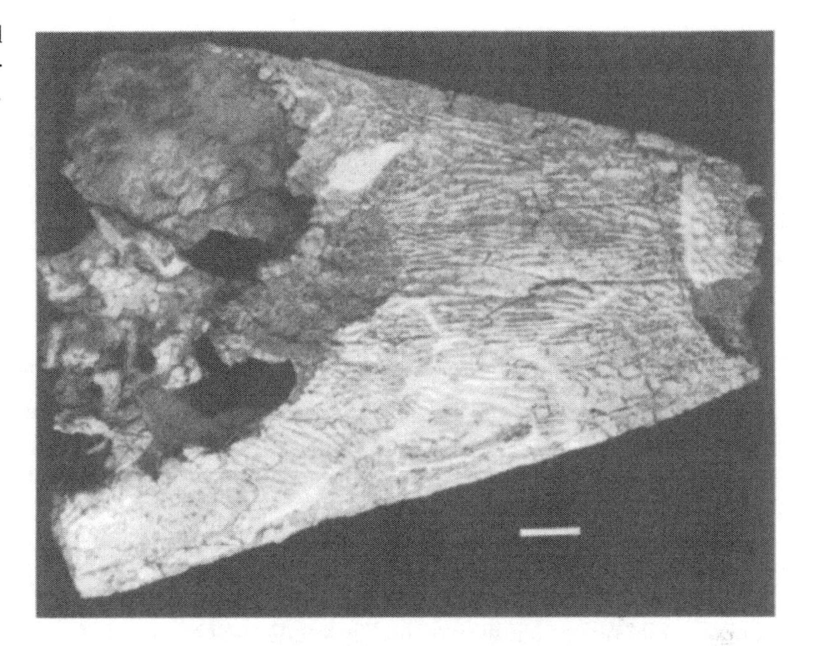

Figure 5.6. Capitosaurid skull from the Gordon Valley, dorsal view (AMNH # 24411). Bar scale = 5 cm.

Figure 5.7. Capitosaurid skull from the Gordon Valley, ventral view (AMNH # 24411). Bar scale = 5 cm.

characters have not been defined for many species, particularly the many open otic notch forms referred to as "*Parotosuchus*.." As Morales (1987a) points out, this genus is based on a plesiomorphic (primitive) character (the open notch) and, thus, represents a grade taxon that is not monophyletic. The Antarctic specimen may represent a new genus; however, a thorough revision of the family is needed before a complete analysis of relationships between this animal and open notch capitosaurids from other continents can be accomplished.

A large snout fragment (AMNH # 24419) of a second genus from the Gordon Valley indicates a total skull length of over 1 m (Fig. 5.8). This animal represents the largest terrestrial vertebrate known from the Antarctic mainland. Its unusual features include an odd-shaped anterior palatal vacuity that is narrow in the middle and widens into two pockets laterally and very large anterior palatal tusks that form a row across the vomers. The nature of the anterior palatal vacuity suggests affinities with the benthosuchids as well as the capitosaurids. Some specimens of *Benthosuchus* (Shishkin and Lozovskiy 1979) and the larger, more derived capitosaurids (Welles and Cosgriff 1965) tend to retain a single vacuity, while others show two smaller vacuities. This specimen (Fig. 5.8) shows a pinching of a single vacuity into two pockets, possibly a transitional state between the single and double forms. Overall, its shape is most similar to that of some specimens of *Benthosuchus;* however, the external nares are not terminal and the snout is not narrow, two features typical of this genus. These characters better match the capitosaurids, hence, provisionally, this animal is assigned to that family.

The row of very large vomerine tusks found on this specimen is not characteristic of either the capitosaurids or the benthosuchids, both of which have a row of relatively small vomerine teeth behind the anterior palatal vacuity. In benthosuchids, this tooth row forms a posteriorly pointed V, in capitosaurids it follows the posterior border of the anterior palatal vacuity. Although the teeth are fewer in number and much larger, the general arrangement of the row of tusks on the Antarctic specimen is again more like that of the capitosaurids, since it follows the posterior border of the vacuity. Future revision of these families and additional, more complete cranial material should clarify the affinities of this specimen; however, it is apparent from its unique features that it does represent a new genus.

At least one fairly large jaw (Fig. 5.9, AMNH # 24415) pertains to the family Benthosuchidae, although this animal is somewhat smaller than either of the capitosaurids.

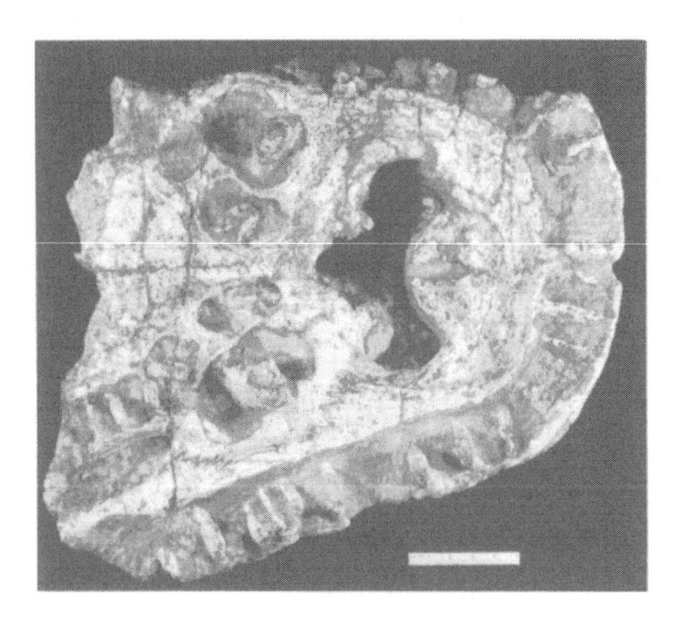

Figure 5.8. Ventral view of large capitosaurid snout tip from the Gordon Valley (AMNH # 24419). Bar Scale = 5 cm.

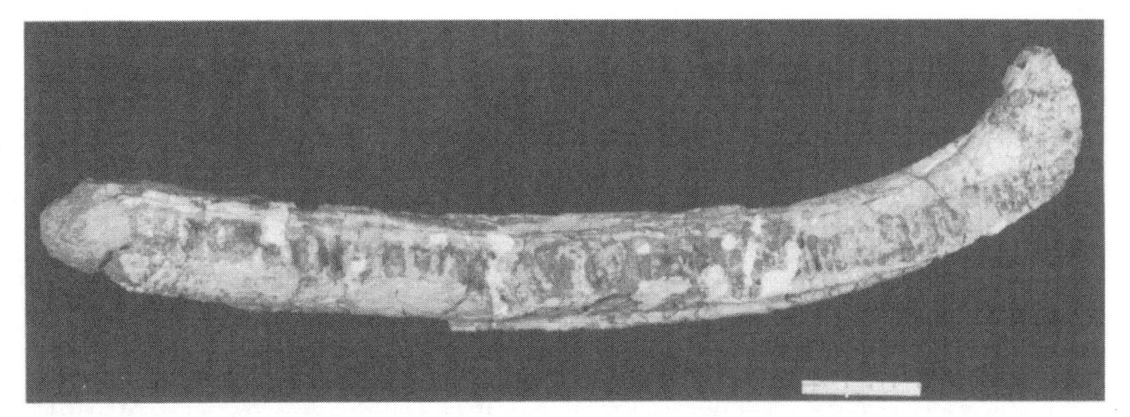

Figure 5.9. Occlusal view of benthosuchid mandible from the Gordon Valley (AMNH # 24415). Bar Scale = 5 cm.

The jaw lacks an anterior meckelian foramen, an apomorphic (derived) character for that family according to Jupp and Warren (1986). Benthosuchids are extremely rare in Gondwana, in fact, the only other southern occurrences are questionable (Romer 1947, Tripathi 1969, Morales 1987b); hence, the Antarctic specimen may represent the only true benthosuchid known from Gondwana.

The Capitosauridae and the Benthosuchidae share a common ancestory, probably within the Rhinesuchidae (Cosgriff 1984) and, consequently, are placed in the same superfamily, the Capitosauroidea. Thus, all three of the temnospondyls from the Gordon Valley assemblage belong to a single evolutionary complex within the Temnospondyli. This superfamily is represented only by indeterminant capitosaurid fragments in the lower Fremouw, yet other typical early Triassic temnospondyl groups, pertaining to separate lineages, are abundant in the lower part of the section. Future collecting should produce taxa of other temnospondyls in the upper Fremouw, but for now, the paucity of capitosauroids in the lower Fremouw coupled with the lack of other temnospondyl groups in the upper Fremouw remains somewhat of an anomaly.

Until the Gordon Valley discovery, efforts of the past 20 years had produced a diversity of age-equivalent Triassic vertebrates from the Transantarctic Mountains. This new assemblage has added the dimension of time to the study of Antarctic vertebrates. It will now be possible to do more than compare Antarctic forms with contemporary animals from other continents, evolution and change within the Antarctic communities can be studied. In addition, it is very probable that in the future additional new discoveries of faunas intermediate or younger in age than the two discussed here will be made, since promising sediments exist throughout the section and many areas remain to be searched. Further collection and study of these faunas may also clarify Antarctica's role as a possible center of origin and distribution for certain Triassic groups. In short, vertebrate paleontological study of the Transantarctic Mountains is still in its infancy, and in the future, Antarctica should continue to make significant contributions to our understanding of Mesozoic life.

References

Barrett PJ (1969) Stratigraphy and petrology of the mainly fluviatile Permian and Triassic Beacon rocks, Beardmore Glacier area, Antarctica. Inst. Polar Studies, Ohio State University, publ. 34, 132 pp.

Barrett PJ, Baillie RJ, and Colbert EH (1968) Triassic amphibian from Antarctica. Science 161:460–462

Barrett PJ, Elliot DH, and Lindsay JF (1986) The Beacon Supergroup and Ferrar Group in the Beardmore area. *In* Turner MD and Splettstoesser JF (eds), Geology of the Central Transantarctic Mountains: Antarctic Research Series, Vol. 36, American Geophysical Union, Washington D.C., pp 339–429

Colbert EH (1972) Antarctic Gondwana Tetrapods. Internat. Union Geol. Sci., Second Gondwana Symp., South Africa, 1970:659–664

Colbert EH (1982) Triassic vertebrates in the Transantarctic Mountains. *In* Turner MD and Splettstoesser JF (eds), Geology of the Central Transantarctic Mountains: Antarctic Research Series, Vol. 36, American Geophysical Union, Washington D.C., pp 339–429

Collinson JW and Elliot DH (1984) Triassic stratigraphy of the Shackleton Glacier area. *In* Turner MD and Splettstoesser JF (eds), Geology of the Central Transantarctic Mountains: Antarctic Research Series, Vol. 36, American Geophysical Union, Washington D.C., pp 103–117

Collinson JW, Stanley KO, and Vavra CL (1981) Triassic fluvial depositional systems in the Fremouw Formation, Cumulus Hills, Antarctica. *In* Cresswell MM and Vella P (eds), Gondwana Five, AA Balkema, Rotterdam, pp 141–148

Cosgriff JW (1983) Large thecodont reptiles from the Fremouw Formation. Antarctic Jour. U.S. 18(5)52–55

Cosgriff JW (1984) The temnospondyl labyrinthodonts of the earliest Triassic. Jour. Vert. Paleon. 4(1):30–46

Cosgriff JW and Hammer WR (1983) The labyrinthodont amphibians of the earliest Triassic from Antarctica, Tasmania and South Africa. *In* Oliver RL, James PR, and Jago JB (eds), Antarctic Earth Science, Australian Academy of Science, Canberra, pp. 590–592

Cosgriff JW and Hammer WR (1984) New material of labyrinthodont amphibians from the Lower Triassic Fremouw Formation of Antarctica. Jour. Vert. Paleon. 4(1):47–56

Cosgriff JW, Hammer WR, Kemp NR, and Zawiskie JW (1978) New Triassic vertebrates from the Fremouw Formation of the Queen Maud Mountains. Antarctic Jour. U.S. 13(4):23–24

Cosgriff JW, Hammer WR, and Ryan WJ (1982) The pangaean reptile, *Lystrosaurus maccaigi*, in the Lower Triassic of Antarctica. Jour. Paleon. 56(2):371–385

Crompton AW (1972) Postcanine occlusion in cynodonts and tritylodontids. Bull. British Mus. Nat. Hist. (Geol.) 21:29–71

Dziewa TJ (1980) Note on a dipnoan fish from the Triassic of Antarctica. Jour. Paleon. 54(2):488–490

Elliot DH, Colbert EH, Breed WJ, Jensen JA, and Powell JS (1970) Triassic tetrapods from Antarctica: evidence for continenal drift. Science 169:1197–1201

Elliot DH, Collinson JW, and Powell JS (1972) Stratigraphy of Triassic tetrapod-bearing beds of Antarctica. *In* Adie RJ (ed), Antarctic Geology and Geophysics, Universitetetsforlaget, Oslo, pp. 387–392

Grine FE, Mitchell D, Gow CE, Kitching JW, and Turner BR (1979) Evidence for salt glands in the Triassic reptile *Diademodon*. Palaeont. Afr. 22:35–39

Hammer WR (1987) Takrouna Formation fossils of Northern Victoria Land. *In* Stump E (ed), Geological Investigations in Northern Victoria Land, American Geophysical Union, Antarctic Research Series 46, pp. 243–247

Hammer WR and Cosgriff JW (1981) *Myosaurus gracilis*, an anomodont reptile from the Lower Triassic of Antarctica and South Africa. Jour. Paleon. 55(2):410–424

Hammer WR and Zawiskie JM (1982) Beacon fossils from northern Victoria Land. Antarctic Jour. U.S. 17(5):13–15

Hammer WR, Ryan WJ, Tamplin JW, and DeFauw SL (1986) New vertebrates from the Fremouw Formation (Triassic) Beardmore Glacier region, Antarctica. Antarctic Jour. U.S. 21(5):24–26

Hammer WR, Ryan WJ, and DeFauw SL (1987) Comments on the vertebrate fauna from the Fremouw Formation (Triassic), Beardmore Glacier region, Antarctica. Antarctic Jour. U.S. 22(5):32–33

Jupp R and Warren AA (1986) The mandibles of the Triassic temnospondyl amphibians. Alcheringa 10:99–124

Kemp TS (1982) Mammal-like reptiles and the origins of mammals. Academic Press, London, 363 p.

Keyser AW and Smith RMH (1978) Vertebrate biozonation of the Beaufort Group with special reference to the western Karoo Basin. Annals Geol. Sur. S. Africa 12:1–35

Kitching JW (1977) The distribution of the Karroo vertebrate fauna. Bernard Price Institute for Palaeontological Research, Memoir 1, 131 p

Kitching JW, Collinson JW, Elliot DH, and Colbert EH (1972) *Lystrosaurus* Zone (Triassic) fauna from Antarctica. Science 175:524–526

Morales M (1987a) Terrestrial fauna and flora from the Triassic Moenkopi Formation of the southwestern United States. Jour. Arizona-Nevada Academy Science 22:1–19

Morales M (1987b) A cladistic analysis of capitosauroid labyrinthodonts: preliminary results (abstract). Jour. Vert. Paleo. 7 (3 supplement):21a

Romer AS (1947) Review of the Labyrinthodontia. Bull. Mus. Comp. Zoo., Harvard, 99(1):1–367

Shishkin MA and Lozovskiy VR (1979) A labyrinthodont from the Triassic deposits in the south of the Soviet Pacific Maritime Province. Doklady Adademii Nauk SSSR 246(1):201–205

Tripathi C (1969) Fossil labyrinthodonts from the Panchet Series of the Indian Gondwanas. Mem. Geol. Sur. India 38:1–45

Welles SP and Cosgriff JW (1965) A revision of the labyrinthodont family Capitosauridae. Univ. California Publ. Geo. Sci. 54:1–148

6—Proterozoic and Paleozoic Palynology of Antarctica: A Review

Geoffrey Playford

Introduction

This chapter provides a summary account of the occurrence and significance of spores, pollen grains, and other palynomorphs that have been recorded from pre-Mesozoic Antarctic sequences ranging in age from Proterozoic (possibly Riphean) to Permian. Occurrences of Paleozoic palynomorphs recycled into younger materials, especially seafloor glacial sediments of the Antarctic continental shelf, are also reviewed. Palynological research on Antarctic pre-Mesozoic palynofloras began some three decades ago and has been directed toward four main ends: (1) stratigraphic correlation and dating of the host sediments that are often devoid of other fossils or at least of age-diagnostic forms; (2) elucidation of the past vegetational and climatic history of the continent; (3) especially for the late Paleozoic, assessment of the extent of floristic similarity with sequences of other southern landmasses to help establish predrift plate juxtapositions within Gondwana; and (4) where encountered as recycled components, allowing inferences to be made concerning the geographic position, age, and metamorphic/maturation level of the (often now hidden) rock sequences from which they derived as erosion products. How far these objectives have presently been realized will be evident from the geochronologically arranged text that follows. Apart from the obviously limited ice-free exposure or accessibility of the Antarctic bedrock, rocks of Triassic and older age have been widely affected by thermal metamorphism from Jurassic volcanic intrusions and extrusions. This has had adverse consequences on palynomorph preservation, as for instance, in upper Paleozoic–Triassic (Victoria Group) strata of the Transantarctic Mountains (Kyle 1977a). Nevertheless, the record shows that, with careful collecting and painstaking laboratory processing, moderately well-preserved or at least identifiable palynological microfossils can be obtained from Antarctic pre-Mesozoic rocks.

While drawing largely upon published work, this chapter includes some additional identifications and illustrations of Permian spores and pollen grains from the Bainmedart Coal Measures (Prince Charles Mountains) samples originally studied by Balme and Playford (1967). The figured specimens are deposited in the micropaleontological collections of the Department of Geology and Mineralogy, University of Queensland, Brisbane, Australia, with registered numbers Y.5758–Y.5809 (see the appendix at the end of this chapter).

Principal Antarctic localities mentioned in the text are shown in Fig. 6.1; more precise details are obtainable from the published sources cited.

Proterozoic–Cambrian

Middle to Upper Proterozoic acritarchs, mostly sparse and poorly preserved, have been reported from scattered localities in

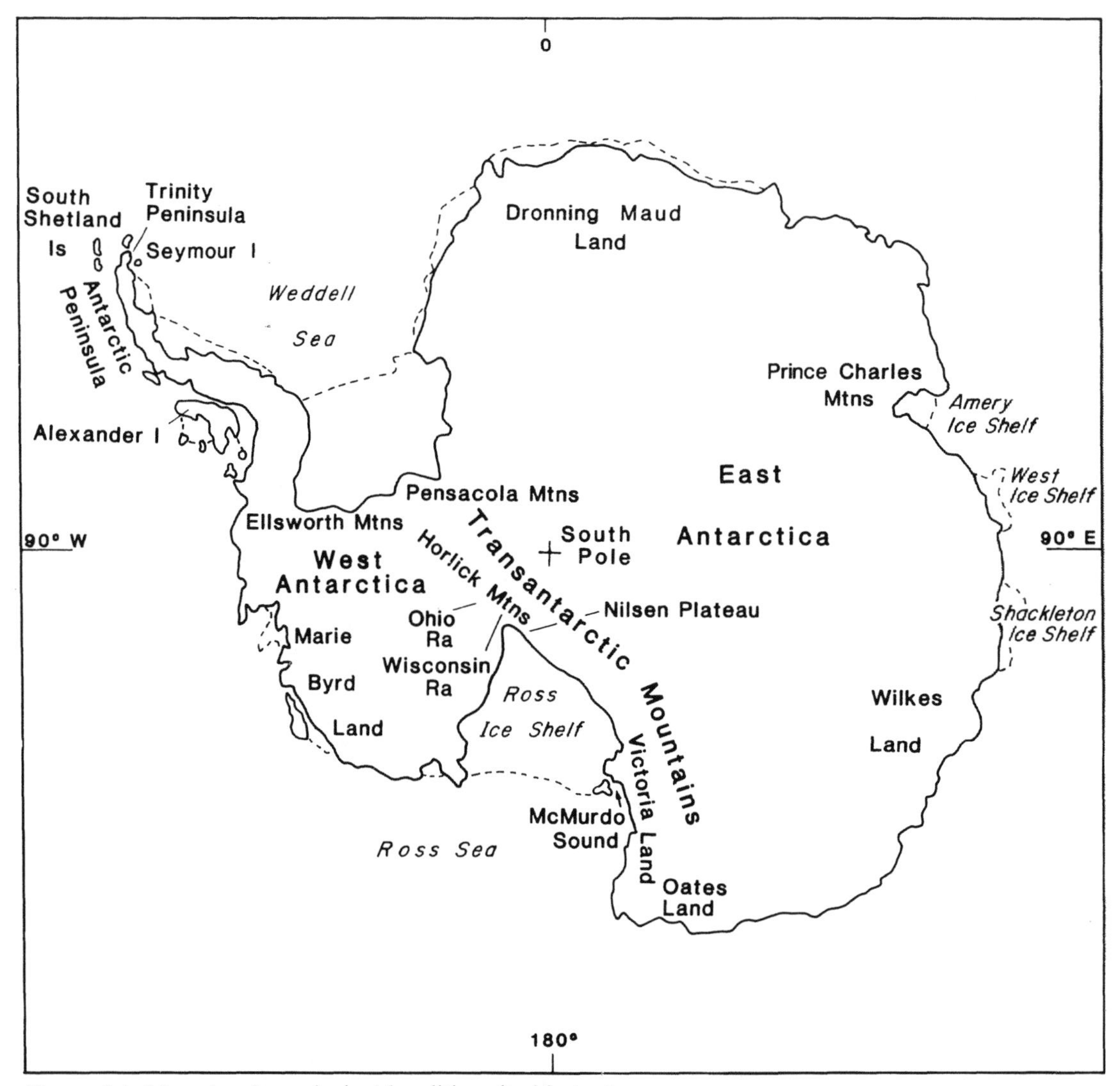

Figure 6.1. Map showing principal localities cited in text.

metasedimentary complexes (of phyllites, slates, metagreywackes, marbles, metaquartzites, argillites) representative of the younger Precambrian basement of Antarctica. Iltchenko's (1972) records are of Riphean acritarchs from the Oates Coast (northern Victoria Land), western Marie Byrd Land, and southern Prince Charles Mountains (Berg Group) and of Vendian forms from western Dronning Maud Land and Clarence Island (South Shetland Islands). However, some caution needs to be exercised with respect to Iltchenko's (1972) data because, as noted by E.M. Truswell (pers. comm.), some of the illustrations (Iltchenko 1972, Figs. 1, 2) appear to be of Mesozoic palynomorphs. Moreover, the Riphean (pre-Vendian) age adduced for the Oates Coast sample, representative of the Robertson Bay Group, could be regarded as anomalously old (cf. Cooper et al. 1982, p. 631). Korotkevich and Timofeev (1959) listed and illustrated acritarch taxa, regarded as Riphean to Early Cambrian in age, from the clastic greenschist-metamorphosed Sandow Group, and from morainic boulders also, in the general Shackleton and West Ice Shelf environs, coastal East Antarctica.

From the northern Victoria Land basement

complex, acritarchs (in places with brachiopods, trilobites, molluscs, and ichnofossils) were reported initially by Vidal in Cooper et al. (1982) from the marine-turbiditic Robertson Bay Group (three localities) and the marine-volcaniclastic Molar Formation (Sledgers Group; several localities), representing the Robertson Bay and Bowers terranes, respectively (Stump et al. 1986). The palynofloral assemblages, though indifferently preserved, were dated as Early Cambrian, possibly extending into the latest Proterozoic (Vendian). However, revised identifications of some of the acritarchs and lithostratigraphic reattribution of one supposed Robertson Bay Group sample site (Mount McCarthy, now Molar Formation) led Cooper et al. (1983) to revised datings, as follows: Robertson Bay Group, Vendian to Middle Cambrian; Molar Formation, Middle Cambrian, supported also by associated trilobite and brachiopod faunas. Further biostratigraphic information is derived from studies by Burrett and Findlay (1984; conodonts) and Wright et al. (1984; conodonts, trilobites) on an olistolith considered to be from within the upper part of the Robertson Bay sequence. The data imply that the group extends as high as the uppermost Cambrian or lowermost Ordovician.

A 10-year study of the acritarch content of 274 (145 productive) upper Precambrian and Cambrian samples, collected by Soviet Antarctic Expeditions to East and West Antarctica, was very briefly summarized by Smirnova (1983). Four acritarch complexes—dated respectively as preRiphean, Riphean, Vendian, and Cambrian—were cited in terms of their taxonomic composition and generalized location. No illustrations of the acritarchs were provided.

Despite the evidently non- or poorly palyniferous nature of many Antarctic Proterozoic–lower Paleozoic rocks (Brown et al. 1982, p. 55), such data as are available suggest that acritarchs offer considerable scope for biostratigraphic utilization and hence for assisting in tectonic and other interpretations of terranes embodied by the Ross orogenic belt (e.g., Stump et al. 1986). There is a clear need for published identifications to be supported by photomicrographs and by normal systematic-descriptive documentation, even in the face of relatively poor preservation of the organic-walled microfossils.

Ordovician–Silurian

Fossil-bearing strata of Ordovician and Silurian ages are as yet unrecorded or unconfirmed from Antarctica (Webby et al. 1981; Rowell et al. 1987), that is, apart from the possibly earliest Ordovician upper part of the Robertson Bay Group mentioned earlier. It is possible, however, that supra-Cambrian rocks in the Bowers Mountains (northern Victoria Land) and in the Ellsworth Mountains (West Antarctica) could be partly Ordovician in age (Webby et al. 1981, p. 46); the same may be true of the Neptune Range (Pensacola Mountains), southern Victoria Land, and Dronning Maud Land. Palynological studies of such unfossiliferous or poorly fossiliferous strata, for example, the Crashsite Quartzite in the Ellsworth Mountains (Webers and Sporli 1983, Fig. 2), could well provide age-definitive data.

Devonian

There have been four published reports of palynomorphs from Antarctic sedimentary rocks ranging in age from Early to Late Devonian (Fig. 6.2). The oldest assemblage is from a sample of the lacustrine Terra Cotta Siltstone, very close to the base of the Beacon Supergroup, at Table Mountain, southern Victoria Land. As noted by Kyle (1977b), the palynoflora is abundant, very poorly preserved, of low diversity, and includes a strongly ribbed form (?*Emphanisporites* McGregor 1961), together with possible annulate and cavate miospores and "an overwhelming abundance of simple smooth-walled forms." On these morphological grounds and the generally diminutive nature of the palynomorphs, Kyle (1977b) proposed a post-Gedinnian, probably Early Devonian, age.

The Horlick Formation, a 50-m-thick, shallow marine unit with an abundant Early Devo-

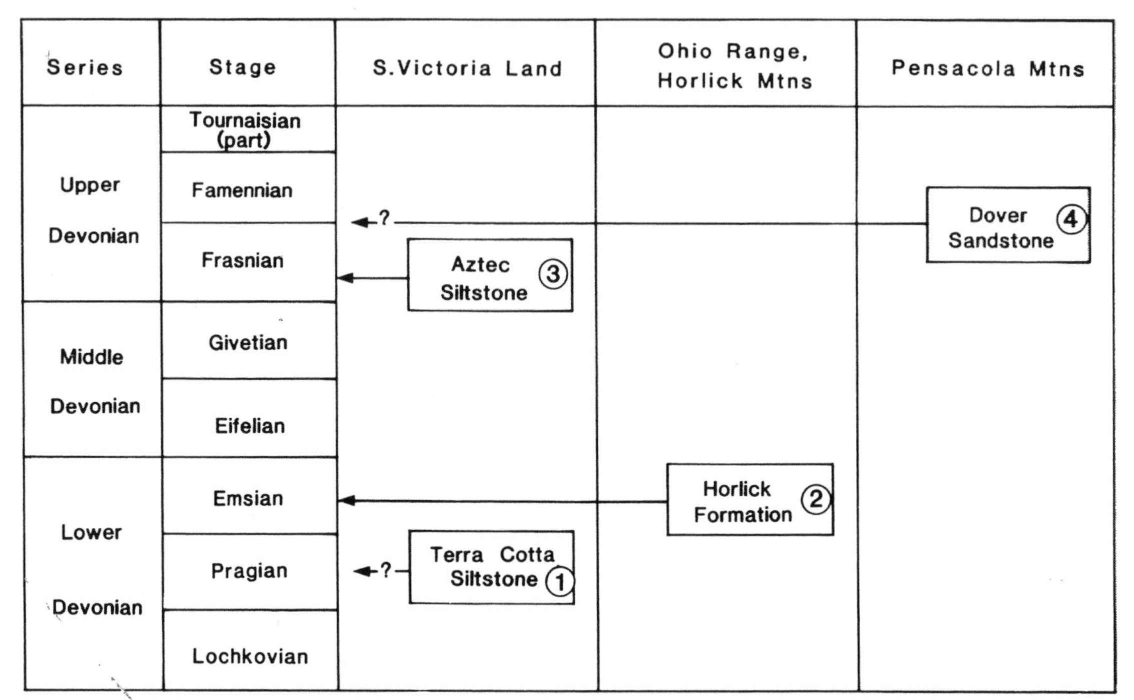

Figure 6.2. Devonian lithostratigraphic units in Antarctica from which palynological data have been published. Key to authors: 1, Kyle (1977b); 2, Kemp (1972c); 3, Helby and McElroy (1969); 4, Schopf in Schmidt and Ford (1969).

nian (early Emsian) invertebrate fauna and less common megaplant remains, was investigated palynologically by Kemp (1972c) from exposures in the Horlick Mountains (Transantarctic Mountains). Kemp described and illustrated 16 species of *sporae dispersae* (all but one trilete) and two chitinozoan specimens, informally called *Angochitina* sp. A and sp. B. Though carbonized and generally not well preserved, Kemp regarded the assemblage as supporting the faunally based age determination (Emsian); Richardson and McGregor (1986, p. 13) suggested a possible equivalence with their *Emphanisporites annulatus-Camarozonotriletes sextantii* Assemblage Zone (early and early late Emsian).

From appreciably higher in the Beacon Supergroup, Helby and McElroy (1969, p. 379; Figs. 3A–D, G) briefly reported and illustrated "a reasonable yield of spores" totaling at least 8 species from near the top of a 46-m-thick exposure of the freshwater alluvial Aztec Siltstone at Aztec Mountain, southern Victoria Land. A Late Devonian (Frasnian) age was adduced mainly from the abundance of the distinctive and widespread *Geminospora lemurata* Balme 1962, which is likewise extremely plentiful in the Frasnian Gneudna Formation of the Carnarvon Basin, Western Australia. *G. lemurata* is a possible progymnosperm derivative, perhaps allied with *Archaeopteris* (Playford 1983).

From carbonaceous interbeds within a coarse sandy sequence in the southernmost Patuxent Range (Pensacola Mountains), Schopf (in Schmidt and Ford 1969) mentioned the presence of plant fossils (microfossils *vide* Kyle 1977b, p. 1147) of probable Late Devonian age. The sequence was considered lithologically correlative with the nonmarine Dover Sandstone of the Neptune Range.

A megafossil flora, consisting principally of the lycopods *Drepanophycus* and *Haplostigma,* was described by Mildenhall (in Grindley et al. 1980) from erratic blocks judged to originate from a nearby deltaic metasedimentary rock sequence (Wilkins Formation) in coastal Marie Byrd Land, West

Antarctica. The metavolcanic rocks (Ruppert Coast Metavolcanics) conformably succeeding the Wilkins Formation had been regarded as young as Jurassic, so the floral dating (late Middle or early Late Devonian) was of considerable interest in providing evidence of the earliest known calcalkaline volcanic activity in this part of the southwest Pacific region (Grindley and Mildenhall 1980). The occurrence of *Haplostigma* connoted floral links for the Middle to Late Devonian with eastern Australia, South Africa, and South America, as well as with East Antarctica; and *Drepanophycus* reinforced the South African alliance. The plant-bearing argillite erratics were evidently too high ranking for palynological analysis.

In summary, the few published palynological studies of Antarctic Devonian strata, both marine and terrestrial in origin, indicate the presence of mainly miospores that appear to have derived from moderately diverse terrestrial vegetation; the marine Horlick Formation additionally contains rare chitinozoans, but (surprisingly) no acritarchs. The palynomorphs are characteristically dark colored (carbonized) and had been extracted with some difficulty from the evidently somewhat metamorphosed host rocks. For instance, the adverse condition of Kyle's (1977b) assemblage was probably caused by local heating from the Jurassic Ferrar Dolerite sill (Plume 1982). At the form-generic level especially, the Devonian spore assemblages imply floral cosmopolitanism, but at the specific level—especially in Kemp's (1972c) comprehensive study—there are suggestions of some degree of endemism.

Carboniferous–Permian

The bulk of Antarctic Paleozoic palynological studies has focused on Late Carboniferous (?)–Permian strata, representative of glacial, periglacial, and postglacial Lower Gondwana sequences; and particularly on stratigraphic-correlative and phytogeographic relationships with better known (and often more suitably palyniferous) successions elsewhere in Gondwana (see, e.g., Kemp 1975; Schopf and Askin 1980).

The occurrence of preglacial rocks of indubitable Carboniferous age has not been authenticated, although Grikurov and Dibner (1968) and Grikurov (1978) reported allegedly Early and mid Carboniferous miospores from the northern Antarctic Peninsula (Hope Bay) and from Alexander Island in rocks considered to belong to the Trinity Peninsula "Series" (now Group: Hyden and Tanner 1981) and to the possibly correlative LeMay Group (Tranter 1987), respectively. The plant microfossils were documented only as unillustrated listings of species known previously from the Lower to mid Carboniferous of the Moscow and Donetz Basins and of Kazachstan. Schopf (1973) and Truswell (in press) have justifiably expressed doubt on the veracity of these findings. The laboratories in which the Antarctic samples were processed would surely have been a potential source of contamination from Soviet Carboniferous samples; and it would seem to be stretching the concept of global Carboniferous floral uniformity (Chaloner and Lacey 1973) rather too far to give unqualified credence to the Antarctic presence of especially the more distinctive Soviet miospore taxa. Elsewhere in the Southern Hemisphere, there is in fact evidence of floral regionalism in at least the species-level complexion of Lower Carboniferous palynofloral assemblages (e.g., Playford 1985). The full age range of the thick, multiply deformed, turbiditic Trinity Peninsula Group remains unresolved at this stage. But marine megafossil and radiolarian faunas reported from the LeMay Group of Alexander Island suggest that it spans at least the mid-Triassic to mid-Cretaceous (Edwards 1982; Burn 1984; Thomson and Tranter 1986).

In East Antarctica, the flat-lying, nonmarine, largely Permian–Triassic Victoria Group constitutes the upper part of the Beacon Supergroup and is well developed throughout the Transantarctic Mountains (see, e.g., Elliot 1975; Barrett 1981). Much of the sequence has been thermally metamorphosed by tholeiitic dolerite intrusions and comagmatic basalt flows, all belonging to the widespread Jurassic

Ferrar Group; this has had deleterious effects on the preservational quality of Victoria Group palynomorphs, which tend to be carbonized and often lacking in fine morphological detail (Kyle and Schopf 1977; Kyle 1977a; Truswell 1980). Most appropriately, Barrett (1981) has paid tribute to the work of Rosemary Kyle (now Askin), "who by perseverance and skill extracted recognisable palynomorphs from enough samples to link the [Victoria Group] stratigraphy from Late Carboniferous to Late Triassic with that of Eastern Australia"; see Kyle (in Kemp et al. 1977), Kyle (1977a), Schopf and Askin (1980), Kyle and Schopf (1982). The generally inimical preservation factor has tended to inhibit systematic-descriptive treatment, that is, apart from Balme and Playford's (1967) account dealing with miospores from the Amery Group (Mond 1972) in the Prince Charles Mountains. The only really well preserved Permian plant microfossils from Antarctica have been mentioned as recycled elements in Shackleton Ice Shelf seafloor sediments (Truswell 1982), thus implying primary derivation from an essentially unmetamorphosed and not deeply buried source rock.

Most of the published upper Paleozoic palynological research has been directed toward dating and correlation of the Victoria Group and equivalent sequences within Antarctica and with palynologically documented (and in places faunistically dated) sequences from elsewhere in the Gondwana region. However, Schopf and Askin (1980) and Truswell (in press) have discussed in fairly general floristic terms the plant microfossil record in relation to that of the plant megafossils.

Useful summaries of previous research on the palynology of the Victoria and Amery Groups have been provided by Kemp et al. (1977), Kyle (1977a), Truswell (1980), and Kyle and Schopf (1982). The last-mentioned work supplemented Kyle's (1977a) palynostratigraphic account of the Victoria Group in southern Victoria Land and extended the study with observations on material from the Ohio Range, Queen Maud Mountains (Nilsen Plateau), and the Shackleton and Beardmore Glacier areas.

Kyle and Schopf (1982) synthesized their own and previously published data from the Victoria Group (principally as sampled in the Transantarctic Mountains) and recognized three informal miospore zonal subdivisions within the Permian to possibly Upper or uppermost Carboniferous interval. In ascending stratigraphic order (but not embracing a stratigraphic continuum), these are the *Parasaccites* zone; the *Protohaploxypinus* zone; and an unnamed unit, regarded as correlative with stage 5 of the Australian palynosuccession (Kemp et al. 1977) and here termed the *Praecolpatites* zone. Kyle and Schopf (1982, Fig. 80.3) illustrated some spore–pollen taxa representative of their listed assemblages and suggested palynostratigraphic correlations with eastern and western Australia (Kyle and Schopf, Fig. 80.4) (see also Fig. 6.3).

Parasaccites Zone

The oldest and least diverse of the informal palynostratigraphic units, the *Parasaccites* zone, is characterized by a preponderance of radiosymmetric monosaccate pollen grains formerly attributed to *Parasaccites* Bharadwaj and Salujha 1964, but now more appropriately assigned to *Cannanoropollis* Potonié and Sah 1960 and *Plicatipollenites* Lele 1964 (see Foster 1975, 1979). Other components include bilaterally symmetrical monosaccate forms (*Potonieisporites* spp.) and rare taeniate and nontaeniate disaccates. The monosaccate grains and the nontaeniate disaccates are probably of cordaitalean and pteridospermous derivation, respectively; the taeniate forms presumably reflect the presence of early glossopterids in the parent flora. Cycadophytic or ginkgo-like pollen grains, specifically the monocolpate *Cycadopites cymbatus* (Balme and Hennelly) Segroves 1970, are reportedly common to abundant, and there is a range of pteridophytic acavate trilete spores with laevigate and sculptured exines. Detailed form-taxonomic listings were given by Kyle (1977a) and Kyle and Schopf (1982). They noted the zone's presence in several glacial and immediately postglacial sequences

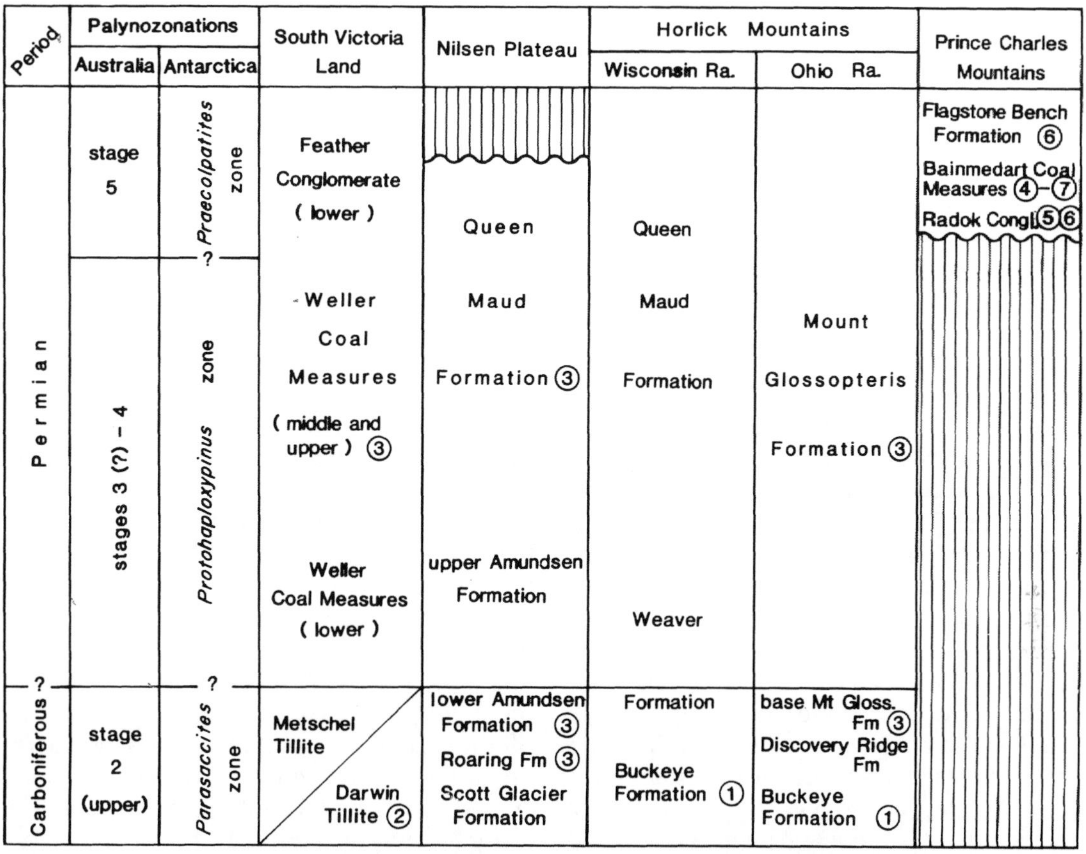

Period	Palynozonations: Australia	Palynozonations: Antarctica	South Victoria Land	Nilsen Plateau	Horlick Mountains: Wisconsin Ra.	Horlick Mountains: Ohio Ra.	Prince Charles Mountains
Permian	stage 5	*Praecolpatites* zone	Feather Conglomerate (lower)				Flagstone Bench Formation (6); Bainmedart Coal Measures (4)–(7); Radok Congl. (5)(6)
— ? —	— ? —	— ? —					
Permian	stages 3 (?) – 4	*Protohaploxypinus* zone	Weller Coal Measures (middle and upper) (3)	Queen Maud Formation (3)	Queen Maud Formation	Mount Glossopteris Formation (3)	
Permian			Weller Coal Measures (lower)	upper Amundsen Formation	Weaver Formation		
— ? —		— ? —					
Carboniferous ?	stage 2 (upper)	*Parasaccites* zone	Metschel Tillite / Darwin Tillite (2)	lower Amundsen Formation (3); Roaring Fm (3); Scott Glacier Formation	Buckeye Formation (1)	base Mt Gloss. Fm (3); Discovery Ridge Fm; Buckeye Formation (1)	

Figure 6.3. Palynological subdivision and correlations of the lower (pre-Mesozoic) part of the Victoria Group in the Transantarctic Mountains (southern Victoria Land, Nilsen Plateau, and Horlick Mountains) and of the Amery Group in the Prince Charles Mountains. Formation names with associated circled numbers signify that one or more sampled horizons within those formations have yielded palynological data. The numbers denote relevant published sources, as follows: 1, Kemp (1975); 2, Barrett and Kyle (1975); 3, Kyle (1977a), Kyle and Schopf (1982); 4, Balme and Playford (1967); 5, Kemp (1973); 6, Dibner (1976, 1978); 7, this chapter. Note that, although the Antarctic zones occur in the sequential order shown, their mutual boundaries have not been established.

of the basal Victoria Group, namely, the Darwin Tillite, the Roaring Formation and possibly the lower Amundsen Formation, the Buckeye Formation (Kemp 1975), and possibly the basal Mount Glossopteris Formation (see Fig. 6.3).

In comparing the palynofloras with those known from Australia (Kemp et al. 1977), Kyle and Schopf (1982) found closest similarity with the stage 2 palynoflora, in particular with its younger part as implied by the presence of such species as *Microbaculispora tentula* Tiwari 1965, *Cycadopites cymbatus*, and *Granulatisporites* (cf.) *micronodosus* Balme and Hennelly 1956. The age of the *Parasaccites* zone has not been established precisely insofar as it is linked to the biostratigraphic placement of the Carboniferous–Permian boundary, which is still controversial in Australia (Kemp et al. 1977; Balme 1980; Archbold 1982, 1984; Dickins 1984) as elsewhere in Gondwana. However, given the range of contention or probability, the Antarctic palynozone surely lies close to the systemic boundary. Kemp (1975) and Schopf and Askin (1980) have noted the virtually pan-Gondwanic distribution of palynofloras akin to stage 2.

Protohaploxypinus Zone

This zone is based primarily on nine palyniferous samples from the middle and upper parts of the *Glossopteris*-bearing Weller Coal measures in southern Victoria Land (samples from the formation's lower part proved barren). Species lists furnished by Kyle (1977a) and Kyle and Schopf (1982) attest to the greater taxonomic diversity of the *Protohaploxypinus* zone compared to the preceding *Parasaccites* zone. Disaccate pollen grains account for 47–55% of the palynoflora; over half of these are taeniate [notably *Protohaploxypinus* spp., including *P. amplus* (Balme and Hennelly) Hart 1964, *P. limpidus* (Balme and Hennelly) Balme and Playford 1967, and *Striatopodocarpites cancellatus* (Balme and Hennelly) Hart 1963] and are presumably of glossopterid derivation. Monosaccate grains ('*Parasaccites*') occur impersistently. Laevigate and ornamented acavate trilete spores are fairly plentiful and varied and imply an expanding pteridophytic component of the vegetation under progressively warmer conditions that also clearly promoted the burgeoning development of the glossopterids. Praecolpate pollen, notably *Marsupipollenites triradiatus* Balme and Hennelly 1956 and *M. striatus* (Balme and Hennelly) Foster 1975, are rare to abundant. Organic-walled microfossils of probable algal derivation are represented sporadically (e.g., *Peltacystia monile* Balme and Segroves 1966) or persistently and sometimes abundantly (*Circulisporites parvus* de Jersey 1962).

Kyle and Schopf (1982) judged their *Protohaploxypinus* zone to be broadly compatible with stage 4 of eastern Australia and with equivalent palynostratigraphic subdivisions of western Australian Permian sequences as documented by Kemp et al. (1977), thus suggesting an early Permian ('Artinskian' or Aktastinian–Baigendzhinian) age. In addition to its presence in the middle and upper Weller Coal Measures, the *Protohaploxypinus* zone may also occur in the lower (but not basal) Mount Glossopteris Formation, where a poorly preserved assemblage carrying an appreciable taeniate-disaccate component has been encountered; this could be stage 3 age or younger (Kyle and Schopf 1982).

Praecolpatites Zone (New) (Figs 6.4–6.55)

This informal biostratigraphic category is proposed to embrace primarily the palyniferous units of the fluviatile-paludal Amery Group in the Prince Charles Mountains (Fig. 6.3); namely (in ascending order), the Radok Conglomerate, the Bainmedart Coal Measures, and the Flagstone Bench Formation (Mond 1972). It provides a convenient, albeit provisional, appellation for host strata of what Kyle (1977a) and Kyle and Schopf (1982) have already recognized as a palynoflora of intermediate stratigraphic/temporal position between their *Protohaploxypinus* and *Alisporites* zones. The *Praecolpatites* zone is proposed with similar intent and qualifications to those stated by Kyle (1977a), when she established her informal zonation covering the palyniferous Victoria Group intervals of southern Victoria Land. As with Kyle's zones, the fact that no bounding biostratigraphic criteria are currently available for the *Praecolpatites* zone emphasizes its informal status in light of essentially reconnaissance-level studies.

The eponym signifies the characteristic presence of the praecolpate pollen species *Praecolpatites sinuosus* (Balme and Hennelly) Bharadwaj and Srivastava 1969 (Figs 6.47–6.49), which is unrecorded from either the *Protohaploxypinus* zone or the *Alisporites* zone and is, moreover, readily recognizable even when poorly preserved (e.g., Kyle and Schopf 1982). The palynoflora contains high frequencies of taeniate-disaccate pollen grains, including such forms as *Protohaploxypinus amplus, P. limpidus, Striatopodocarpites cancellatus, S. fusus* (Balme and Hennelly) Potonié 1958, *S. rarus* (Bharadwaj and Salujha) Balme 1970, *S. solitus* (Bharadwaj and Salujha) Foster 1979, *Striatoabieites multistriatus* (Balme and Hennelly) Hart 1964, and *Guttulapollenites hannonicus* Goubin 1965 (Figs 6.29–6.44). Nontaeniate disaccates include *Scheuringipollenites ovatus* (Balme and Hennelly) Foster 1975 and *S. maximus*

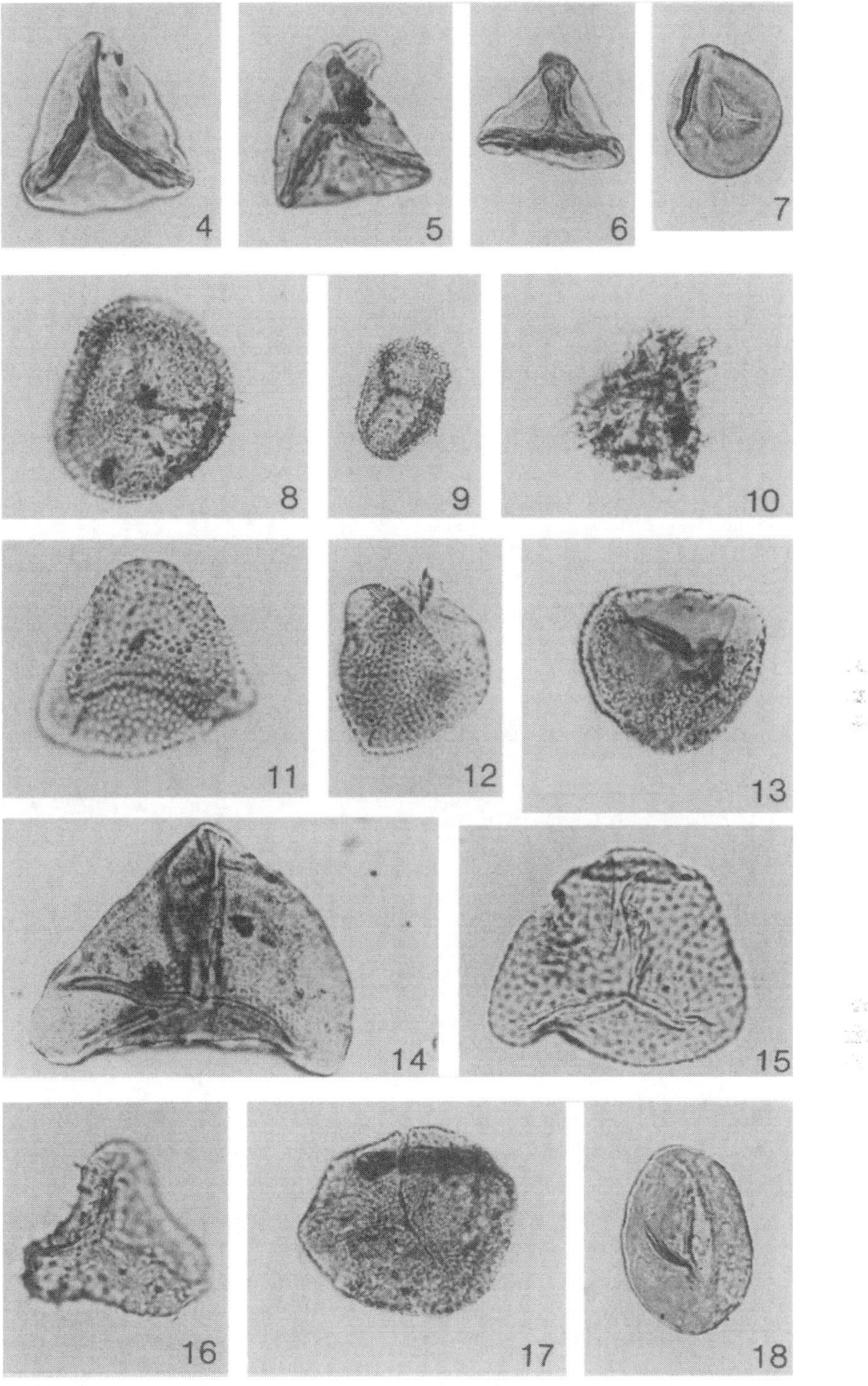

Figures 6.4–6.18. Permian spores of the *Praecolpatites* zone. Magnifications ×750 unless stated otherwise. 6.4–6.6, *Leiotriletes directus* Balme and Hennelly 1956; 6.6, ×500. 6.7, *Retusotriletes nigritellus* (Luber) Foster 1979; ×500. 6.8, 6.9, *Granulatisporites absonus* Foster 1979; 6.9, ×500. 6.10, *Horriditriletes ramosus* (Balme and Hennelly) Bharadwaj and Salujha 1964. 6.11–6.13, *Granulatisporties micronodosus* Balme and Hennelly 1956; 6.12, 6.13, ×500. 6.14, *Granulatisporites trisinus* Balme and Hennelly 1956. 6.15, *Granulatisporites quadruplex* Segroves 1970. 6.16, cf. *Acanthotriletes tereteangulatus* Balme and Hennelly 1956. 6.17, *Pseudoreticulatispora pseudoreticulata* (Balme and Hennelly) Bharadwaj and Srivastava 1969; ×500. 6.18, *Laevigatosporites vulgaris* (Ibrahim) Ibrahim 1933; ×500.

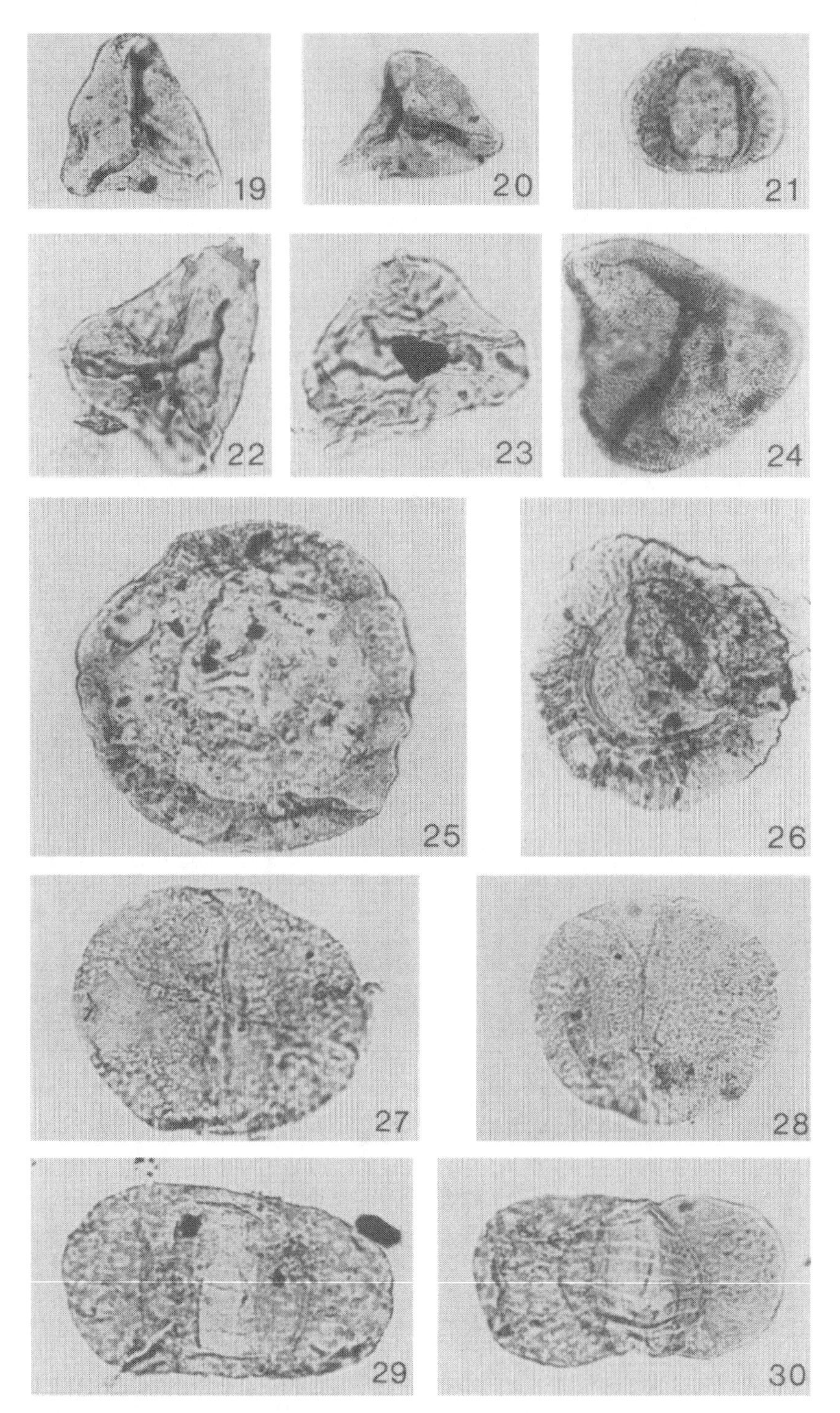

Figures 6.19–6.30. Permian spores and pollen grains of the *Praecolpatites* zone. Magnifications ×500 unless stated otherwise. 6.19, 6.20, *Microbaculispora tentula* Tiwari 1965. 6.21, *Potonieisporites* sp. 6.22, 6.23, *Indospora clara* Bharadwaj 1962; ×750. 6.24, *Microbaculispora villosa* (Balme and Hennelly) Bharadwaj 1962. 6.25, *Barakarites rotatus* (Balme and Hennelly) Bharadwaj and Tiwari 1964. 6.26, *Cannanoropollis mehtae* (Lele) Bose and Maheshwari 1968. 6.27, *Scheuringipollenites maximus* (Hart) Tiwari 1973. 6.28, *Scheuringipollenites ovatus* (Balme and Hennelly) Foster 1975; ×750. 6.29, 6.30, *Protohaploxypinus amplus* (Balme and Hennelly) Hart 1964.

(Hart) Tiwari 1973 (Figs 6.27–6.28). Other saccates are represented, albeit sparingly, by forms such as *Barakarites rotatus* (Balme and Hennelly) Bharadwaj and Tiwari 1964, *Densipollenites indicus* Bharadwaj 1962, and *Bascanisporites undosus* Balme and Hennelly 1956. In addition to *Praecolpatites sinuosus*, there is persistent praecolpate representation by *Marsupipollenites triradiatus* and *M. striatus* (Figs. 6.50–6.53). Acavate trilete and monolete spores with variously sculptured exines include such species as *Microbaculispora villosa* (Balme and Hennelly) Bharadwaj 1962, *M. tentula, Didecitriletes ericianus* (Balme and Hennelly) Venkatachala and Kar 1965, *Acanthotriletes tereteangulatus* Balme and Hennelly 1956, *Granulatisporites micronodosus, G. quadruplex* Segroves 1970 (*sensu* Foster 1979, p. 31), *G. trisinus* Balme and Hennelly 1956, *Pseudoreticulatispora pseudoreticulata* (Balme and Hennelly) Bharadwaj and Srivastava 1969, and *Indospora clara* Bharadwaj 1962 (Figs. 6.4–6.24). For a more comprehensive inventory of this reasonably diverse and (by Antarctic pre-Mesozoic standards) well-preserved palynoflora, the reader is referred to Balme and Playford (1967), Kemp (1973), and Dibner (1976, 1978). In the Dibner publications, some quantitative variations of broad miospore categories through the Amery Group sequence were presented. Selected miospores illustrated herein (Figs. 6.4–6.55) are from original residues and from reprocessed Balme and Playford (1967) samples representative of the Bainmedart Coal Measures.

Floristically, the *Praecolpatites* zone's palynological complexion signifies a fairly varied parental vegetation, dominated by glossopterids, which are plentiful also as megafossils at four collecting sites in the Bainmedart Coal Measures (White 1973).

Kyle and Schopf (1982, p. 653) noted what may be perceived as *Praecolpatites* zone occurrences in the Victoria Group of the Transantarctic Mountains: (1) in the Queen Maud Formation of the Nilsen Plateau, where assemblages are dominated by disaccates (60–70%, almost half of these being taeniate grains) and include common *Marsupipollenites,* together with *Praecolpatites sinuosus* and *Bascanisporites undosus,* and (2) less definitively, in the upper part of the Mount Glossopteris Formation of the Ohio Range, assemblages of which are dominated by taeniate and nontaeniate disaccates (up to 80%).

From its high content of diverse taeniate disaccates associated with certain other distinctive forms (e.g., *Praecolpatites sinuosus, Microbaculispora villosa, Bascanisporites undosus*), Balme and Playford (1967) considered what is termed here the *Praecolpatites* zone to be correlative with younger Permian sequences of eastern and western Australia classified palynologically (in Kemp et al. 1977) as stage 5 and unit VII–?VIII, respectively. That opinion was endorsed subsequently, for example, by Kemp (1973), Kyle (1977a, and in Kemp et al. 1977). Kemp (1973) added *Guttulapollenites hannonicus* and *Indospora clara* to the chronologically significant species list. As originally conceived, stage 5 is a unit of fairly broad stratigraphic scope; it has been subdivided (Kemp et al. 1977; Price 1983) into as many as 6 finer units on the basis of successive species entrance levels, notably of representatives of *Dulhuntyispora* Potonié emend. Price 1983. Unfortunately, *Dulhuntyispora*-type spores are unknown, in situ at least, from Antarctica and hence are stratigraphically inapplicable there. It should be noted that, although particulary characteristic of stage 5/unit VII, *Praecolpatities sinuosus* is known also from older eastern Australian strata (upper stage 4, Paten 1969; and even stage 3, Rigby and Hekel 1977). If the first appearance of *Didecitriletes ericianus* (base lower stage 5b, Truswell 1980; Price 1983) can be extrapolated to Antarctica, then the *Praecolpatites* zone might be regarded as attributable to the lower stage 5b–upper stage 5 interval. However, as noted by Balme and Playford (1967), *D. ericianus* is extremely rare and was unrecorded by Kemp (1973); moreover, Dibner's (1978) specimen is almost certainly misidentified. No indices of post-stage 5 zones (e.g., Foster 1982; Bowen Basin, Queensland) have been encountered in the zone. Considering all available evidence, a broad assignment to stage 5 remains appro-

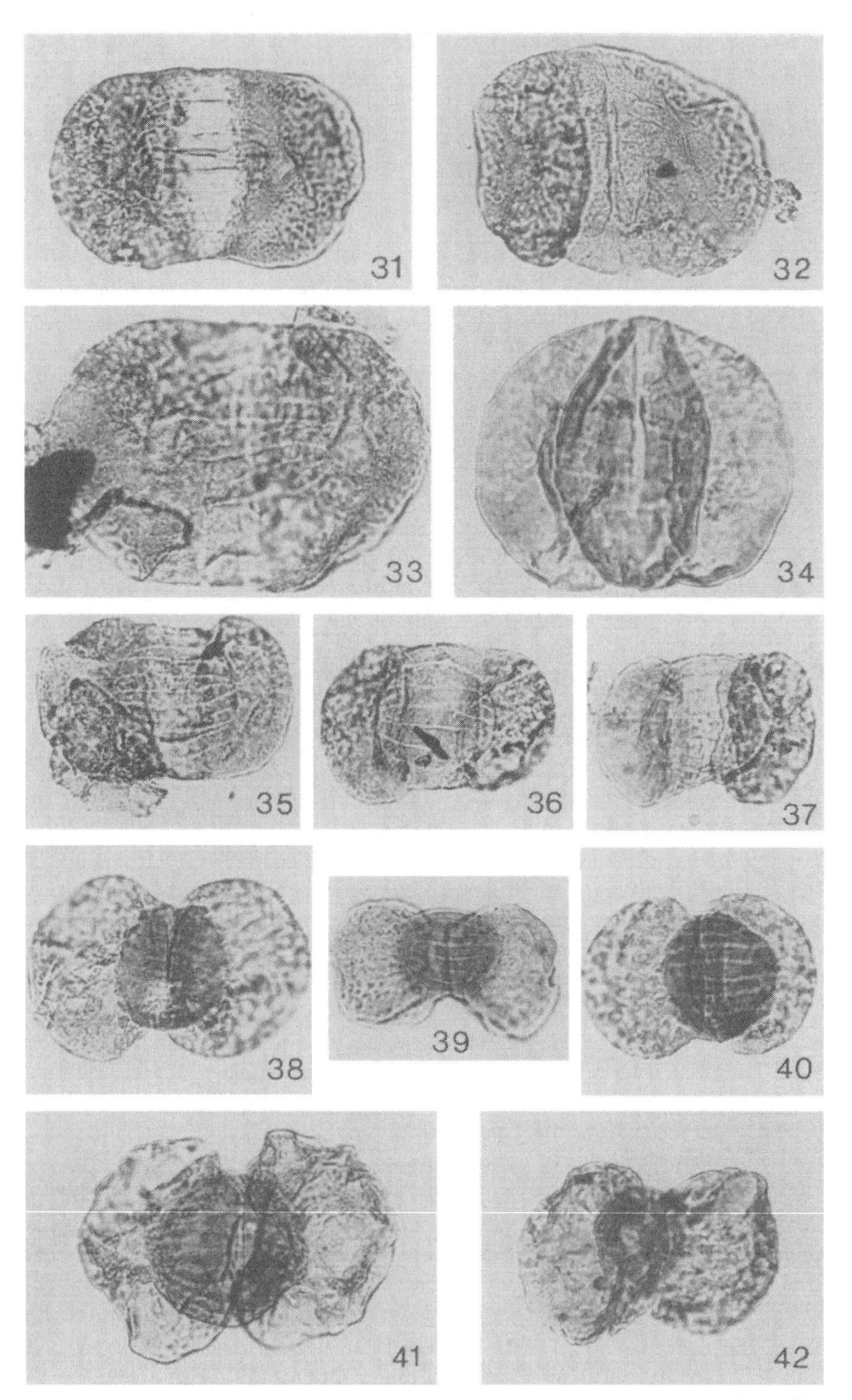

Figures 6.31–6.42. Permian pollen grains of the *Praecolpatites* zone. All magnifications ×500. 6.31 –6.33, *Protohaploxypinus amplus* (Balme and Hennelly) Hart 1964. 6.34, *Protohaploxypinus bharadwajii* Foster 1979. 6.35–6.37, *Protohaploxypinus limpidus* (Balme and Hennelly) Balme and Playford 1967. 6.38–6.40, *Striatopodocarpites cancellatus* (Balme and Hennelly) Hart 1963. 6.41, 6.42, *Striatopodocarpites fusus* (Balme and Hennelly) Potonié 1958.

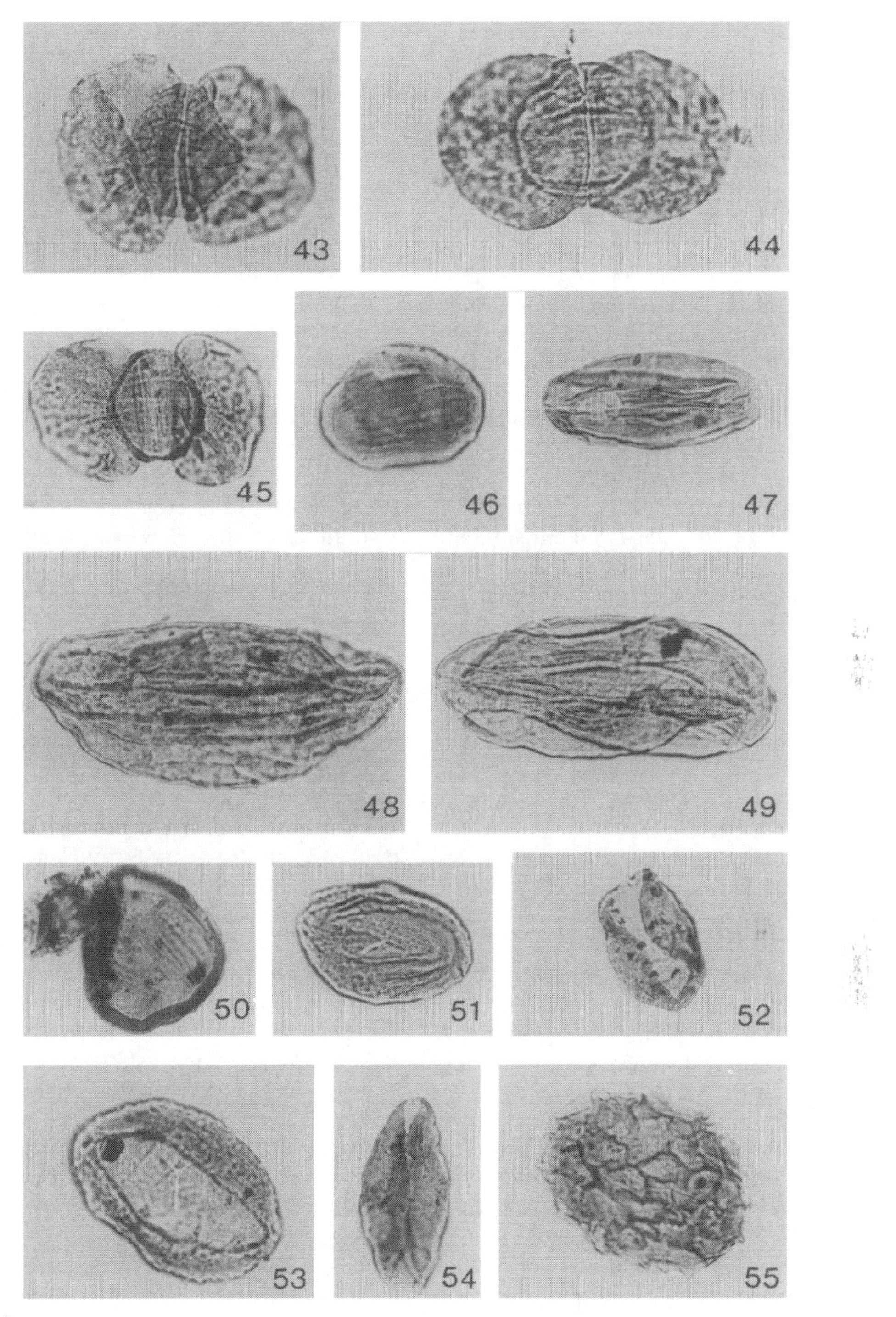

Figures 6.43–6.55. Permian pollen grains (6.43–6.54) and *incertae sedis* palynomorph (6.55) of the *Praecolpatites* zone. Magnifications ×500 unless stated otherwise. 6.43, *Striatopodocarpites* sp. A. 6.44, *Striatopodocarpites* sp. B. 6.45, *Striatopodocarpites* sp. C. 6.46, *Vittatina fasciolata* (Balme and Hennelly) Bharadwaj and Srivastava 1969; ×750. 6.47–6.49, *Praecolpatites sinuosus* (Balme and Hennelly) Bharadwaj and Srivastava 1969. 6.50, 6.51, *Marsupipollenites striatus* (Balme and Hennelly) Foster 1975. 6.52, 6.53, *Marsupipollenites triradiatus* Balme and Hennelly 1956; 6.53, ×750. 6.54, *Cycadopites cymbatus* (Balme and Hennelly) Segroves 1970. 6.55, cf. *Greinervillites* sp.

priate for the *Praecolpatites* zone. The latter resembles palynologically a similarly, if somewhat arbitrarily, assigned Tasmanian interval analyzed by Truswell in Calver et al. (1984; upper part of Upper Marine Sequence, Douglas River corehole). Truswell regarded her assemblages as relatively impoverished—as in Antarctica, notably lacking in *Dulhuntyispora* and related forms—and suggested that the higher paleolatitudes of both regions were responsible (i.e., compared to the 40–50° paleolatitudinal situation of the Canning and Carnarvon Basins, Western Australia, where contemporaneous assemblages are much more diverse).

Assessment of the stage 5 Permian time span obviously depends on integrating, where feasible and in critical regions, the faunal and palyno-successions and relating these to an internationally accepted version of the Permian time scale. The eastern Australian reality is that stage 5, and by inference the *Praecolpatites* zone, can presently be regarded as falling within the broad range of (?late) Baigendzhinian to Chhidruan/early Tatarian (e.g., Truswell 1980).

Recycled Occurrences of Paleozoic Palynomorphs

Studies of the recycled Paleozoic–Paleogene palynological contents of Antarctic seafloor sediments, collected during dredge and grab sampling and coring operations on or adjacent to the continental shelf, have proved to be rewarding in a range of geological and paleobotanical contexts; see, for example, Truswell (1983), Truswell and Drewry (1984), and Truswell and Anderson (1984). The studies assume particular significance in that the often profuse remanié suites are without doubt entirely, or at least predominantly, Antarctic derived: wind- and water-circulation patterns emanate consistently from Antarctica, and there is a numerical decrease in palynomorph content concomitant with increasing distance from the Antarctic coastline (Truswell and Drewry 1984). Redeposited palynomorphs have also been documented in lithified onshore deposits (e.g., Askin and Elliot 1982). Brief mention is made below of known occurrences of recycled Paleozoic palynomorphs (which are generally subordinate to palynomorphs of younger, post-Triassic age with which they are commonly associated).

Seafloor sediments from areas of the East Antarctic continental shelf, between about 75°E and 148°E, have been investigated by Truswell (1982, 1987; Kemp 1972a). In the more westerly parts (West and Shackleton Ice Shelves), the recycled palynomorph admixtures comprise Permian spores–pollen associated with Late Jurassic–early Tertiary palynomorphs. The Permian element evidently derived from erosion of *Praecolpatites* zone strata (as in the Prince Charles Mountains) and possibly also of older Permian rocks. The variable quality of miospore preservation—normally poor to mediocre, but of an unusually high standard in the Shackleton samples according to Truswell (1982)—would suggest a varying degree of metamorphism or maturation level of the eroding Permian source terrain. Surprisingly, no acritarchs were encountered in the recycled palynofloras; these might have been expected in view of the Riphean–Early Cambrian acritarch assemblages reported by Korotkevich and Timofeev (1959) from the onshore Sandow Group and from morainic boulders at Farr Bay, west of the Shackleton Ice Shelf.

Reconnaissance sampling of the Weddell Sea floor (Kemp 1972b; Truswell and Anderson 1984) yielded relatively rich Upper Jurassic–Upper Cretaceous palynomorphs and much less profuse early Tertiary forms. The presence of Permian spores–pollen was mentioned by Kemp (1972b), mainly taeniate-disaccate grains, together with specimens of the distinctive trilete scutulate genus *Dulhuntyispora*. The latter was a notable discovery; still unrecorded as a primary (*in-situ*) component of Antarctic Permian palynofloras, *Dulhuntyispora* and its constituent species are stratigraphically important within Australia's later Permian (stage 5) assemblages (Price 1983).

In bottom sediments of the Ross Sea, initially studied by Wilson (1968) and much more extensively and intensively by Truswell and

Drewry (1984), remanié palynomorphs of Late Cretaceous to early Tertiary age generally overshadow those of greater age. Among the latter category, however, Permian spores–pollen are persistently present; and are more prevalent in the western sector of the Ross Sea, as would be expected from the proximity of southern Victoria Land with its eroding Victoria Group sequence. The Permian elements conform with contents of Kyle's (1977a) *Parasaccites* and *Protohaploxypinus* zones. A single Devonian miospore specimen, belonging to *Emphanisporites,* was recorded by Truswell and Drewry (1984) and likened to a species recorded from the Horlick Formation of the Ohio Range by Kemp (1972c).

From core hole samples of MSSTS-1, drilled at a near-shore site from sea ice in the McMurdo Sound (southwest extremity of the Ross Sea), Truswell (1986) has reported common Permian spores–pollen; these attest to significant input from the nearby Victoria Group during deposition of all but the upper part of the Late Oligocene–Early Miocene sequence penetrated.

Palynological analysis of the marine Campanian–Maastrichtian and the succeeding nonmarine lower Tertiary sequence of Seymour Island (near the tip of the Antarctic Peninsula, West Antarctica) has disclosed a diverse admixture of Permian, Triassic, and pre-Campanian Cretaceous palynomorphs associated with well-preserved, in-situ palynofloras dominated by podocarpacean and *Nothofagus* pollen (Askin and Elliot 1982). The recycled forms are especially common (up to 9%) in the upper part of the sequence. The Permian assemblage is dominated by taeniate disaccates (*Protohaploxypinus*); and, among the trilete spores identified, a single broken specimen of *Dulhuntyispora* (*D. dulhuntyi* Potonié 1956; Askin and Elliot 1982) adds to Kemp's (1972b) hitherto unique Antarctic record from the Weddell Sea floor, just east of Seymour Island. A closer relationship of the recycled Permian palynoflora of Seymour Island with those known from Australia and South Africa, rather than South America, was intimated by Askin and Elliott (1982). Cognizant of the seeming absence of likely source rocks on the Antarctic Peninsula, and using various lines of evidence, Askin and Elliot hypothesized derivation from a topographic high to the west or northwest, in the forearc terrain of the Gondwana plate margin. Baldoni and Barreda (1987) analyzed the palynology of samples from evidently the same Seymour Island sequence as Askin and Elliot (1982), albeit adopting a different lithostratigraphic terminology, and noted, incidentally, the presence of remanié Permian pollen (*Protohaploxypinus*).

Summary and Conclusions

Studies to date indicate that, generally speaking, Antarctica's pre-Mesozoic rocks do not yield very advantageously preserved assemblages of palynological microfossils; indeed, many samples have proven totally unproductive. This situation is chiefly a consequence of metamorphism of the Precambrian and older Paleozoic basement and the well-documented baking/carbonizing effects of extensive Jurassic volcanic activity (e.g., on the virtually undeformed Devonian–Triassic Beacon Supergroup in the Transantarctic Mountains). Hence, palynological coverage of the pre-Mesozoic rocks tends to be very intermittent and fortuitious; especially for the older sequences, it is largely unsupported by photographic documentation. Proterozoic (Riphean?)–Middle Cambrian acritarchs have been reported from scattered localities; and Devonian spores from four units developed in different parts of the Transantarctic Mountains (one unit, the Emsian Horlick Formation, has also yielded rare Chitinozoa). There is one dubious record of Early and mid-Carboniferous spores from the Antarctic Peninsula and several, relatively detailed accounts of Late Carboniferous(?)–Permian spore–pollen floras from East Antarctica (from the Victoria Group of the Transantarctic Mountains and Amery Group of the Prince Charles Mountains).

A provisional and informal zonal scheme, synthesized from what is known of the composition and distribution of Late Carboniferous(?)–Permian palynofloras, comprises (in ascending order) the *Parasaccites, Protohaploxypinus,* and *Praecolpatites* zones (the latter newly designated herein). This scheme can be broadly related to upper Paleozoic palynozonations developed in Australia and also connotes some floristic links with other Gondwanan regions. The Antarctic palynofloras, more diverse than the preserved megafloras, imply the development of a progressively more varied postglacial vegetation in which the taeniate pollen-producing glossopterid plants steadily expanded and in which other gymnosperms and pteridophytes were clearly represented. The flora appears to have been less diverse, however, than those represented, for instance, in mainland Australia, presumably because of appreciably higher Antarctic paleolatitudes.

Recycling of Paleozoic palynomorphs (mainly Permian spores–pollen) into upper Mesozoic–Paleogene sequences—and, more particularly, into Holocene sediments of the Antarctic continental shelf—has received considerable attention. Such studies are valuable in enabling predictions of the age, nature, and position of source rocks, many of which may now be concealed, and are enhanced where patterns of sediment-distributing ice flow are known.

Unquestionably, palynological analyses of Antarctic pre-Mesozoic sequences, though beset with problems of productivity and preservation quality, offer unique opportunities for resolving a variety of geological problems and, in the paleobotanical sphere, for helping to reconstruct the continent's past floras.

Acknowledgments. Dr. Basil Balme (University of Western Australia) generously provided raw samples and palynological residues of the Prince Charles Mountains (Bainmedart Coal Measures) material (Balme and Playford 1967). Dr. Elizabeth Truswell (Bureau of Mineral Resources, Canberra) kindly allowed reference to her manuscript on Antarctic vegetation history (Truswell in press) and provided helpful advice. Peter Price (APG Consultants, Brisbane) generously assisted with information on aspects of Australian Permian palynostratigraphy. In the Department of Geology and Mineralogy, University of Queensland, Elvira Burdin and Barbara Reiss, respectively, drafted the text figures and typed the manuscript and Marshall Butterworth provided laboratory assistance. Last, but not least, sincere thanks are extended to Drs. Thomas and Edith Taylor (Ohio State University) for organizing the Workshop on Antarctic Paleobotany/Palynology at which this review was presented.

References

Archbold NW (1982) Correlation of the Early Permian faunas of Gondwana: implications for the Gondwanan-Carboniferous–Permian boundary. J Geol Soc Aust 29:267–276

Archbold NW (1984) Early Permian faunas from Australia, India and Tibet: an update on the Gondwanan Carboniferous-Permian boundary. Bull Ind Geol Assoc 17:133–138

Askin RA, Elliot DH (1982) Geologic implications of recycled Permian and Triassic palynomorphs in Tertiary rocks of Seymour Island, Antarctic Peninsula. Geology 10:547–551

Baldoni AM, Barreda V (1987) Estudio palinológico de las Formaciones López de Bertadano y Sobral, Isla Vicecomodor Marambio, Antartida. Bol IG-USP Sér Cient 17:89–98

Balme BE (1980) Palynology and the Carboniferous–Permian boundary in Australia and other Gondwana continents. Palynology 4:43–55

Balme BE, Playford G (1967) Late Permian plant microfossils from the Prince Charles Mountains, Antarctica. Rev Micropaléont 10:179–192

Barrett PJ (1981) History of the Ross Sea region during the deposition of the Beacon Supergroup 400-180 million years ago. J Roy Soc N Z 11:447–458

Barrett PJ, Kyle RA (1975) The Early Permian glacial beds of south Victoria Land and the Darwin Mountains, Antarctica. *In* Campbell KSW (ed) Gondwana Geology. Aust Nat Univ Press, Canberra, pp 333–346

Brown AV, Cooper RA, Corbett KD, Daily B, Green GR, Grindley GW, Jago J, Laird MG, VandenBerg AHM, Vidal G, Webby BD, Wilkinson HE (1982) Late Proterozoic to Devonian sequences of southeastern Australia, Antarctica and New Zealand and their correlation. Spec Publs Geol Soc Aust 9:i–ii, 1–103

Burn RW (1984) The geology of the LeMay Group of Alexander Island. Scient Rept Br Antarct Surv 109:1–65

Burrett CF, Findlay RH (1984) Cambrian and Ordovician conodonts from the Robertson Bay Group, Antarctica and their tectonic significance. Nature 307:723–726

Calver CR, Clarke MJ, Truswell EM (1984) The stratigraphy of a late Palaeozoic borehole section at Douglas River, eastern Tasmania: a synthesis of marine macro-invertebrate and palynological data. Pap Proc Roy Soc Tasm 118:137–161

Chaloner WG, Lacey WS (1973) The distribution of late Palaeozoic floras. Spec Pap Palaeontology 12:271–289

Cooper RA, Jago JB, MacKinnon DI, Shergold JH, Vidal G (1982) Late Precambrian and Cambrian fossils from northern Victoria Land and their stratigraphic implications. *In* Craddock C (ed) Antarctic Geoscience. Univ Wisconsin Press, Madison, pp 629–633

Cooper RA, Jago JB, Rowell AJ, Braddock P (1983) Age and correlation of the Cambrian–Ordovician Bowers Supergroup, northern Victoria Land. *In* Oliver RL, James PR, Jago JB (eds) Antarctic Earth Science. Aust Acad Sci, Canberra, pp 128–131

Dibner AF (1976) Late Permian palynoflora of sedimentary deposits of the Beaver Lake area (East Antarctica). Antarctika 15:41–52 (Russian)

Dibner AF (1978) Palynocomplexes and age of the Amery Formation deposits, East Antarctica. Pollen et Spores 20:405–422

Dickins JM (1984) Late Palaeozoic glaciation. BMR Jl Aust Geol Geophys 9:163–169

Edwards CW (1982) New paleontologic evidence of Triassic sedimentation in West Antarctica. *In* Craddock C (ed) Antarctic Geoscience. Univ Wisconsin Press, Madison, pp 325–330

Elliot DH (1975) Gondwana basins of Antarctica. *In* Campbell KSW (ed) Gondwana Geology. Aust Nat Univ Press, Canberra, pp 493–536

Foster CB (1975) Permian plant microfossils from the Blair Athol Coal Measures, central Queensland, Australia. Palaeontographica Abt B 154:121–171

Foster CB (1979) Permian plant microfossils of the Blair Athol Coal Measures, Baralaba Coal Measures, and basal Rewan Formation of Queensland. Publs Geol Surv Qd 372:1–244

Foster CB (1982) Spore–pollen assemblages of the Bowen Basin, Queensland (Australia): their relationship to the Permian/Triassic boundary. Rev Palaeobot Palyn 36:165–183

Grikurov GE (1978) Geology of the Antarctic Peninsula. Amerind Publ Co, New Delhi Bombay Calcutta New York

Grikurov GE, Dibner AF (1968) New data on the Trinity Series (C_{1-3}) of West Antarctica. Dokl Akad Nauk 179:410–412 (Russian)

Grindley GW, Mildenhall DC (1980) Geological background to a Devonian plant fossil discovery, Ruppert Coast, Marie Byrd Land, West Antarctica. *In* Cresswell IM, Vella P (eds) Gondwana Five. Balkema, Rotterdam, pp 23–30

Grindley GW, Mildenhall DC, Schopf JM (1980) A mid-late Devonian flora from the Ruppert Coast, Marie Byrd Land, West Antarctica. J Roy Soc N Z 10:271–285

Helby RJ, McElroy CT (1969) Microfloras from the Devonian and Triassic of the Beacon Group, Antarctica. N Z Jl Geol Geophys 12:376–382

Hyden G, Tanner PWG (1981) Late Palaeozoic–early Mesozoic fore-arc basin sedimentary rocks at the Pacific margin in Western Antarctica. Geol Rundsch 70:529–541

Iltchenko LN (1972) Late Precambrian acritarchs of Antarctica. *In* Adie RJ (ed) Antarctic Geology and Geophysics. Universitetsforlaget, Oslo, pp 599–602

Kemp EM (1972a) Reworked palynomorphs from the West Ice Shelf area, East Antarctica, and their possible geological and palaeoclimatological significance. Mar Geol 13:145–157

Kemp EM (1972b) Recycled palynomorphs in continental shelf sediments from Antarctica. Antarct Jl U S 7(5):190–191

Kemp EM (1972c) Lower Devonian palynomorphs from the Horlick Formation, Ohio Range, Antarctica. Palaeontographica Abt B 139:105–124

Kemp EM (1973) Permian flora from the Beaver Lake area, Prince Charles Mountains, Antarctica. 1. Palynological examination of samples. Bull Bur Miner Resour Geol Geophys Aust 126:7–12

Kemp EM (1975) The palynology of late Palaeozoic glacial deposits of Gondwanaland. *In* Campbell KSW (ed) Gondwana Geology. Aust Nat Univ Press, Canberra, pp 397–413

Kemp EM, Balme BE, Helby RJ, Kyle RA, Playford G, Price PL (1977) Carboniferous and Permian palynostratigraphy in Australia and Antarctica: a review. BMR Jl Aust Geol Geophys 2:177–208

Korotkevich ES, Timofeev BV (1959) On the age of the rocks of East Antarctica (from spore analysis). Bull Soviet Antarct Exped 12:41–46 (Russian)

Kyle RA (1977a) Palynostratigraphy of the Victoria Group of south Victoria Land, Antarctica. N Z Jl Geol Geophys 20:1081–1102

Kyle RA (1977b) Devonian palynomorphs from the basal Beacon Supergroup of south Victoria Land, Antarctica. N Z Jl Geol Geophys 20:1147–1150

Kyle RA, Schopf JM (1977) Palynomorph preservation in the Beacon Supergroup of the Transantarctic Mountains. Antarct Jl U S 12(4): 121–122

Kyle RA, Schopf JM (1982) Permian and Triassic palynostratigraphy of the Victoria Group, Transantarctic Mountains. *In* Craddock C (ed) Antarctic Geoscience. Univ Wisconsin Press, Madison, pp 649–659

Mond A (1972) Permian sediments of the Beaver Lake area, Prince Charles Mountains. *In* Adie RJ (ed) Antarctic Geology and Geophysics. Universitetsforlaget, Oslo, pp 585–589

Paten RJ (1969) Palynologic contributions to petroleum exploration in the Permian formations of the Cooper Basin, Australia. APEA Jl 9:79–87

Playford G (1983) The Devonian miospore genus *Geminospora*: a reappraisal based upon topotypic *G. lemurata* (type species). Mem Ass Australas Palaeontols 1:311–325

Playford G (1985) Palynology of the Australian Lower Carboniferous: a review. C R 10th Congr Internat Strat Géol Carb (Madrid 1983) 4:247–265

Plume RW (1982) Sedimentology and paleocurrent analysis of the basal part of the Beacon Supergroup (Devonian [and older?] to Triassic) in south Victoria Land, Antarctica. *In* Craddock C (ed) Antarctic Geoscience. Univ Wisconsin Press, Madison, pp 571–580

Price PL (1983) A Permian palynostratigraphy for Queensland. *In* Permian Geology of Queensland. Geol Soc Aust Qd Div, Brisbane, pp 155–211

Richardson JB, McGregor DC (1986) Silurian and Devonian spore zones of the Old Red Sandstone continent and adjacent regions. Bull Geol Surv Can 364:1–79

Rigby JF, Hekel H (1977) Palynology of the Permian sequence in the Springsure Anticline, central Queensland. Publs Geol Surv Qd 363:1–76

Rowell AJ, Rees MN, Braddock P (1987) Silurian marine fauna not confirmed from Antarctica. Alcheringa 11:137

Schmidt DL, Ford AB (1969) Geology of the Pensacola and Thiel Mountains (Sheet 5, Pensacola and Thiel Mountains) Pl. V. *In* Bushnell VC, Craddock C (eds) Geologic maps of Antarctica. Antarctic Map Folio Series 12

Schopf JM (1973) Plant material from the Miers Bluff Formation of the South Shetland Islands. Rept Inst Polar Studies Ohio State Univ 45:1–45

Schopf JM, Askin RA (1980) Permian and Triassic floral biostratigraphic zones of southern land masses. *In* Dilcher DL, Taylor TN (eds) Biostratigraphy of Fossil Plants. Dowden, Hutchinson & Ross, Stroudsburg, Pa, pp 119–151

Smirnova LN (1983) Acritarchs of the late Precambrian and Cambrian in Antarctica. *In* Papulov GN (ed) Stratigraphy and Correlation of Sediments by Palynological Methods. Akad Nauk SSSR Ural Nauchn Tsentr Sverdlovsk, pp 31–34

Stump E, White AJR, Borg SG (1986) Reconstruction of Australia and Antarctica: evidence from granites and recent mapping. Earth Planet Sci Letters 79:348–360

Thomson MRA, Tranter TH (1986) Early Jurassic fossils from central Alexander Island and their geological setting. Bull Br Antarct Surv 70:23–39

Tranter TH (1987) The structural history of the LeMay Group of central Alexander Island, Antarctic Peninsula. Bull Br Antarct Surv 77:61–80

Truswell EM (1980) Permo-Carboniferous palynology of Gondwanaland: progress and problems in the decade to 1980. BMR Jl Aust Geol Geophys 5:95–111

Truswell EM (1982) Palynology of seafloor samples collected by the 1911–14 Australasian Antarctic Expedition: implications for the geology of coastal East Antarctica. J Geol Soc Aust 29:343–356

Truswell EM (1983) Geological implications of recycled palynomorphs in continental shelf sediments around Antarctica. *In* Oliver RL, James PR, Jago JB (eds) Antarctic Earth Science. Aust Acad Sci, Canberra, pp 394–399

Truswell EM (1986) Palynology. *In* Barrett PJ (ed) Antarctic Cenozoic History from the MSSTS-1 Drillhole, McMurdo Sound. DSIR Bull 237, Wellington, pp 131–134

Truswell EM (1987) The palynology of core samples from the S.P. Lee Wilkes Land cruise. *In* Eittreim SL, Hampton MA (eds) The Antarctic Continental Margin: Geology and Geophysics of Offshore Wilkes Land. CPCEMR Earth Sci Ser 5A, Houston, pp 215–218

Truswell EM (in press) Antarctica: a history of terrestrial vegetation. *In* Tingey RJ (ed) Geology of Antarctica. Oxf Univ Press, Oxford

Truswell EM, Anderson JB (1984) Recycled palynomorphs and the age of sedimentary sequences in the eastern Weddell Sea. Antarct Jl U S 19(5):90–92

Truswell EM, Drewry DJ (1984) Distribution and provenance of recycled palynomorphs in surficial sediments of the Ross Sea, Antarctica. Mar Geol 59:187–214

Webby BD, VandenBerg AHM, Cooper RA, Banks MR, Burrett CF, Henderson RA, Clarkson PD, Hughes CP, Laurie J, Stait B, Thomson MRA, Webers GF (1981) The Ordovician system in Australia, New Zealand and Antarctica. Publ Int Union Geol Sci 6:1–64

Webers GF, Sporli KB (1983) Palaeontological and stratigraphic investigations in the Ellsworth Mountains, West Antarctica. *In* Oliver RL, James PR, Jago JB (eds) Antarctic Earth Science. Aust Acad Sci, Canberra, pp 261–264

White ME (1973) Permian flora from the Beaver Lake area, Prince Charles Mountains, Antarctica. 2. Plant fossils. Bull Bur Miner Resour Geol Geophys Aust 126:13–18

Wilson GJ (1968) On the occurrence of fossil microspores, pollen grains, and microplankton in bottom sediments of the Ross Sea, Antarctica. N Z Jl Mar Freshw Res 2:381–389

Wright TO, Ross RJ Jr, Repetski JE (1984) Newly discovered youngest Cambrian or oldest Ordovician fossils from the Robertson Bay terrane (formerly Precambrian), northern Victoria Land, Antarctica. Geology 12:301–305

Appendix

Register of illustrated specimens from Bainmedart Coal Measures (Amery Group), Beaver Lake area, Prince Charles Mountains, East Antarctica. Note: All palynomorph specimens are lodged in the micropaleontological collections of the Department of Geology and Mineralogy, University of Queensland, Brisbane, Australia. Sample numbers are those cited in Balme and Playford (1967, p. 180). Slide coordinates are derived from standard "England Finder" slide.

Species	Figure number	Catalog number	Sample number	Slide number	Coordinates
Leiotriletes directus					
	6.4	Y.5758	44212	B1314/16	X40/0
	6.5	Y.5759	44210	B1279/6	R38/0
	6.6	Y.5760	44212	B1314/11	R46/4
Retusotriletes nigritellus					
	6.7	Y.5761	44213	293/40	M37/0
Granulatisporites absonus					
	6.8	Y.5762	44210	B1279/4	035/3
	6.9	Y.5763	44212	B1314/21	L45/2
Horriditriletes ramosus					
	6.10	Y.5764	44210	B1279/37	H34/2
Granulatisporites micronodosus					
	6.11	Y.5765	44210	B1279/27	J27/2
	6.12	Y.5766	44213	293/44	P36/4
	6.13	Y.5767	44212	B1314/25	Q33/3
Granulatisporites trisinus					
	6.14	Y.5768	44212	D1314/32	R39/1
Granulatisporites quadruplex					
	6.15	Y.5769	44213	293/10	L36/2
cf. *Acanthotriletes tereteangulatus*					
	6.16	Y.5770	44210	B1279/29	M38/4
Pseudoreticulatispora pseudoreticulata					
	6.17	Y.5771	44212	B1314/27	R44/1
Laevigatosporites vulgaris					
	6.18	Y.5772	44213	293/39	L36/2
Microbaculispora tentula					
	6.19	Y.5773	44212	B1314/8	M30/0
	6.20	Y.5774	44210	B1279/17	J35/0
Potonieisporites sp.					
	6.21	Y.5775	44212	B1314/26	S27/2
Indospora clara					
	6.22	Y.5776	44213	293/28	P34/2
	6.23	Y.5777	44213	293/27	M30/4
Microbaculispora villosa					
	6.24	Y.5778	44210	B1279/34	M35/0
Barakarites rotatus					
	6.25	Y.5779	44210	B1279/25	M32/0
Cannanoropollis mehtae					
	6.26	Y.5780	44212	B1314/4	H37/0
Scheuringipollenites maximus					
	6.27	Y.5781	44213	293/24	L31/0
Scheuringipollenites ovatus					
	6.28	Y.5782	44213	293/6	N33/1
Protohaploxypinus amplus					
	6.29	Y.5783	44213	293/34	N39/0
	6.30	Y.5784	44213	293/26	N31/4
	6.31	Y.5785	44213	293/36	Q37/2
	6.32	Y.5786	44213	293/16	N28/2
	6.33	Y.5787	44213	293/16	N30/0

Species	Figure number	Catalog number	Sample number	Slide number	Coordinates
Protohaploxypinus bharadwajii					
	6.34	Y.5788	44213	293/25	N38/0
Protohaploxypinus limpidus					
	6.35	Y.5789	44213	293/5	T36/0
	6.36	Y.5790	44213	293/41	U37/0
	6.37	Y.5791	44213	293/18	S34/3
Striatopodocarpites cancellatus					
	6.38	Y.5792	44213	293/9	S35/3
	6.39	Y.5793	44213	293/42	K43/2
	6.40	Y.5794	44213	293/14	K31/1
Striatopodocarpites fusus					
	6.41	Y.5795	44212	B1314/29	Q42/4
	6.42	Y.5796	44210	B1279/39	E32/3
Striatopodocarpites sp. A					
	6.43	Y.5797	44213	293/20	K41/2
Striatopodocarpites sp. B					
	6.44	Y.5798	44213	293/31	J29/4
Striatopodocarpites sp. C					
	6.45	Y.5799	44213	293/30	M39/4
Vittatina fasciolata					
	6.46	Y.5800	44210	B1279/18	M39/2
Praecolpatites sinuosus					
	6.47	Y.5801	44212	B1314/24	O39/0
	6.48	Y.5802	44212	B1314/13	L31/4
	6.49	Y.5803	44212	B1314/23	M35/4
Marsupipollenites striatus					
	6.50	Y.5804	44210	B1279/24	M25/0
	6.51	Y.5805	44213	293/8	K34/3
Marsupipollenites triradiatus					
	6.52	Y.5806	44210	B1279/9	O40/1
	6.53	Y.5807	44210	B1279/36	R47/3
Cycadopites cymbatus					
	6.54	Y.5808	44210	B1279/30	L41/0
cf. *Greinervillites* sp.					
	6.55	Y.5809	44213	293/29	N34/3

7—Cretaceous and Tertiary Vegetation of Antarctica: A Palynological Perspective

Elizabeth M. Truswell

Introduction

The data base on which the Cretaceous and Tertiary vegetation history of Antarctica can be built remains very slender; it consists of a few sequences onshore that have been sampled for their spore and pollen content, and an equally limited number of drillsites offshore, on the continental shelf, that have provided palynological data (Fig. 7.1). To these records of in situ pollen and spores, there can be added information available from the palynomorphs that are recycled by glacial processes and incorporated into surficial sediments on the seafloor around Antarctica.

This data base, meager as it is, can provide at least partial answers to various questions: What was the nature of the Cretaceous and Tertiary vegetation of Antarctica? Did it differ significantly from that of other southern continents? How long did the vegetation persist with the increasing development of the Cenozoic icecap? How does the available record fit in with the record of glacial history? What can be deduced of the climates that supported the Cretaceous and Tertiary vegetation? How does the history derived from the pollen record accord with that from macrofossil sources?

This review attempts to answer some of these issues by treating known pollen floras sequentially, beginning with the Early Cretaceous.

Cretaceous

Palynological data of Cretaceous age are known in greatest detail from the Antarctic Peninsula, where outcrop sequences have been studied in the South Shetlands and in the James Ross Island Basin. Data are also available from drill sites offshore, on the East Antarctic continental margin. The stratigraphic distribution of these sites is shown in Fig. 7.2. In addition, spores and pollen recognizably of Cretaceous age are common in surficial seafloor muds.

Sequence of Spore and Pollen Assemblages

Tithonian to Barremian

The oldest Cretaceous spore and pollen assemblages known are those reported from Livingston and Snow Islands, in the South Shetlands (Askin 1983). Localities include marine mudstones and nonmarine plant-bearing beds associated with volcanic debris. Palynomorphs from the area reflect a flora with a diversity of ferns, with other herbaceous taxa, including lycopods and bryophytes, probably as an understory to a canopy of conifers. The fern flora was diverse and included, as common elements, Cyatheaceae, Osmundaceae, and Gleicheniaceae. Present too were schizaeaceous elements [reflected in the dispersed spores *Cicatricosisporites aus-*

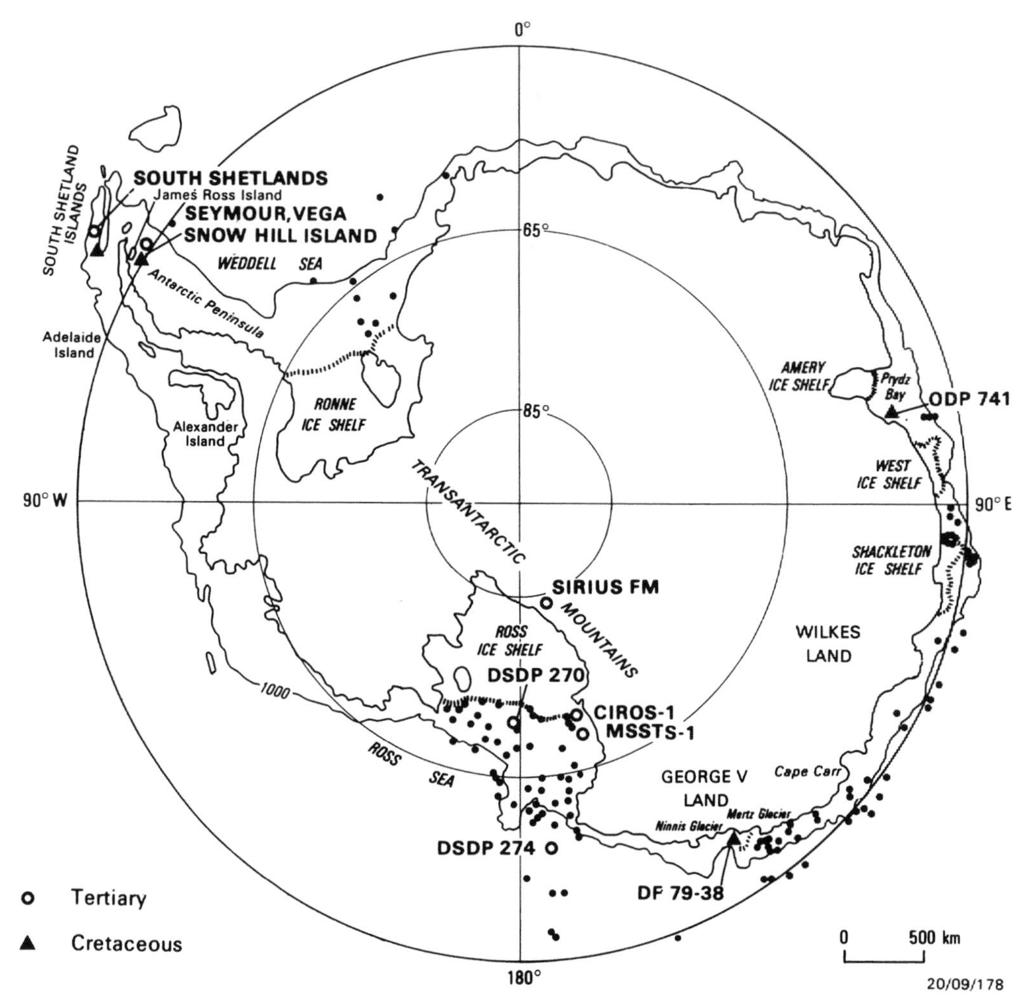

Figure 7.1. Locality map showing Cretaceous and Tertiary *in situ* palynological localities. Small dots show sites with recycled palynomorphs in surficial sediments offshore.

tralis, and *C. ludbrooki,* some of which may reflect *Ceratopteris* rather than Schizaeaceae; see Dettmann (1986) and *Klukisporites*].

Some assemblages from the nonmarine beds are dominated by the distinctive spore *Cyatheacidites annulatus* (referred to as *C. tectifera*? conspecific with *C. annulatus*). This dispersed spore is very similar to those of *Lophosoria quadripinnata,* a monotypic genus of tree ferns now inhabiting cool, moist habitats in central and South America, where it is an opportunistic invader of disturbed sites. In its present range, it occurs in forest and grassland habitats in the highland tropics, but further south, in Chile, Argentina, and the Juan Fernandez Islands, it grows from sea level to 1000 m. *C. annulatus* spores are widely distributed in Cretaceous and Tertiary sediments in Australia, southern South America, the Falklands, and southern Africa. The Berriasian record from the South Shetlands is the oldest from within this region; the Antarctic Peninsula may thus be part of a "cradle area" from whence later radiations occurred, but data are too sparse to trace migration paths.

Tree taxa within the South Shetlands assemblages include, as common elements, pollen referable to *Podocarpus;* less common is that of *Microcachrys* and *Araucaria.* The Cheirolepidiaceae or related groups are re-

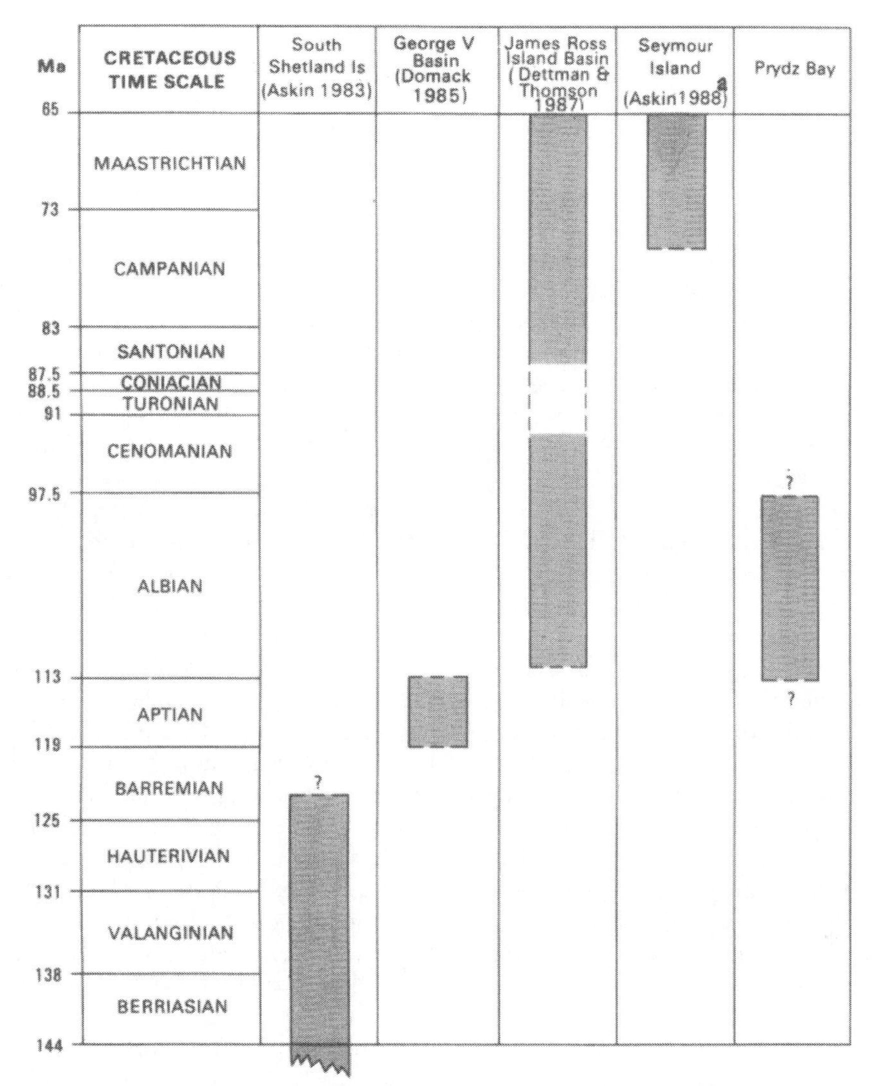

Figure 7.2. Columns showing stratigraphic extent of Cretaceous palynofloras from Antarctica.

flected by infrequently occurring *Classopollis* species. Lycopod and bryophyte spore taxa include *Ceratosporites equalis* (Selaginellaceae), *Foraminisporis dailyi,* and *Aequitriradites spinulosus* (Hepaticae).

Aptian

The palynology of Aptian sediments is known with confidence only from East Antarctica, a wide region generally poor in palynological data. On the continental margin offshore from the George V/Adelie coast, pollen and spores were recovered from brecciated siltstones drilled in the Mertz-Ninnis Trough, an inner shelf depression lying parallel to the coast (Domack et al. 1980). Similar assemblages have also been reported from brown mudstone clasts in diamictites overlying the siltstones (Domack 1985). Comparison of the spore and pollen assemblages with sequences from southeastern Australia suggests assignment to the *Cicatricosisporites hughesii* subzone of the *Dictyotosporites speciosus* zone; equivalence to Burger's (1980) *Osmundacidites dubious* zone, established in intracratonic sequences further north in Australia, was also suggested. This correlation suggests an Aptian age for the siltstones (see Dettmann 1986 for a chronological calibration of current zonal schemes).

The probable affinities of the spore and pollen species described by Domack et al. (1980) were listed by Truswell (1983, Table 4). A parent vegetation is suggested with diverse ferns, including Cyatheaceae, Schizaeaceae, and Osmundaceae, lycopods (Selaginellaceae and Lycopodiaceae), bryophytes, and conifers, the last including Podocarpaceae and Cheirolepidiaceae. Recently, palynomorph-rich residues from the core DF79-38 were reexamined (Truswell unpublished data) and the pollen species *Vitreisporites pallidus* of possible pteridosperm origin (?Caytoniales) and *Cycadopites nitidus,* representing Cycadales, Bennettitales, or Ginkgoales, were added to the known list. Quantitatively, the pollen and spore assemblages suggest an abundance of podocarpaceous conifers (45% of the spore and pollen sum) and of cyatheaceous/dicksoniaceous ferns (40%), although without figures for production and dispersal of propagules in these groups it is not easy to interpret these pollen frequencies in terms of relative importance in the contemporary vegetation. *Classopollis* occurs in frequencies of less than 1%. No angiosperm pollen was recorded.

New information on Aptian spore and pollen assemblages is likely to be available in the near future, from organic-rich mudstones drilled during 1987 in the eastern Weddell Sea off Dronning Maud Land (Barker et al. 1987). No details of the spore–pollen content of these are yet published. Recycled palynomorphs in recent muds in the Weddell Sea were used to predict the occurrence of the *in situ* sequences (Truswell and Anderson 1985). Off Dronning Maud Land, the recycled suite contains, as the most common elements, probable schizaeaceous ferns (*Cicatricosisporites, Contignisporites*), lycopods (*Leptolepidites, Retitriletes*), and podocarpaceous conifers.

Albian to Cenomanian

Palynomorph assemblages of Albian to Cenomanian age are best known from the James Ross Island Basin of the Antarctic Peninsula. The following account of the main vegetation features that they reflect is summarized from Dettmann and Thomson (1987). The sequence from James Ross Island is predominantly marine, with faunas that allow reasonably accurate dating in terms of an international time scale. The stratigraphic sequence and the taxonomic composition, in terms of major plant groups, are shown in Fig. 7.3 (from Dettmann and Thomson 1987).

Early Albian assemblages are distinguished by high frequencies of podocarpaceous pollen—*Podocarpus* and *Microcachrys* types—and by abundant *Araucaria* and pteridosperms. Pollen of cheirolepidiaceous conifers is present in low frequencies, a contrast with contemporaneous sediments in South America. In the proportions of major taxa, which suggest that the vegetation was one of temperate podocarp–araucarian rainforests, the assemblages resemble those of Australia and New Zealand. Differences, however, are to be found in understory communities. Fern spores in the James Ross Island basin suggest rainforest associates—Osmundaceae, Hymenophyllaceae, Dicksoniaceae, and Cyatheaceae (*Lophosoria*). In New Zealand, the component of lycopodiaceous spores is higher, perhaps reflecting open moorland or rain forest fringe habitats. In southeastern Australia, the presence of rain forest and drier zone ferns, together with lycopods, suggests diverse habitats perhaps reflecting slope topography associated with the Australia/Antarctica rift. The distribution of these variants on a temperate rain forest theme is shown in Fig. 7.4.

The pollen taxon *Clavatipollenites* shows angiosperms to have been established in the Antarctic Peninsula by the Early Albian. This appearance date is slightly younger than the first confidently dated angiospermous pollen in Australia. In the James Ross Island sequence, the diversity of angiosperm pollen increased through the later part of the Albian and into the Cenomanian. The chloranthaceous stock, which *Clavatipollenites* probably represents, was increased by the addition of the trichotomosulcate *Asteropollis*. Tricolpate, tricolporate, and monosulcate forms,

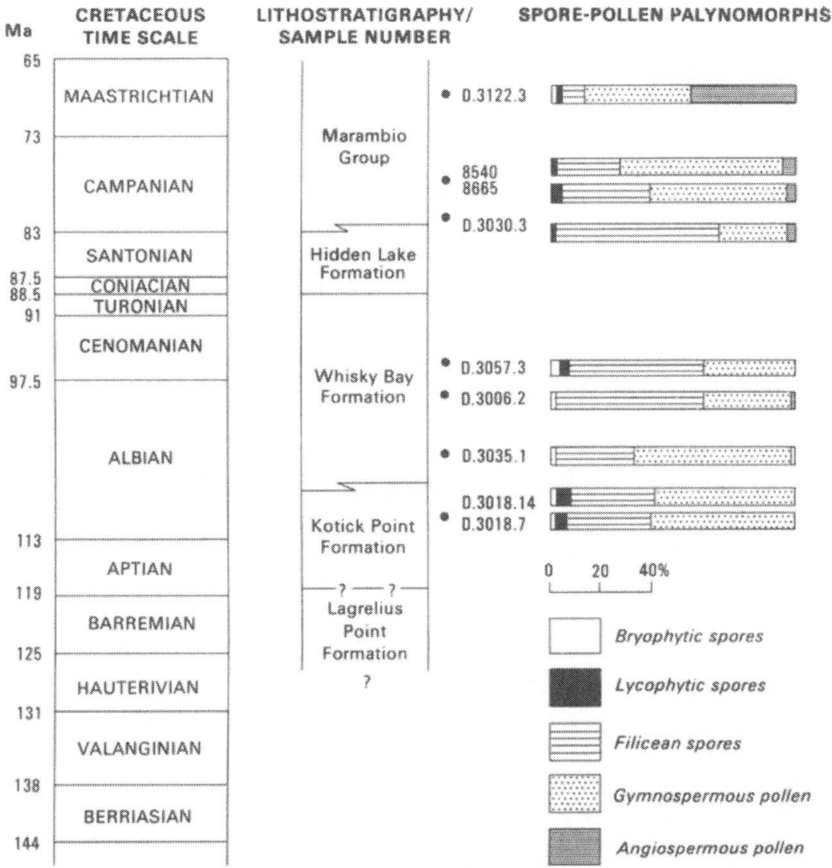

Figure 7.3. Stratigraphy of James Ross Island, with palynomorph samples on right of lithostratigraphy column, and composition of assemblages in major plant groups (modified from Dettmann and Thomson 1987).

most of unknown affinity, also appear in this interval, although representation of podocarpaceous conifers remains high.

On the East Antarctic continental margin, palynological assemblages probably of Albian age have been obtained from siltstones drilled during the Ocean Drilling Program leg 119 cruise to Prydz Bay. The palynology of seaward-dipping sequences on the continental shelf there has been examined in only a preliminary way (Truswell in prep.), but assemblages from site 741 are probably of Albian age. They are of low diversity, but fern spores of probable schizaeaceous affinity (as *Appendicisporites* sp., suggesting *Anemia* or *Mohria* types) are unusually common. There is a high frequency of podocarpaceous pollen and a notable component of hepatics (*Aequitriradites, Foraminisporis, Coptospora*). Fungal hyphae and resting spores are of unusually high abundance. In keeping with other Antarctic sites of this age, *Classopollis* is rare.

In the Weddell Sea, the presence of sequences of Aptian through Cenomanian age in the eastern sector has already been noted. Recycled indicators of Albian age in recent seafloor muds (Truswell and Anderson 1985) include the fern *Appendicisporites distocarinatus* (?Schizaeaceae), the hepatic *Coptospora paradoxa,* and the megaspore *Balmeisporites glenelgensis*.

Campanian to Maastrichtian

In situ palynological assemblages representing this time interval are known only from the Antarctic Peninsula area. Information from James Ross, Seymour, and Vega islands has been summarized by Dettmann and Thomson

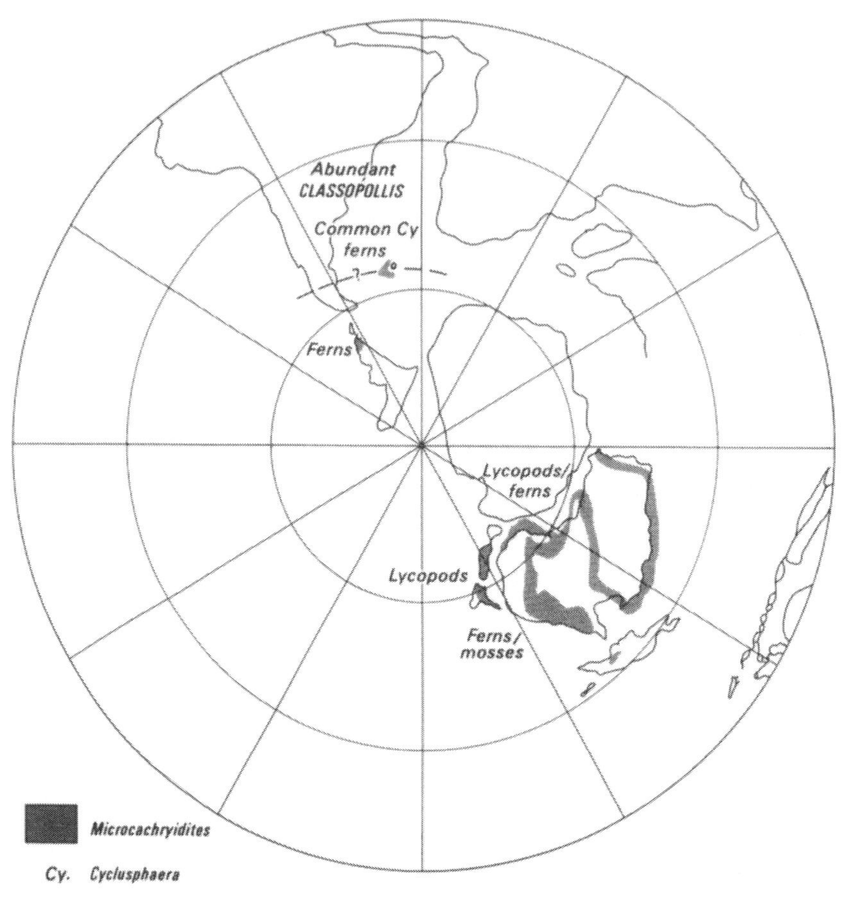

Figure 7.4. Regional variations of Cretaceous (Albian) palynofloras (modified from Dettmann and Thomson 1987).

(1987) and Askin (1988 a and b). There is presently no information on the Turonian and Coniacian.

In the late Santonian or early Campanian, the podocarp/araucarian rain forests were modified by the introduction of a number of taxa characteristic of later, Tertiary austral rain forests. The ancestral complex of the extant Huon Pine, *Lagarostrobus franklinii,* appeared in the early Campanian, as witnessed by the pollen type *Phyllocladidites mawsonii;* in Australia and New Zealand, a major radiation of this lineage occurred in the earliest Tertiary. About the mid-Campanian, pollen of *Dacrydium* (Podocarpaceae), *Nothofagus,* Proteaceae, and Myrtaceae appears, all of which play an increasingly important role in the latest Cretaceous and Tertiary forests. The earliest *Nothofagus* pollen, according to Dettmann and Thomson (1987), is distinct from all extant morphological groups and may be viewed as an ancestral type.

The vegetation of the Santonian and Campanian, as interpreted from the pollen data, shows close parallels between the Antarctic Peninsula, New Zealand, Australia, and the Falklands Plateau. The greatest similarities are evident among the gymnosperms, with most pollen taxa in that group being widely distributed throughout the region; there are similarities (but some differences), too, in the *Nothofagus* element. The major regional differences are focused on the cryptogam floras and on some of the angiosperms.

Further vegetation change occurred in the Maastrichtian. Askin (1988b) notes an increasing diversity in spore and pollen species through the Campanian–Maastrichtian, particularly among the angiosperms. Dettmann and Thomson (1987) suggest that change occurred during the Maastrichtian, as evident in the material from Vega Island, with angiosperm pollen reaching peak diversity then, suggesting that angiosperms matched gymnosperms in importance in the vegetation. Twelve species of *Nothofagus* pollen are apparent, including representatives of all three extant morphotypes. Tricolpate, polyporate, and triporate pollen types are frequent. The affinities of most of the pollen types are obscure or speculative. Apart from *Nothofagus,* types related to the Proteaceae can be identified; the common species *Peninsulapollis gillii* shares features with extant *Beauprea* (Dettmann and Jarzen 1988), but further work is required to confirm this affinity. Many morphotypes clearly reflect extinct groups, for instance the *Gambierina* types, which have some morphological affinities with the northern hemisphere Normapolles group.

Provincialism in Late Cretaceous Floras

It is within the latest Cretaceous pollen suites, especially among the angiosperms, that regionalism is most apparent within the southern floras. Askin (1988a), in her study of Seymour and nearby islands, distinguished five levels of provincialism. There are species endemic to the James Ross Island Basin; species confined to Antarctica; others occurring only within the Weddellian Province (encompassing southern South America, Antarctica, New Zealand, and southeastern Australia); species with an austral, or wider southern distribution; and those that are cosmopolitan in their geographic range.

Dettmann and Jarzen (1988) have described a variety of angiosperm pollen, with a variety of provincial connotations. Province boundaries have been recognized on the basis of in situ localities for pollen and spores, but the recycled suites of palynomorphs that occur in recent muds on the Antarctic continental shelf provide an additional source of information on the former distribution of species. Species assigned to *Beaupreaidites,* possibly with affinities to extant *Beauprea,* appear to have an Antarctic/southern Australian distribution [Fig. 7.5(a)], with the species *B. orbiculatus* being more southern in its range than *B. elegansiformis*. *Peninsulapollis gillii,* also with possible affinities to *Beauprea,* is Weddellian in its distribution, with records on the Antarctic Peninsula, in southern South America, New Zealand, southeastern Australia, and on the East Antarctic continental margin [Fig. 7.5(b)]. Only in the Antarctic Peninsula, however, is it associated with morphologically similar species, suggesting that that area was an evolutionary center. *P. askiniae,* on present data, has a much narrower range, being confined to the James Ross Island Basin and to the eastern Ross Sea.

Notable among species that are markedly southern in their distribution are taxa morphologically similar to the heterogenous northern hemisphere Normapolles group. Distribution of three recognized southern species is shown in Fig. 7.5(c). The fact that the southern forms appear later, in the Senonian, and show some distinct morphological differences from their northern counterparts, suggests that evolution of their parent taxa may have proceeded independently.

Relation of Pollen Floras to Macrofossils

Macrofossil floras of Cretaceous age are known only from the Antarctic Peninsula, with records from the South Shetlands, Alexander Island, and, largely undescribed, Snow Hill Island. The flora at Hope Bay from the tip of the Antarctic Peninsula, discovered during the Swedish South Polar Expedition of 1901–1903, was considered initially to be Jurassic in age (Nathorst 1904, 1907; Halle 1913), but more recent assessments, based on floristic comparisons with South American floras and on lithostratigraphic correlations, suggest that it may be as young as Neocomian (Stipanicic and Bonetti 1970; Farquarson 1984; Baldoni 1986; Gee 1987).

Only the macrofossil flora from the South Shetlands has palynomorphs reported from the same area and stratigraphic section. From

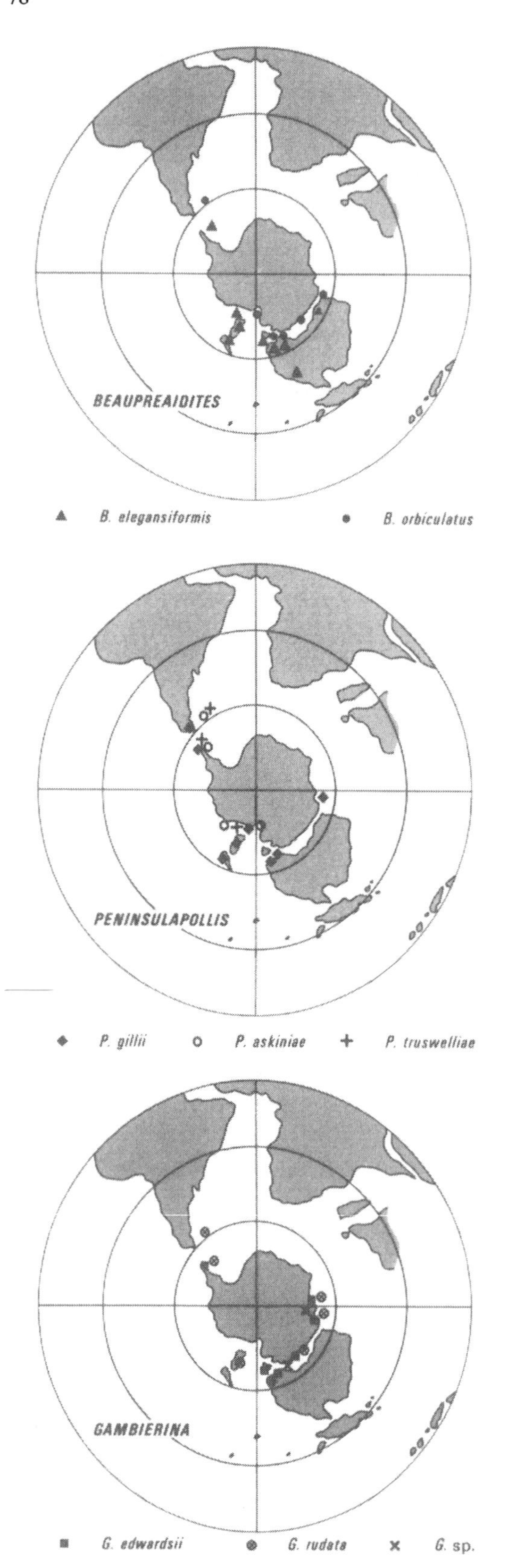

Livingston Island, Hernandez and Azcarate (1971) reported a flora of ferns, cycadophytes (including *Williamsoniella*), and conifers from a locality west of Cerro Negro; from President Head, Snow Island, Fuenzalida et al. (1972) listed leaf floras with cycadeoid fronds, including species assigned to *Dictyozamites*, *Ptilophyllum*, and *Otozamites*, and others that could be sterile foliage of pteridosperms (*Pachypteris*, *Stenopteris*) and conifers (*Elatocladus*). In some ways, the plant microfossil suites reflect the macrofossil, but differences are apparent. The spore record shows a diversity of ferns unparalleled as yet in the frond record; conversely, the frond associations include a diversity of cycadophytes that has no palynological expression.

Attempts to recover palynomorphs from the other macrofossil localities, namely, from Mt. Flora, and the ?Barremian to Albian Fossil Bluff Formation of Alexander Island, have proven unsuccessful because of levels of alteration of the containing sediments. Nevertheless, the macrofossil assemblages reflect in a most general way the composition of the Early Cretaceous vegetation as this is understood from the wider palynological record of Antarctica. Jefferson (1981, 1982) used the macrofossil data to reconstruct a variety of plant communities growing within a fluvially dominated delta; this environment included flood plain associations dominated by ferns (with leaf forms suggestive of Osmundaceae, Matoniaceae, and Dicksoniaceae) and lake margin communities of variable diversity, ranging from fern-dominated to diverse fern–conifer–ginkgoalean–cycad communities. The dominant vegetation was podocarp forest, with trees widely spaced, as witnessed by the distribution of standing stumps, and with an understory of taeniopterids and ferns. Trees were fast growing, with high annual increments of wood and pronounced sensitivity to environmental change. Rare angiosperm

Figure 7.5. Distribution of selected angiospermous pollen taxa in the Late Cretaceous and Early Tertiary (from Dettmann and Jarzen 1988, with additional data from R.A. Askin).

leaves occur in the upper (Albian) part of the sequence.

The Hope Bay flora also contains diverse ferns and abundant conifer foliage, both features which are echoed in the palynology from a range of Antarctic sites. There is, however, as in the South Shetland floras, an abundance of cycadophytic material, chiefly leaf fronds but including a *Williamsonia* "flower"; this group is underrepresented palynologically.

The prevalence of forests of podocarps and araucarians, which is suggested by the palynological record, is supported by fossil wood studies. Data on Early Cretaceous growth patterns are available from the South Shetlands, from Mt Flora, and from Whisky Bay on James Ross Island; Late Cretaceous wood data include sites at Lachman Crags and the Naze (James Ross Island) and Cape Lamb, Vega Island (Francis 1986). Angiosperm wood, recovered from the Late Cretaceous only, is dominantly that of *Nothofagus,* in parallel with palynological evidence for radiation of this genus in the Late Cretaceous. Growth rings in wood from the northern end of the Antarctic Peninsula are again wide, suggesting favorable conditions; rates were, however, more uniform from year to year than on the southern end of the peninsula.

Tertiary

The Cretaceous–Tertiary Transition

The sequence on Seymour Island allows examination of vegetation events across the Cretaceous–Tertiary boundary. In palynological terms, the boundary there has been narrowed to the changes occurring between successive dinoflagellate zones (Askin 1988b), a change that occurs above a laterally persistent glauconite bed. This bed, which also marks the last appearance of ammonites and other groups of shelly fossils, occurs within the uppermost Lopez de Bertodano Formation. Within the spore and pollen flora, reflecting the land vegetation, no abrupt extinction events can be recognized, but there is a continual sequence of first and last appearances over the transition interval. In general, Paleocene assemblages show a reduction in species diversity from that of the Late Cretaceous, suggesting a period of cooling.

The Tertiary Sequence of Palynological Assemblages

The following describes the sequence of vegetation changes that can be discerned through the Tertiary of Antarctica on the basis of evidence presently available. The data are scarce, however, so the picture remains hazy. The discussion centers on specific regions of Antarctica, as sequences there provide insights into particular parts of the geological column; there is, however, some time overlap between areas. Stratigraphic distribution of the known sequences is shown in Fig. 7.6.

Paleocene through Eocene: The Antarctic Peninsula

Askin's (1988a,b) account of the palynostratigraphy of Seymour Island, through the Lopez de Bertodano to the Sobral Formation, devotes only brief comment to the land-derived component. The broad picture through this interval is one in which conifer pollen—mainly of araucarians and podocarps—predominates, with abundant and diverse *Nothofagus* and fern spores. Other angiosperm groups include Proteaceae, Loranthaceae, Myrtaceae, Casuarinaceae, Ericales (?Epacridaceae), and Liliaceae. There is a diversity increase in the Eocene, with increases in abundance of *Nothofagus* and Proteaceae. A similar picture, using frequency data for the major plant groups, was portrayed by Baldoni and Barreda (1986).

The earliest account of Seymour Island palynology was that of Cranwell (1959) of pollen from a tuffaceous limestone collected on the Swedish South Polar Expedition, probably from the Paleocene Cross Valley Formation. She recorded a dominance and a diversity of *Nothofagus,* with subsidiary podocarps and *Araucaria.* Among subsidiary angiosperms, she reported Cruciferae, (the only report to date from Antarctica), Myrtaceae, Proteaceae, (aff. *Lomatia*), Lo-

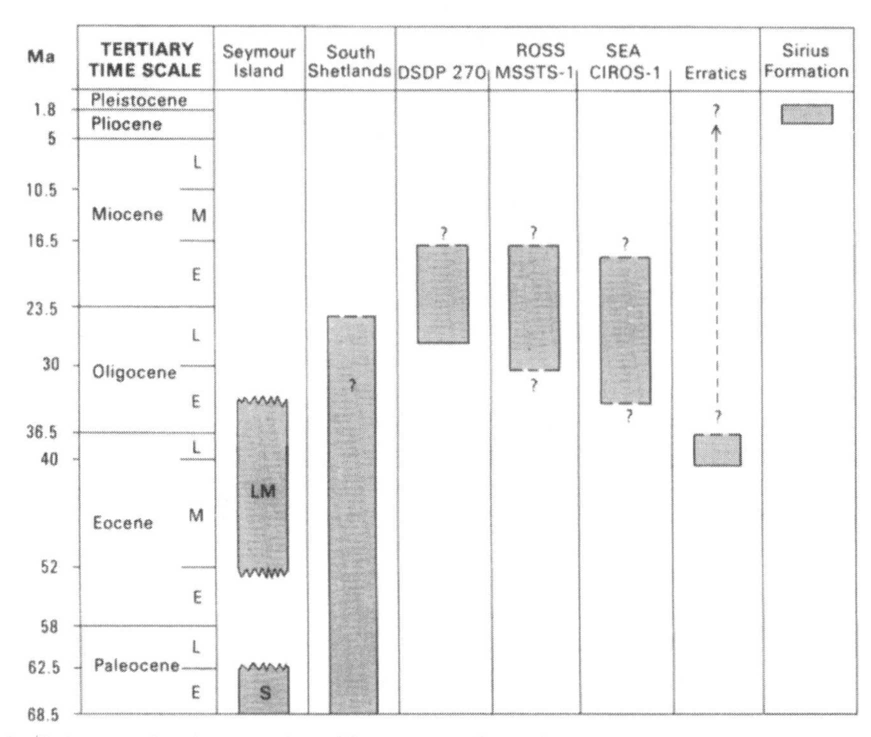

Figure 7.6. Columns showing stratigraphic extent of Tertiary palynofloras from Antarctica. LM, LaMeseta Formation, S, Sobral Formation.

ranthaceae, *Fuchsia*-like grains (not subsequently reported), and possible Winteraceae. She also noted small grains possibly of Cunoniaceae (cf. *Weinmannia*) or Elaeocarpaceae; these are of particular interest because pollen of these has recently been reported as abundant in deposits of Late Paleocene to Eocene age in central Australia, and the vegetation there is described as "Cunoniaceae-dominated vine forests with a significant conifer component" (Sluiter in press). This vegetation type seems likely to have been part of the mosaic of Australian Early Tertiary vegetation; Cranwell's data hint that it may have been a type with a wider southern distribution.

A coal seam, possibly within the Cross Valley Formation (Fleming and Askin 1982), may reflect a specialized swamp forest flora. Pollen assemblages in it are dominated by *Phyllocladidites,* representing the lineage incorporating the extant *Lagarostrobus franklinii,* the Tasmanian Huon Pine. This is also particularly abundant in Tertiary coal seams in the Gippsland and Murray Basins of southeastern Australia (Stover and Partridge 1973; Martin 1984).

Other comments on the Tertiary part of the Seymour Island sequence include Hall's (1977) observation of grass pollen in the La Meseta Formation. The only subsequent record of grass in Antarctica is that of Mildenhall (in press) for the Ross Sea, a report that is qualified by doubts concerning the in situ nature of the occurrence.

Elsewhere in the Antarctic Peninsula, pollen and spores have been recovered in the South Shetlands, from the Ezcurra Inlet Group exposed at Admiralty Bay, King George Island (Stuchlik 1981). The Ezcurra Inlet Group occurs about midway in the series of calc-alkaline volcanics that make up the King George Island Supergroup (Birkenmajer 1980); dates on the plant-bearing Point Hennequin Group indicate an age of 24.5 ± 0.5 Ma (just below the Oligo–Miocene boundary). The pollen suites are older than that, but other age constraints are poor. The assemblages show a co-dominance of *Nothofagus* and ferns. Nine morphotypes of *Nothofagus* pollen were identified; *fusca* types predominate, and species comparable to pollen of extant *N. betuloides* and *N. antarcticus* are most common. Rhamnaceae pollen is recorded as a minor component. Among the ferns, cyatheaceous types predominate. Stuchlik suggests a *Nothofagus* forest with a fern understory; conifers do not seem to have been a significant feature.

To complete the record of palynomorph discoveries on the Antarctic Peninsula, brief mention should be made of a record from Adelaide Island (Jefferson 1980), where carbonized pollen grains suggestive of those of *Nothofagus* occur in association with poorly preserved leaves of the same genus.

?Eocene through Miocene: The Ross Sea

When glacial processes became a significant factor in sedimentation, the palynological record becomes difficult, almost impossible, to interpret in terms of vegetation communities and cannot provide a clear record of the vegetation growing at the time the sediments were deposited. It can, however, furnish an inventory of taxa that grew within a region, but no record of when a particular species might have flourished there. Nowhere is this problem more apparent than in the Ross Sea. In this region, all available palynological data are dogged by the specter of recycling, by the fact that some, if not all of the observed pollen does not represent vegetation contemporaneous with sedimentation. From the area, data are available from core holes drilled during Deep Sea Drilling Project cruises, from cores drilled from sea-ice platforms (the New Zealand MSSTS-1 and CIROS-1 boreholes) or through the Ross Ice Shelf itself (J9), from erratic boulders in morainal debris, and from material clearly recycled into surficial sediments.

The earliest known sources of pollen were in erratic boulders from Black Island and Minna Bluff, in the McMurdo Sound area, and thought to be Eocene on the basis of their dinoflagellate content (McIntyre and Wilson 1966). Pollen assemblages in these sources are dominated by *Nothofagus* (up to 83%), with all morphological groups present. Other angiosperms suggest Proteaceae and Casuarinaceae, or else they are of unknown affinity. Gymnosperms were referred to *Podocarpus*, cf. *Libocedrus*, and *Dacrydium*. Until recently, the flora in these erratics has been considered to represent a standard for the Eocene of Antarctica; however, doubt has been cast on the in situ nature of the palynomorphs in the erratics with the identification by Stott et al. (1983) of Miocene or Pliocene diatoms in the erratics from Minna Bluff.

The alleged rain forest vegetation represented by the palynomorph suite in the erratics persisted into the Late Oligocene, according to records in DSDP site 270, drilled off the edge of the Ross Ice Barrier (Kemp 1975; Kemp and Barrett 1975). Sedimentological evidence indicates that glaciers had reached sea level by this time, so the Oligocene vegetation, it was argued, persisted into these early phases of continental glaciation. The Late Oligocene was again dominated by *Nothofagus;* proteaceous pollen was represented by 4 or 5 species; also present were Myrtaceae, *Podocarpus,* and rare, mainly cyatheaceous ferns. All species identified are known to be long ranging in Australia and New Zealand, so a recycled origin for them could not be precluded.

Pollen recovered from a drill core penetrating bottom sediments beneath the Ross Ice Shelf was dated as Miocene on the basis of associated fauna and was thought to represent a continuation of the earlier *Nothofagus*/podocarp forests (Brady and Martin 1979). However, diatoms in the core have been established as Pleistocene in age, so the palynomorphs have a recycled origin and do not necessarily reflect Miocene vegetation.

Data from the two New Zealand core holes, MSSTS-1 and CIROS-1, are similarly insecure, but they do provide some insights into vegetation history. The MSSTS-1 hole, dated by diatoms and magnetic reversal stratigraphy as being as old as Late Oligocene, was drilled entirely in glacigene strata. Recovery of pollen and spores was sparse, assemblages were of low diversity and of long-ranging taxa. Dominant was *Nothofagus,* with subsidiary *Dacrydium, Podocarpus, Microcachrys, Casuarina,* Liliaceae, and Proteaceae. In CIROS-1, 715 m of mudstones, muddy sandstones, and diamictites were penetrated; foraminifera at the base of the hole suggest an Early Oligocene age, the oldest record of sea-level ice around Antarctica. Although pollen recovery was sparse in the core, and diversity low, Mildenhall (in press) has argued that the number and variety of *Nothofagus* pollen, and the fact that in some intervals they cohere in clumps of monospecific grains, suggesting anthers, indicate that they are in situ, and that forests containing a variety of *Nothofagus* species must have flourished nearby in the Oligocene. The presence in the core at 215 m of a leaf resembling those of the deciduous Tasmanian alpine species *Nothofagus gunnii* is clear evidence for the presence of such forests (Hill in press). Pollen of Liliaceae and of some unknown angiosperms is also reported to occur in clumps. The distinctiveness of these forests is evident from the fact that many of the *Nothofagus* pollen types that occur in Antarctica differ from those of Tertiary Australia and New Zealand. Other elements in CIROS-1 that may reflect pollen production from contemporary forests include Podocarpaceae (*Podocarpus, Dacrydium, Microcachrys, ?Phyllocladus,* and *Lagarostrobus*), rare Casuarinaceae, Proteaceae, and Myrtaceae, as well as a variety of unknown angiosperms. The site provided additions to the summary list of Antarctic Tertiary taxa given in Truswell (1983, Table 5), including *Epilobium* (Onagraceae), Chenopodiaceae, possibly Convolvulaceae, and *Coprosma* (Rubiaceae).

In both MSSTS-1 and CIROS-1, there is a gradual diminution of taxa throughout the Oligocene and Early Miocene, suggesting increasing stress on the vegetation with increasing glaciation. Bearing in mind the uncertainty imposed on the reconstructed vegetation history because of the uncertain stratigraphic position of many of the pollen grains, Mildenhall (in press) suggested that the coastal vegetation in the Ross Sea region during the Oligocene was essentially a *Nothofagus* forest with podocarps, Proteaceae, and other shrubby angiosperms. The prominence in the pollen spectrum of the *fusca* morphotype of *Nothofagus,* morphologically similar to the pollen of *N. dombeyi,* suggests similarity to modern Chilean forests in which that species is prominent. The Antarctic material, however, suggests a richer variety of *Nothofagus* species than in the modern *N. dombeyi* forest.

There is as yet little evidence to suggest that the Early Tertiary forests of greater Antarctica declined through a tundra or scrubland phase with increased glaciation. Herbaceous taxa are poorly represented in the pollen spectrum, although Mildenhall reports chenopods and grasses (?contaminants) in CIROS-1. It remains possible that these represent an open vegetation phase beginning in the Late Oligocene.

Post-Miocene Vegetation: The Sirius Formation

The decrease in abundance of pollen into the Miocene, from already depauperate Late Oligocene assemblages, means that there is really no credible record of Miocene vegetation. The available record of glaciation, from sedimentological and isotopic data (Mercer 1986; Murphy and Kennett 1985) suggests major ice cap development in the midMiocene, so it may be that the coastal forests fringing the Ross Sea were largely eliminated in that interval. That being the case, the evidence for vegetation presented by fossil assemblages in the Sirius Formation is the more remarkable. Wood fragments, reflecting in situ woody shrubs, and pollen have been recovered from glacial deposits at Oliver Bluffs near the head of the Beardmore Glacier at elevations of 1800–1900 m (Webb et al. 1987). Recycled marine diatoms in the unit suggest an age as young as 2.5 Ma. The wood fragments (Carlquist 1987) are of a single species of *Nothofagus,* matching in cell detail the features of the extant *N. betuloides* of South America and *N. gunnii* of Tasmania. These two species have the southernmost ranges of living members of the genus. Pollen recovered from the formation similarly suggests floristic poverty: according to Askin and Markgraf (1986), the sediments contain pollen of the *N. fusca* group (to which both *N. gunnii* and *N. betuloides* belong), with rare podocarps and tricolpate pollen referable to Labiateae or Polygonaceae.

Relation of Pollen Floras to Macrofossils

Apart from the single *Nothofagus* leaf from the Ross Sea and the Sirius Formation wood, macrofossil remains of Tertiary age are known only from the Antarctic Peninsula. There they have been described from Seymour Island, the South Shetlands, and Adelaide and Alexander islands. Descriptive works published in the early part of this century (e.g., Dusèn 1908, Gothan 1908) need critical revision in terms of leaf architecture, and the assignment of many fossils to living genera needs reevaluation.

The leaf floras from Seymour Island described by Dusèn probably came from the Cross Valley Formation. They contain abundant ferns, *Araucaria,* and a diversity of angiosperms, including leaves identified as *Nothofagus* (also *Fagus,* with species recently transferred to *Nothofagus* by Tanai 1986), *Drimys, Caldcluvia, Laurelia, Lomatia,* and *Knightia*—these Dusèn considered to be a cool temperate element related to the living flora of southern Chile and Patagonia. He also identified what he considered to be a warmer element—leaves referred to *Mollinedia* (Monimiaceae) and *Miconiphyllum*—which has suggested affinity with *Miconia* (Melastomataceae). The recent study by Case (1988), describing leaf assemblages from Paleocene and Middle and Late Eocene horizons on Seymour Island includes only the cool temperate element. The Paleocene flora is dominated by ferns, with some *Nothofagus* and podocarps, the Middle Eocene by a large leaved *Nothofagus,* and the Late Eocene by two smaller leaved species of the same genus. Published pollen data are as yet too few to compare with leaf floras, but both fossil sources agree in a general way, with *Araucaria, Nothofagus,* and Podocarpaceae prominent in the pollen flora. Of interest in the comparative context is Cranwell's (1959) record of common pollen of Cunoniaceae; this may reflect the same parentage as leaves Dusèn assigned to *Caldcluvia.*

The fossil woods from the Cross Valley and La Meseta Formations described by Francis (1986) provide, through their annual ring widths, a record of the climates that sustained tree growth. A decrease in ring widths from the Late Cretaceous to the Paleocene suggests climatic cooling; increased width in the Eocene may reflect subsequent warming. Specimens are, however, very few. There are parallel shifts in pollen diversity, which increased from the Paleocene into the Eocene (Askin 1988b).

In the South Shetlands, fossil floras are known from a number of levels in the volcanic–sedimentary sequence on King George Island, including the Dufayel Island, Barton Peninsula, Fildes Peninsula, and Point Hennequin groups. The flora from the Dufayel Island Group, bracketed by K–Ar dates on associated basalt as being in the 51–56 Ma age range (Late Paleocene to Early Eocene), is entirely angiospermous, containing impressions of *Nothofagus,* leaves of lauraceous and myrtaceous affinities, and a variety of other taxa doubtfully referred to as *?Dodonea, ?Sterculia, Tetracera,* Cochlospermaceae, Dilleniaceae, Leguminoseae, Sapindaceae, and Verbeniaceae (Birkenmajer and Zastawniak 1986). Some analogies to the Valdivian rain forest of Patagonia have been suggested, but uncertainty of taxonomic assignment in the fossil flora makes comparison very tenuous. The Fildes Peninsular flora (Orlando, 1964) contains an apparent mix of dicotyledonous taxa, with some, including Anacardiaceae, Dilleniaceae, Monimiaceae, of supposed subtropical distribution, with gymnospermous taxa, including those assigned to *Fitzroya.*

The youngest fossil floras from the South Shetlands are those from the Point Hennequin Group, where plant beds occur in the Mt Wawel Formation and in the Dragon Glacier moraine, not far above a lava flow dated at 24 ± 0.5 Ma (Zastwniak 1981; Zastawniak et al. 1985). Both assemblages are similar except for the greater richness of ferns at the Dragon Glacier site. *Nothofagus* leaves predominate, and resemblances have been noted among these to southern South American species, including *N. alessandri, N. alpina, N. dombeyi, N. glauca,* and *N. procera.* Podocarpaceae are represented in the floras by shoots and seeds. Angiosperm leaves other than *Nothofagus* are rare, but fragments possibly attributable to Rhamnaceae occur, as do palmately lobed leaves referred to the tropical *Cochlospermum.* Leaf size classes show a predominance (75%) of microphylls. The Point Hennequin floras immediately predate the onset of continental glaciation in the King George Island area, an event that Birkenmajer (1987) considers to be latest Oligocene in age.

Palynological assemblages from lower in the South Shetlands sequence (Stuchlik 1981) differ little in composition from the macrofossils, except that they are dominated by ferns and the podocarpaceous element is missing, probably because of local factors. The Point Hennequin floras are of particular interest in that they are similar in composition to the *Nothofagus*/podocarp forests suggested by contemporaneous pollen assemblages from the Ross Sea, even to the fact that *Nothofagus* species are diverse in them. This low-diversity vegetation thus seems to have been widespread, covering both low-latitude Antarctica and at least coastal areas closer to the pole during the early phases of Cenozoic glaciation.

Summary

The palynological record from the Cretaceous and Tertiary of Antarctica is still meager when compared with that from other continents. The available data consist of information from a very few outcrop sequences on land and from a small number of borehole sequences drilled offshore, augmented by what can be gleaned from pollen recycled into recent marine muds on the continental shelf. The available data are sufficient to provide, in sketchy outline, a history of the vegetation of at least coastal Antarctica during the Late Mesozoic and Tertiary. The picture presented accords with the information (also sparse) from plant macrofossils.

The Cretaceous vegetation was not markedly different from that of other southern continents. The Early Cretaceous vegetation was probably a conifer forest, rich in podocarps and Araucariaceae, with a fern understory, whose compositon varied from region to region. Variable, too, was the regional representation of lycopods and bryophytes. The importance of Cheirolepidiaceae appears to have been significantly less in Antarctica than in regions further to the north.

Palynological assemblages from the James Ross Island Basin record the vegetation history of peninsular, low-latitude Antarctica after the advent of the angiosperms, which appears to have occurred in Antarctica in the Early Albian. The Late Cretaceous saw radiations within gymnosperm and angiosperm components. In approximately the Early Campanian, the podocarp/araucarian rain forests were modified by the introduction, among the conifers, of the *Lagarostrobus* (Huon Pine) type and by *Dacrydium;* among angiosperms, *Nothofagus,* Proteaceae, and Myrtaceae appeared. A further increase in variety seems to have characterized the Maastrichtian, when angiosperm pollen reached its peak diversity, suggesting that the flowering plants were then of comparable importance to the gymnosperms in the Antarctic vegetation. In the Late Cretaceous, provinciality is most apparent in the southern floras, with regional distributions clearest among angiosperms.

The Cretaceous/Tertiary transition, as known from Seymour Island, saw no abrupt extinction events; rather, there was a series of extinctions and first appearances. There was a general reduction in diversity from the Cretaceous to the Paleocene, which accords with cooling. In the Paleocene, the picture continues to be one of an Antarctic vegetation dominantly of podocarp/araucarian forests, with a marked diversity of *Nothofagus* and other angiosperms including Proteaceae, Myrtaceae, Casuarinaceae, Liliaceae, and Cunoniaceae. Eocene floras are as yet published in rough outline only, and data on in situ assemblages are available only from the Antarctic peninsula.

Late Tertiary information comes from Ross Sea sites, where glaciers at sea level in the Early Oligocene complicate interpretation of the palynological record by recycling palynomorphs from older sequences. Pollen suites from the Ross Sea have been cautiously interpreted as reflecting a Late Oligocene coastal vegetation comprising forests with a variety of *Nothofagus* species; Podocarpaceae were important, but other angiosperm groups less so. There is a general decrease in diversity of pollen species through the Oligocene and Miocene, but no evidence yet of tundra with increasing glaciation. The picture of forests with diverse *Nothofagus* in the Late Oligocene is paralleled by assemblages of leaf fossils of the same age from the South Shetlands, with resemblances to extant South American species. There are no clear data for Miocene vegetation.

The evidence for Pliocene vegetation presented by the fossil material from the Sirius Formation suggests a continuation of the trend toward depauperization of high-latitude Antarctic floras. Species poverty among fossil wood fragments is matched by that of the pollen flora, which includes mainly a single species of *Nothofagus*.

Some Paleoclimatic Considerations

Available evidence suggests that forest vegetation was present on Antarctica through the Cretaceous and Paleogene, and that it persisted well into the Neogene, coexisting with the Antarctic ice cap for much of its Cenozoic history. Palaeogeographic reconstructions of the southern continents based on seafloor magnetic anomalies and on the fitting of continental fragments according to geologic constraints indicate a high-latitude position for Antarctica throughout the Late Mesozoic and Tertiary. The problems that these palaeolatitudes must have posed for plant growth have been discussed by many authors, among them Jefferson (1982), Creber and Chaloner (1984, 1985), Axelrod (1984), Francis (1986), and Truswell (in press).

There appears to be a developing consensus that the extreme photoperiod conditions imposed by such high latitudes need not have inhibited plant growth, provided that temperatures remained high enough to sustain growth and allow provision of moisture. Creber and Chaloner (1985) pointed out that ambient temperature, rather than available light, is the factor limiting tree growth in high latitudes today, citing data that show the total light energy available within Arctic and Antarctic circles is not significantly less than that at temperate latitudes. The efficiency of utilization of the light energy depends on spacing and structure of the vegetation, in order that the low-angle sun's rays are not obstructed by mutual shading. Such analyses appear to render superfluous explanations for high-latitude vegetation that depend on changing physical factors, such as the obliquity of the earth's rotational axis.

There is an increasing body of information relating to past temperatures in high latitudes. Climatic inferences have been drawn from physical sources—from sedimentological and isotopic records—or from the fossil remains of the biota itself. For Antarctica, pertinent sedimentological data include the record of ice-sheet initiation and expansion contained in sequences offshore. Isotopic records, utilizing mainly foraminifera, are available from a limited number of high-latitude sites and from sites in lower latitudes where water masses retain an imprint of events in polar regions.

Information from plant fossils provides an independent record of climatic fluctuations, but needs to be cross-checked against other sources. In this context, it is instructive to note that recently published isotopic records relate well to the suggested vegetation history. Oxygen isotopic analyses of Late Campanian to Early Paleocene foraminifera from Seymour Island suggest deep shelf-water temperatures of 4–10°C and surface water temperatures of about 9–10.5°C (Barrera et al. 1987). These relatively warm waters—which presumably would be linked with high precipitation and equable conditions on nearby land areas—coincide with an interval of high diversity in land floras as evinced from the pollen record. Cooling from the Late Cretaceous into the Tertiary is evident from sediment sequences drilled recently in the Weddell Sea (Barker et al. 1987), and accords with on-land diversity decreases shown by palynomorphs. However, in some parts of the record, problems remain in equating isotopic temperature data with the interpreted vegetation record: for instance, surface water temperatures in the Ross Sea probably fell to 0°C at the Eocene–Oligocene boundary, as extrapolated from the isotopic record in the Tasman Sea (Murphy and Kennett 1985). Such freezing surface waters, which would have resulted in severe conditions on land, are difficult to reconcile with the pollen (and leaf) record, which suggests the persistence of a coastal forest into the Late Oligocene and beyond.

As more information on the Antarctic vegetation record accumulates, there is a need for more quantitative techniques for estimating former climatic conditions. For instance, the installation in Australia of an extensive data base detailing the climatic conditions under which extant rain forests, both tropical and temperate, flourish provides an opportunity to quantify palaeoclimates using pollen data. The computer program, BIOCLIM (Nix 1984 and in press) provides climatic profiles for any rain forest taxon, species, or genus, as well as for structural rain forest categories, and has already been applied to Australian Paleocene climates.

Sluiter (in press) produced rainfall and mean annual temperature figures for rain forests in central Australia based on the assumption that the ecophysiological characteristics that determine climatic responses of the vegetation are unlikely to have changed greatly with time. If the vegetation of the Oligocene Ross Sea margins was indeed rain forest, then it would be instructive to utilize the bioclimatic profiles available for forests growing now under microtherm climatic regimes—in Australia these are the categories microphyll fern forest and microphyll/nanophyll mossy forest, both of which have *Nothofagus* species. Currently, they flourish under mean annual temperatures of 4–14°C, with coldest month mean minimum temperatures of −4 to 1°C. The extant Australian formations, however, are evergreen. To realistically apply this kind of analysis to the Antarctic data, it would be necessary to have a bioclimatic data base that encompassed deciduous microtherm formations. The application of a taxon profile to *Nothofagus gunnii* would be instructive, given the similarity of the fossil leaf in Ross Sea cores to this species.

Acknowledgments. Publication is by permission of the Director, Bureau of Mineral Resources, Canberra. I an indebted to Rosemary Askin for provision of much information, and for critically reviewing the manuscript. Thanks are due to Tom and Edie Taylor for making possible my attendance at the meeting at which this material was presented.

References

Askin RA (1983) Tithonian (Uppermost Jurassic)—Barremian (Lower Cretaceous) spores, pollen and microplankton from the South Shetland Islands, Antarctica. In Antarctic Earth Sciences (4th International SCAR Symposium, Adelaide, 1982). Oliver RL, James PR, Jago JB (eds), Australian Academy of Science, 295–297

Askin, RA (1988a) The Campanian to Paleocene palynological succession of Seymour and adjacent islands, northeastern Antarctic Peninsula. Geological Society of America, Memoir 169, 131–153

Askin RA (1988b) The palynological record across the Cretaceous/Tertiary transition on Seymour Island, Antarctica. Geological Society of America, Memoir 169, 155–162

Askin RA, Markgraf V (1986) Palynomorphs from the Sirius Formation, Dominion Range, Antarctica. Antarctic Journal of the United States 21(5), 34–35

Axelrod D (1984) An interpretation of Cretaceous and Tertiary biota in polar regions. Palaeogeography, Palaeoclimatology, Palaeoecology 45, 105–147

Baldoni AM (1986) Características generales de la megaflora, especialmente de la especie *Ptilophyllum antarcticum,* en el Jurásico superior-Cretácico inferior de Antartida y Patagonia, Argentina. Boletin IG-USP, Série Científica, Instituo Universidad de São Paulo, 17, 77–87

Baldoni AM, Barreda V (1986) Estudio palinológico de las formaciones López de Bertodano y Sobral, Isla Vicecomodoro Marambio, Antartida. Boletin IG-USP, Série Científica Instituo Geociencias, Universidad de São Paulo, 17, 89–98

Barker PF and Leg 113 shipboard scientific party

(1987) Glacial history of Antarctica. Nature 328, 115-116
Barrera E, Huber BT, Savin SM, Webb PN (1987) Antarctic marine temperatures: late Campanian through early Palaeocene. Paleoceanography 2, 21–47
Birkenmajer K (1980) Tertiary volcanic–sedimentary succession at Admiralty Bay, King George Island (South Shetland Islands, West Antarctica). Studia Geologica Polonica 64, 8–65
Birkenmajer K (1987) Oligocene–Miocene glaciomarine sequences of King George Island (South Shetland Islands), Antarctica. Palaeontologia Polonica 49, 9–36
Birkenmajer K, Zastawniak, E (1986) Plant remains of the Dufayel Island Group (Early Tertiary?), King George Island, South Shetland Islands (West Antarctica). Acta Palaeobotanica 26, 33–54
Brady H, Martin HA (1979) Ross Sea region in the Middle Miocene: a glimpse into the past. Science 203, 437–438
Burger D (1980) Palynology of the Lower Cretaceous in the Surat Basin. Bulletin, Bureau of Mineral Resources, Australia, 189, 1–106
Carlquist S (1987) Pliocene *Nothofagus* wood from the Transantarctic Mountains. Aliso 11(4), 571–583
Case JA (1988) Paleogene floras from Seymour Island, Antarctica Peninsula. Geological Society of America, Memoir 169, 523–530
Cranwell LM (1959) Fossil pollen from Seymour Island, Antarctica. Nature 184, 1782–1785
Creber GT, Chaloner WG (1984) Climatic indications from growth rings in fossil wood. In Brenchley P (ed) Fossils and Climate, John Wiley & Sons, pp 49–74
Creber GT, Chaloner WG (1985) Tree growth in the Mesozoic and early Tertiary and the reconstruction of palaeoclimates. Palaeogeography, Palaeoclimatology, Palaeoecology 52, 35–60
Dettmann ME (1986) Early Cretaceous palynoflora of subsurface strata correlative with the Koonwarra Fossil Bed, Victoria. Memoirs, Association of Australasian Palaeontologists 3, 79–110
Dettmann ME, Jarzen DM (1988) Angiosperm pollen from uppermost Cretaceous strata of southern Australia and the Antarctic Peninsula. Memoirs, Association Australasian Palaeontologists 5, 217–237
Dettmann ME, Thomson MA (1987) Cretaceous palynomorphs from the James Ross Island area, Antarctica—a pilot study. British Antarctic Survey, Bulletin 77, 13–59
Domack EW (1985) Glacial erosion on the George V/Adélie continental margin, East Antarctica. Antarctic Journal of the United States 20(5), 76–78
Domack EW, Fairchild WW, Anderson JB (1980) Lower Cretaceous sediment from the East Antarctic continental shelf. Nature 287, 625–626
Dusèn P (1908) Über die tertiäre Flora der Seymour Insel. Wissenschaftliche Ergebnisse der schwedischen Südpolarexpedition 1901–1903. Bd3(3), 1–127
Farquarson GW (1984) Late Mesozoic non-marine conglomeratic sequences of northern Antarctic Peninsula (The Botany Bay Group). British Antarctic Survey Bulletin 65, 1–32
Fleming F, Askin RA (1982) An early Tertiary coal bed on Seymour Island, Antarctic Peninsula. Antarctic Journal of the United States 17(5),67
Francis JE (1986) Growth rings in Cretaceous and Tertiary wood from Antarctica and their palaeoclimatic implications. Palaeontology 29, 665–684
Fuenzalida H, Araya R, Herve F (1972) Middle Jurassic flora from north-eastern Snow Island, South Shetland Islands. In Adie RJ (ed), Antarctic Geology and Geophysics, Universitetetsforlaget, Oslo, pp 93–97
Gee CT (1987) Revision of the Early Cretaceous flora from Hope Bay, Antarctica. XIV International Botanical Congress Abstracts, Berlin, 24 July-1 August, 1987 p 333
Gothan W (1908) Die fossilien Hölzer von der Seymour and Snow Hill Insel. Wissenschaftliche Ergebnisse der Schwedischen Südpolar-Expedition 1901–1903, 3(8), 1–33
Hall SA (1977) Cretaceous and Tertiary dinoflagellates from Seymour Island, Antarctica. Nature 267, 239–241
Halle TG (1913) The Mesozoic flora of Graham Land. Wissenschaftliche Ergebnisse der schwedischen Südpolar-Expedition, 1901–1903, 3(14), 3–124
Hernandez, P, Azcarate V (1971) Estudio paleobotánico preliminar sobre restos de una tafoflora de las Peninsula Byers (Cerro Negro), Isla Livingston, Islas Shetland del Sur, Antártida. Instituto Antártico Chileno, Ser. Cientifica 2, 15–50
Hill RS (in press) Fossil leaf with affinities to *Nothofagus gunnii*. Report on CIROS-1 drillhole, McMurdo Sound, Antarctica. DSIR New Zealand
Jefferson TH (1980) Angiosperm fossils in supposed Jurassic volcanogenic shales, Antarctica. Nature 285, 157–158
Jefferson TH (1981) Palaeobotanical contributions to the geology of Alexander Island, Antarctica. Ph.D. Thesis, University of Cambridge (unpublished)
Jefferson TH (1982) Fossil forests from the Lower Cretaceous of Alexander Island, Antarctica. Palaeontology 25, 681–708
Kemp EM (1975) Palynology of Leg 28 drillsites, Deep Sea Drilling Project. In Hayes DE, Frakes LA et al (eds), Initial Reports of the Deep Sea Drilling Project, 28, Washington (United States Govt Printing Office) pp 599–623
Kemp EM, Barrett PJ (1975) Antarctic glaciation and early Tertiary vegetation. Nature 258, 507–508

Martin HA (1984) The use of quantitative relationships and palaeoecology in stratigraphic palynology of the Murray Basin in New South Wales. Alcheringa 8, 253–272
McIntyre DJ, Wilson GJ (1966) Preliminary palynology of some Antarctic Tertiary erratics. New Zealand Journal of Botany 4, 314–321
Mercer JH (1986) Southernmost Chile: A modern analog of the southern shores of the Ross Embayment during Pliocene warm intervals. Antarctic Journal of the United States 21(5), 103–105
Mildenhall DC (in press) Terrestrial palynology. In Report on the CIROS drillhole, McMurdo Sound, Antarctica. DSIR New Zealand
Murphy MG, Kennett JP (1985) Development of latitudinal thermal gradients during the Oligocene: oxygen isotope evidence from the southwest Pacific. In Kennett JP, von der Borch CC et al. (eds), Initial Reports of the Deep Sea Drilling Project 90, (US Government Printing Office) Washington, pp 1347–1360
Nathorst AG (1904) Sur la flore fossile des régions antarctiques. Comptes Rendus, Académie Science, Paris, 138, 1447–1540
Nathorst AG (1907) On the upper Jurassic flora of Hope Bay. Comptes Rendus, 10th Int. Geol. Congr., Mexico, 2, 1269–1270
Nix HA (1984) An environmental analysis of Australian rainforests. In GL Werren & AP Kershaw (eds), Australian National Rainforest Study Report Vol. 1. Proceedings of a workshop on the past, present and future of Australian rainforests, Griffith University, December 1983. Department of Geography, Monash University, pp 421–425
Nix HA (in press) An environmental analysis of Australian rainforests. J. Biogeography
Sluiter IR (in press) Early Tertiary vegetation and paleoclimates, Lake Eyre region, northeastern South Australia. In De Deckker P, Williams MAJ (eds), The Cainozoic of the Australian Region. Geological Society of Australia, Special Publication
Stipanicic PN, Bonnetti MIR (1970) Posiciones estratigráficas y edades de la principales floras jurásicas argentinas II. Floras doggerianas y málmicas. Ameghiniana 7(2), 101–118
Stott LD, McKelvey BC, Harwood DM, Webb PN (1983) A revision of the ages of Cenozoic erratics at Mount Discovery and Minna Bluff, McMurdo Sound. Antarctic Journal of the United States, 18(5), 36–38
Stover LE, Partridge AD (1973) Tertiary and late Cretaceous spores and pollen from the Gippsland basin, southeastern Australia. Proceedings of the Royal Society of Victoria 85, pp 237–286
Stuchlik L (1981) Tertiary pollen spectra from the Ezcurra Inlet Group of Admiralty Bay, King George Island, (South Shetland Islands, Antarctica). Studia Geologica Polonica 72, 109–132
Tanai T (1986) Phytogeographic and phylogenetic history of the genus *Nothofagus* Bl. (Fagaceae) in the southern hemisphere. Journal of the Faculty of Science, Hokkaido University, 21, Series IV 505–582
Truswell EM (1983) Recycled Cretaceous and Tertiary pollen and spores in Antarctic marine sediments: a catalogue. Palaeontographica Abt B186, 121–174
Truswell EM (in press) Antarctica: a history of terrestrial vegetation. In Tingey RJ (ed), The Geology of Antarctica. Oxford University Press
Truswell EM, Anderson JB (1985) Recycled palynomorphs and the age of sedimentary sequences in the eastern Weddell Sea. Antarctic Journal of the United States 19(5), 90–92
Webb PN, McKelvey BC, Harwood DM, Mabin MCG, Mercer JH (1987) Sirius Formation of the Beardmore Glacier region. Antarctic Journal of the United States 21(1/2), 8–13
Zastawniak E (1981) Tertiary leaf flora from the Point Hennequin Group of King George Island (South Shetland Islands, Antarctica). Preliminary Report. Studia Geologica Polonica 72. Geological Results of the Polish Antarctic Expeditions, K. Birkenmajer (ed), pp 97–108
Zastawniak E, Wrona R, Gozdick A, Birkenmajer K (1985) Plant remains from the top part of the Point Hennequin Group (Upper Oligocene), King George Island (South Shetland Islands, Antarctica). Studia Geologica Polonica 81, Geological results of the Polish Antarctic Expeditions, K. Birkenmajer (ed), pp 143–164

8—Silurian–Devonian Paleobotany: Problems, Progress, and Potential

Dianne Edwards

By tradition the first land plants originated in the North Atlantic Region—particularly in Britain (Boucot and Gray 1982).

It is now a matter for history that studies on early pteridophytes began on fossils from the Old Red Continent, that the finest preservation has been found there and that the land mass has been the subject of some classic sedimentological studies (Edwards and Fanning 1985).

The second quotation was a somewhat defensive response to the first, which was Gray and Boucot's expression of criticism, admittedly partly justified, of overemphasis on megafossils from the present northern hemisphere in hypotheses relating to the early evolution of land plants. However, it cannot be denied that most of the relevant data derive from assemblages in Europe, Asia, and North America, and that the resulting history of early vascular plants was satisfying not the least because it conformed to theoretical anticipation. Complacency was shattered by two developments—first, the discovery of a Ludlow assemblage in Australia (Garratt et al. 1984) containing plants with almost all the lycophyte characteristics and, second, the description of microfossils (tetrads and cuticle fragments) of land-plant origin in Late Ordovician and Early Silurian sediments prior to the appearance of vascular plants (Gray 1985). The earliest records from Libya (Gray et al. 1982) come from extremely high paleolatitudes (Cocks and Fortey 1988) and focused attention on the need for more information on assemblages in the present southern hemisphere, and in particular from the Siluro–Devonian Gondwana continent (Figs. 8.4–8.7).

Progress: The Record in 1988

In this section, I list and comment with some apprehension on the very fragmentary Devonian fossils that were found and described by others and that I have not seen. I feel able to do so only because I have some experience with similar poorly preserved material and have the benefit of acquaintances with recent and extensive literature on early land plants and on the geology of Antarctica (Boucot et al. 1968) not available to earlier researchers and reviewers. My task is not onerous. There are just four records where the fossils were named and two more where plants were mentioned (Fig. 8.1).

Transantarctic Mountains

South Victoria Land

The basal part of the Beacon Supergroup comprises some of the most well-researched Devonian rocks of Antarctica (Barrett et al. 1972). The succession in the Beacon Heights area (McElroy and Rose 1987) is particularly important paleobotanically (Table 8.1). In addition to the Early and Late Devonian palynomorph assemblages (Kyle 1977, Helby and McElroy 1968, Playford, Chapter 6, this volume) megafossils have been found in the

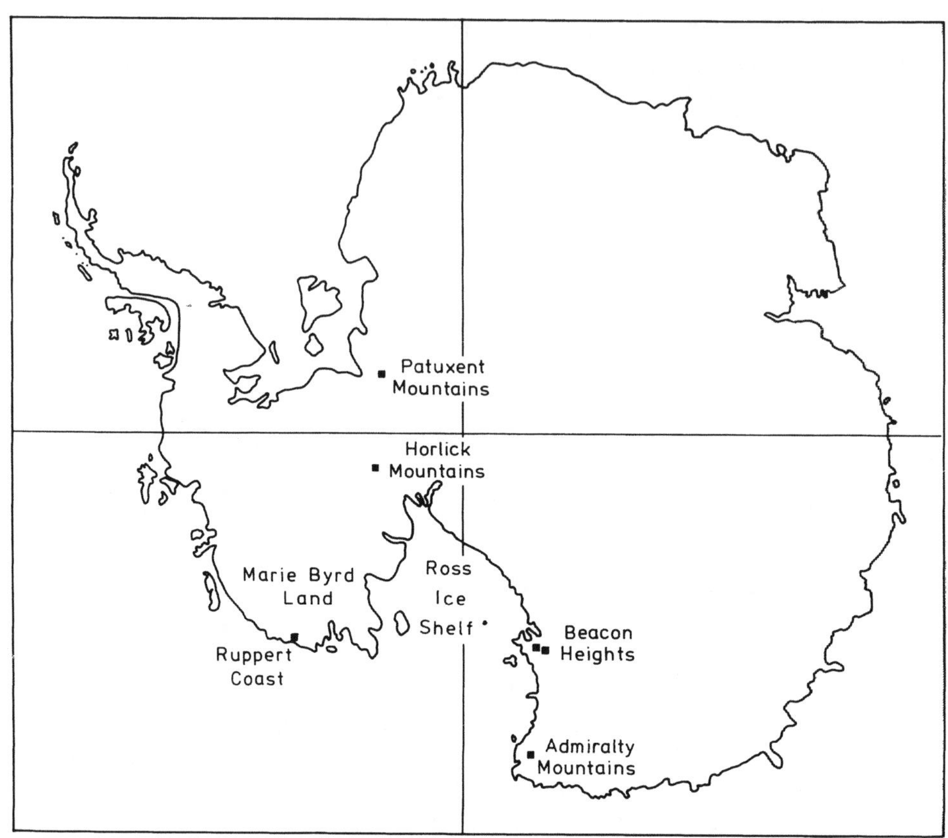

Figure 8.1. Map of Antarctica with positions of Devonian localities.

Beacon Heights Orthoquartzite (Harrington and Speden 1962) believed to be late Middle Devonian. They were described by Plumstead (1962). A single specimen of *Haplostigma irregularis* (Schwarz) Seward was collected from the scree on the west side of the Beacon Heights West area on the southern margin of the upper Taylor Glacier. It is an impression in

Table 8.1. Stratigraphic units in the lower part of the Beacon Supergroup (after McElroy and Rose 1987) Beacon Heights area.

Age				Stratigraphy
Devonian	Late	Beacon Supergroup	Taylor Group	Aztec Siltstone
				Beacon Heights Orthoquartzite
	?			Arena Sandstone
	Middle			Altar Mountain Formation
				Ashtray Sandstone Member Odin Arkose Member Heimdall Erosion Surface New Mountain Sandstone
	?			Terra Cotta Siltstone
	Early			Windy Gully Sandstone Kukri Erosion Surface
Early Paleozoic and/or Precambrian				Basement Complex

a quartzite, a hollow mold of a stem, 50 mm long and 15 mm wide, characterized by spirally arranged circular or oval depressions marking the attachment points of leaves. Laminae were not preserved. A number of axes (ca 15) identified as cf. *Protolepidodendron lineare* were found on a loose block of coarse carbonaceous grit on scree on the south side of Beacon Heights. Up to 40 mm wide, they are sometimes twisted and described as fluted (Plumstead 1962) because they possess uneven longitudinal ridges and grooves. Where a thin film of graphite persists, its surface is minutely and longitudinally striated. The striae "occasionally appear to bend around flat round scars" (Plumstead 1962).

The *Haplostigma* stem closely resembles similarly named fragments in the Falklands and South Africa where such fossils are particularly abundant. They are probably the stems of herbaceous lycophytes, but to date, neither fertile nor anatomically preserved specimens have been found. It seems likely that the squat, conical, acute-tipped appendages interpreted as leaves (Anderson and Anderson 1985) are the remains of leaf bases. In South Africa, *H. irregularis* occurs in the Middle Devonian upper Bokkeveld and Upper Devonian lower Witteberg Groups. This implies a Middle Devonian or basal Late Devonian age for the Beacon Heights quartzite as it is overlain by the Aztec Siltstone with Upper Devonian (?Frasnian) fish and spores (Helby and McElroy 1968). The identification of the sulcate stems with *Protolepidodendron lineare* Walkom is less convincing (as implied by Plumstead's use of cf.). In addition, the use of that genus for the Australian specimens is misleading. The Antarctic specimens are probably better left as indeterminate, but possibly decorticated lycophyte stems. Finally, mention should be made of the plants included in the descriptions of paleosols in the Aztec Siltstone (McPherson 1979). Casts of roots, some with branching rootlets are said sometimes to possess the remains of "woody or cell-like structure," while "primitive plants (*Rhynia*-type), stems and spores" (McPherson 1979) were recorded but not illustrated. Clearly, more information is needed about this assemblage; a more advanced kind of vegetation would be anticipated both from the sediments (indications of roots) and from their age.

Finally, Plumstead (1964) included a record of unidentified Devonian plants (silicified wood fragment with growth rings and branching stems) from Mt. Crean in the Lashly Mountains. This record should be discounted because field relationships of the strata in that area indicate that they are certainly not Devonian but probably Triassic (Dr. J. Collinson, pers. comm.).

Ohio Range: Horlick Mountains

The Horlick Formation in the Discovery Ridge type area comprises 45 m of alternating sandstones and dark carbonaceous siltstones (Long 1964). It has yielded an in situ Emsian assemblage of marine shelly invertebrates (Doumani et al. 1965) and a restricted palynomorph assemblage (Kemp 1972) possibly belonging to the Emsian *Emphanisporites annulatus–Camarozonotriletes sextanti* zone of Richardson and McGregor (1986). Also recorded are sterile psilophytic axes of *Hostinella* aspect (Rigby and Schopf 1969). In Schopf's collection of prints and negatives in the Byrd Polar Research Center, Columbus, Ohio is a photograph of a specimen traversed by numerous, aligned, parallel-sided, smooth axes [Fig. 8.2(e)]. They cannot be identified, but their mode of preservation and abundance are reminiscent of specimens from Pragian Lower Devonian localities in the Gaspé and South Wales.

Pensacola Range: Patuxent Mountains

Plants were found in the Okanogan nunatak at the southwest limit of the Patuxent Mountains. The sediments, from evidence of similar lithology, are considered the equivalent of the Dover Sandstone in the Neptune Range (Schmidt and Ford 1969) and consist of thick-bedded, coarse sandstones showing red weathering, sometimes interbedded with carbonaceous layers. Schopf (1968) recorded a specimen of *Haplostigma* preserved as a mold in a quartzitic sandstone and, in the carbonaceous layers, spores and compressions of *Cyclostigma*-type stems. In the 1968 paper he gave a Middle Devonian age to the locality, although later he was quoted as considering it Late Devonian (Schmidt and Ford 1969). The fossils were not illustrated, but are probably those figured in Fig. 8.2(a)–(e), the negatives having been found in the Schopf collection. Figure 8.2(a) is of a typical *Haplostigma* with leaf attachment points marked by depressions. Figures 8.2(b)–(d) show impressions of axes with circular to oval ?leaf bases arranged in low spirals. They are probably not *Cyclostigma* and differ from typical *Haplostigma* in the relative dimensions of bases and axes and in the distance apart of successive gyres. In the latter character, they resemble defoliated *Colpodexylon cashiriense* from Venezuela (Edwards and Benedetto 1985), but such superficial similarity does not allow assignment to that species, and these axes are better left as *incertae sedis*. Mr. J. Isbell (Byrd Polar Research Center) has recently collected some impressions in a very coarse carbonaceous sandstone from the Dover Sandstone, but again these cannot be identified.

Northern Victoria Land: Central Admiralty Mountains

Poorly preserved impressions were collected in the siltstones or sand-sized volcaniclastics interbedded with the Black Prince volcanics. These overlie unconformably the deformed Robertson Bay Group sediments and Devonian Admiralty Granite. The specimen illustrated by Findlay and Jordan (1984) was found at 3050 m on the northern ridge of Mt. Black Prince. The small impression bears contiguous vertically elongate diamond-shaped areas (from scale, 2.3 mm long and ca 1 mm wide), each with a further raised area at the presumed proximal end. There is no evidence of vascular strand, leaf, or parichnos scars. Leistikow and Schweitzer (pers. comm.) identified the specimens as stems of *Protolepidodendropsis pulchra* syn *Bergeria mimerensis* (*sic Protolepidodendropsis pultor* syn *Bergeria minensis*), which also occurs in the upper Middle Devonian of Spitsbergen (Høeg 1942, Schweitzer 1965), the stem patterning being closer to *Protolepidodendropsis* than to *Bergeria*, where the outlines are rhombic with width equal to height but usually considerably more. It is of some interest that when Høeg (1942) first described *Bergeria mimerensis* he compared it with the genus *Leptophloeum*, which occurs in the Upper Devonian of Australia and South Africa as *L. australe* and in the Permian of South Africa as *L. sanctae–helenae* (Anderson and Anderson 1985). However, there is no justification for placing the Robertson Bay plants in *Leptophloeum* on available evidence. Their identification as *P. pulchra* is consistent with a Middle Devonian age for the sediments, but as its preservation is poor, such an age determination should be treated with great caution. As far as I am aware, *Protolepidodendropsis* is not recorded from the Carboniferous.

Thus, erring on the cautious side, at most this new plant record indicates the existence of lycophytes with swollen and extended leaf bases, that the rocks are probably not older than Middle Devonian (but see records of Lower Devonian lycophytes in Libya: Lejal-Nicol 1975), and that the Robertson Bay plants are quite distinct from the lycophytes recorded from South Victoria Land and Marie Byrd Land. However, it would be unwise to attach too much significance to this isolated record in considering the tectonic history of the Robertson Bay area.

Marie Byrd Land: Ruppert Coast

The most numerous and most informative plant fossils in the Antarctic Devonian come from dark carbonaceous slaty argillite erratics in the most southernly nunatak (Milan rock) at Mt. Hartkopf. Similar lithologies with coalified plants are recorded from below a calc-alkaline metavolcanic sequence at the Ruppert Coast (Grindley and Mildenhall 1980). The plants include *Drepanophycus schopfii*, *Haplostigma irregularis*, ?*Protolepidodendron*, and some indeterminate ?psilophyte axes. (Grindley et al. 1980). The new *Drepanophycus* species was first investigated by Schopf and then named after him. Its axes are described as longitudinally corrugated with tear-shaped leaf bases in relief or as depres-

Figure 8.2. (*1*)*Haplostigma,* sp (2058), mold in quartzitic sandstone. Okanogan nunatak, Patuxent Mountains, ?Upper Devonian, × 1.3.(*2*)-(*4*) Indeterminate impressions of lycophyte axes with circular to oval leaf attachment sites arranged in a low spiral. Locality details as in (*1*). (*2*). 2092, ×2.5; (*3*). 2068, ×1.9; (*4*). part of (*3*) enlarged, ×4.8.(*5*) Indeterminate smooth axes (2180), Horlick Formation, Ohio Range, Lower Devonian, ×1.2.

sions. For the first time in an Antarctic Devonian lycophyte, leaves, although incomplete, were found attached to the stems. They are ca 2.5 mm long and a little over 1 mm wide just above the attachment point. The leaves are sickle shaped (presumably in profile), with a relatively long attenuate or acicular point of an arrowhead-shaped base (Grindley et al. 1980). Each leaf contains a single vascular strand. The tips of one row of leaves overlap the bases of those above.

Specimens compared with *Haplostigma irregularis* are superficially similar to *D. schopfii* in the position and type of preservation of leaf bases and in the longitudinal ribbing of axes. They differ in that the stems are larger and each leaf base comprises a raised central area surrounded by a depressed ring. Grindley et al.'s Fig. 7 shows short, rigid, curved leaves said to be broader (? in profile) than those in *D. schopfii*. They contain a single vascular strand. The authors considered the possibility that the two kinds of axes were different preservation states of the same plants, but decided to separate them.

Grindley et al. compared their new species with those in *Drepanophycus*. This genus is one of the most confusing and puzzling among Devonian taxa (Grierson and Banks 1963), and following the discovery of information on fertile parts (Schweitzer 1980) and anatomy (Rayner 1984) of *D. spinaeformis*, its use is perhaps best confined to the type species. A good example of reassignment was the transfer of *D. colophyllous* to *Haskinsia colophylla*, a herbaceous lycophyte with triangular leaves (Grierson and Banks 1983). A second species, *H. sagittata*, Edwards and Benedetto (1985), lacked anatomy but differed morphologically from *H. colophylla* in its possession of sagittate leaves, a feature now demonstrated in the type (H.P. Banks, personal communication). Leaves of Antarctic and Venezuelan specimens (Fig. 8.3) suggest that they come from the same genus, the most obvious difference being the conspicuous midvein in the Antarctic specimens. The range of variation described for American and Venezuelan specimens matches that illustrated by Grindley et al. for *D. schopfii* and *Haplostigma* and may even encompass the indeterminate lycopod axis, ?*Protolepidodendron* (their figure 16). Their tear-shaped leaf cushions may have been produced by thick-walled cells of the outer cortex arranged in a fusiform pattern around leaf bases, a pattern similar to that described for *H. colophylla*.

Smooth stems were tentatively described as psilophyte axes. The one illustrated branches dichotomously, with a granular appearance and a number of small outgrowths, "pustules," identified as leaf bases. The affinities of such axes are obscure: they are almost certainly not psilophytes and should be left as indeterminate.

The occurrence of *Haplostigma* and *Dre-*

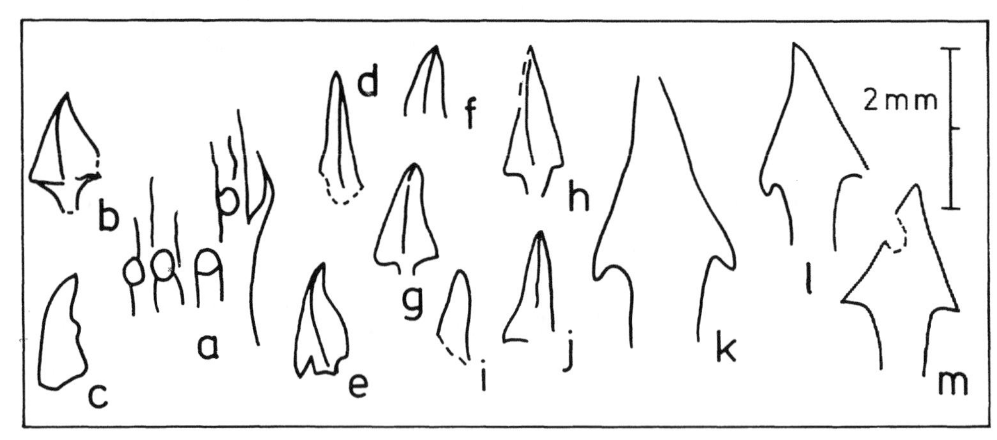

Figure 8.3. Outlines of leaves (b–j) and shoot (a) of *Drepanophycus schopfii* redrawn from Grindley et al. (1980) for comparison with leaves (k–m) of *Haskinsia sagittata* (Edwards and Benedetto 1985).

panophycus led the authors to postulate a late Middle to early Late Devonian age for the plant horizon. This is broadly consistent with the Late Devonian–Carboniferous age assigned from radiometric dating to the overlying Ruppert Coast metavolcanics in eastern Marie Byrd Land. *Haskinsia colophylla* ranges from Givetian to Lower Frasnian strata in New York state and again supports the radiometric data.

Paleobiogeographical Consideration

The role of Antarctic assemblages in such analysis is restricted by the paucity of records, the uncertainty over the identification and affinities of the plants (all are probably lycophytes; none are fertile), and the lack of reliable, independent dating. Similar limitations also often apply when analyzing other Gondwana assemblages. Their distribution is shown on the Upper Devonian paleocontinental reconstruction in Fig. 8.7. Most occurrences are in South Africa in the Middle Devonian upper Bokkeveld Group and the lowermost Upper Devonian lower Witteberg Group of Cape Province (Anderson and Anderson 1985). *Haplostigma irregularis* is abundant, usually as monospecific assemblages particularly near the base of the Witteberg Group. Toward the top, it occurs with *Archaeosigillaria caespitosa* or with *A. caespitosa* and *H. kowiensis*. In the Upper Bokkeveld, it is associated with yet another presumed lycophyte, *Palaeostigma sewardi*, and species of *Archaeosigillaria*. Vegetation was thus dominated by herbaceous lycopods, although Anderson and Anderson described *H. irregularis* as herbaceous to semi-arboreal, a description not substantiated by the small size of their stems.

Haplostigma irregularis is also recorded from the Falklands, but the age of the sediments is uncertain (Seward and Walton 1923). In Argentina, the only independently dated, and hence unequivocal, Devonian record is that of *Haplostigma furquei* (Middle Devonian from trilobites: Menendez 1965a,b). *Haplostigma* is also found in the Givetian upper Porta Grossa beds of Brazil (Chaloner et al. 1974) but is absent from the more extensive lycophyte assemblage (*Palaeostigma sewardi, Protolepidodendron kegeli*, and *Archaeosigillaria picosensis*) in the Middle Devonian of the Maranhao Basin (Bär and Riegel 1974 but see Edwards and Benedetto 1985). The exact age (?Givetian/Frasnian) of the Venezuela assemblage is unknown. In addition to the herbaceous lycophytes *Colpodoxylon* and *Haskinsia*, it contains characteristic Laurentian taxa, such as *Pseudosporochnus* and *Tetraxylopteris*, as well as Australian *Astralocaulis*. Similarities between Australian and Antarctic assemblages are via ?*Protolepidodendron lineare* and *Haplostigma*, but it should be noted that the former are mixed because they also contain Laurentian taxa, such as *Leclercqia complexa* (Fairon-Demaret 1974; ?Givetian of Queensland).

This very limited evidence suggests that in Middle–early Late Devonian times, Antarctic vegetation had most in common with that paleogeographically closest (Table 8.2, Argentina, Falklands, South Africa) and that there was greater diversity in assemblages at lower latitudes. Earlier reviewers of Antarctic fossils point to similarities between Antarctic assemblages and those elsewhere. Thus, for example, Rigby and Schopf (1969, p. 92) stated "Nothing has been discovered that seems at variance with the concept of a Devonian flora that was surprisingly cosmopolitan." In fact, it is the herbaceous lycophyte elements in the flora that were cosmopolitan. However, before discussing this further, taphonomic aspects must be considered. The most productive and informative clastic lithologies for plant fossils and palynomorphs are grey siltstones and sandstones, preferably in unaltered states. Thus, the best preservation of megafossils in the Devonian of Antarctica is in the siltstones from Marie Byrd Land, while the best palynomorph assemblages have been extracted from the Horlick Formation and Aztec Siltstone (Playford, Chapter 6, this volume). Such sediments appear not to be very common. Records of *Haplostigma* are in coarse-grained, thick-bedded sandstones,

Table 8.2. Geographic and stratigraphic occurrences of taxa present in Antarctica.

Taxa present in Antarctica	S. Africa	S. America	Australia	Laurentia
Haplostigma irregularis	M, U	M		
Haskinsia sp (*Drepanophycus schopfi*)		? M/U		M/U
? *Protolepidodendron lineare*			M/U	
? *Protolepidodendropsis pulchra*				M

(M = Middle; U = Upper Devonian).

both in Antarctica and South Africa (Plumstead 1962), and it seems very likely that sorting occurred during transport. Further it is possible that lycophytes are more resistant to degradation and disintegration during high-energy transport than other plants and thus remain intact to be eventually buried. However, it is also likely that the woody progymnosperms would have survived transport and that at least vegetative parts would be

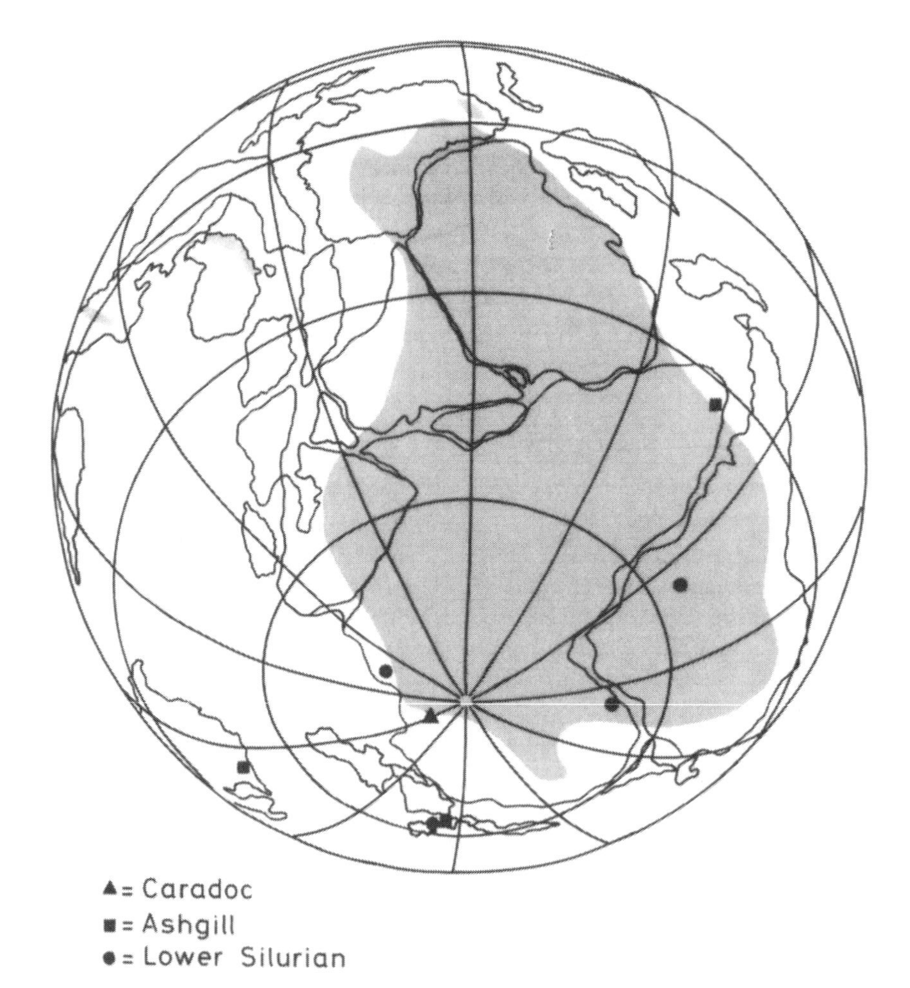

Figure 8.4. Distribution of certain Ordovician and basal Silurian assemblages of microfossils derived from land plants on a base map produced by Denham and Scotese (1987) for the Late Ordovician.

preserved. I therefore very tentatively suggest that the high-latitude vegetation was dominated by herbaceous lycophytes in the Middle to early Late Devonian and was of lower diversity than the vegetation of lower latitudes. As a corollary, it would seem that such cosmopolitan herbaceous lycopods were tolerant of extremes of climate (Edwards and Benedetto 1985).

Similar problems relating to quality of preservation and actual presence or absence occur when analyzing the significance of dispersed spores. Thus, for example, the Lower Devonian Terra Cotta siltstone assemblage and that from the Horlick Formation are both poorly preserved and of relatively low diversity. Unfortunately, there is little information on the producers of most of the spores, and we are particularly ignorant of the spores of herbaceous lycophytes, such as *Drepanophycus* and *Baragwanathia*. This prevents the testing of the above hypothesis.

Potential

The meagerness of the record outlined above emphasizes the need for more collecting in Devonian sediments. In the Early Devonian, the Horlick Formation seems particularly promising, especially as the associated invertebrate fauna characteristic of the Malvinokaffric faunal province is indicative of a cold climate. In younger sediments, the single find of fairly abundant plants in an erratic on a nunatak in Marie Byrd Land augers well for future collecting in similar lithologies in situ elsewhere in the Wilkins Formation, while the plants reported associated with the Upper Devonian paleosols require detailed description.

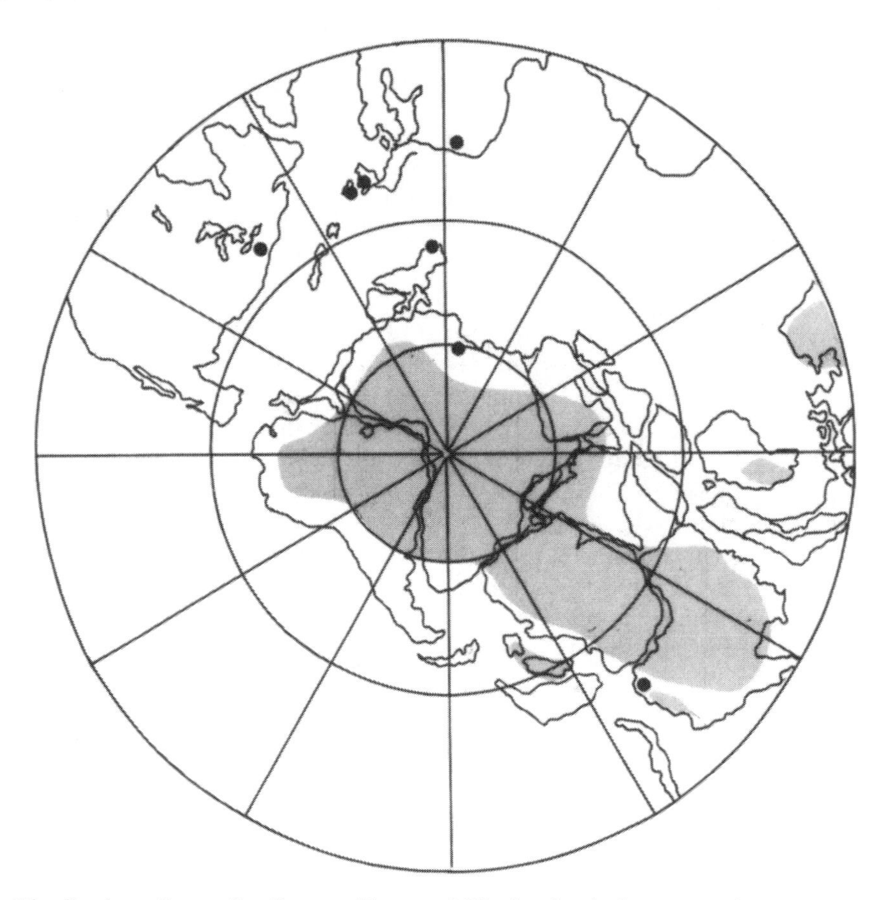

Figure 8.5. Distribution of megafossil assemblages of Silurian land plants on a base map produced by Denham and Scotese (1987) for the Late Silurian.

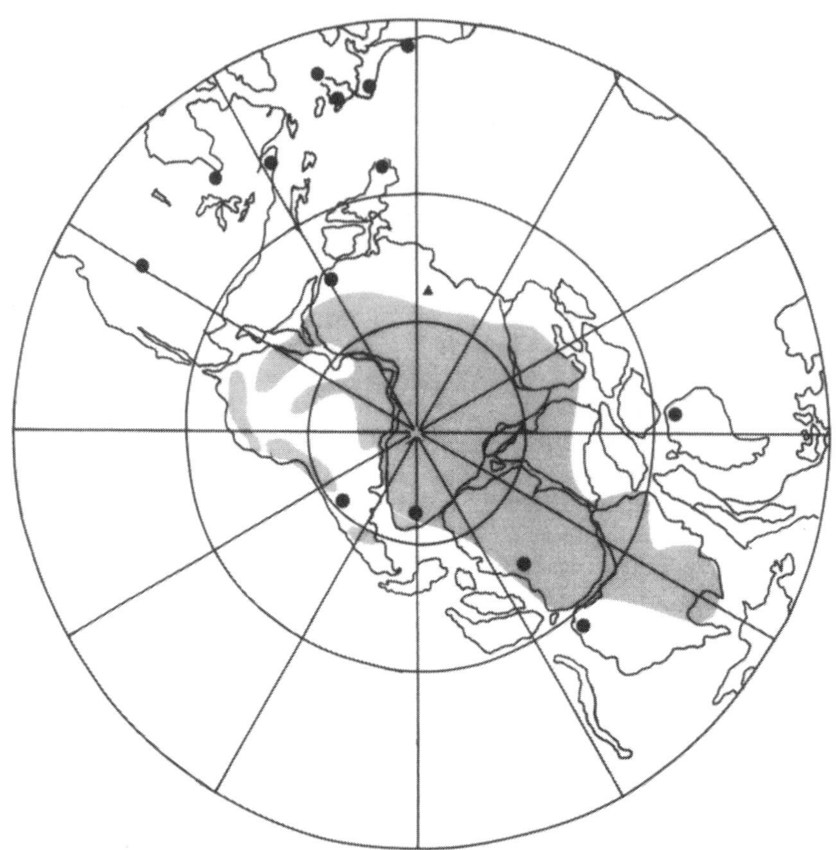

Figure 8.6. Distribution of Early Devonian megafossil assemblages on a base map produced by Denham and Scotese (1987).

Further information is also required on the poorly preserved plant stems recorded in black shales of probable mid-Late Devonian age at Buttress Peak in the Gallipoli Heights area, northern Victoria Land (Grindley and Oliver 1983); but exploration should not be confined to the Devonian.

In the present northern hemisphere, there is evidence for fertile erect pteridophyte-like plants in the Wenlock, while spores and cuticles extend the record of land plants into the Ordovician (Caradoc of Libya: Gray et al. 1982, Gray 1985). Until recently, there was no direct evidence for Ordovician or Silurian sediments in Antarctica, although it must have formed quite a substantial part of the Gondwana landmass at that time (Cocks and Fortey 1988) (Fig. 8.4).

Comparatively unexplored are the Ellsworth Mountains, where sediments at least 13,000 m thick extend from the Precambrian into the Permian (Craddock 1969) and where, as yet, no unconformities have been detected. The Crashsite Quartzite has Cambrian fossils at its base and probably Devonian brachiopods near the top (Webers and Spörli 1983), so that there is a good chance of Ordovician and Silurian sediments between. Unfortunately, in contrast to the Paleozoic rocks of the Transantarctic Mountains, the sedimentary volcanic Ellsworth rocks are strongly folded and slightly to moderately metamorphosed and hence are not particularly promising for the extraction of palynomorphs or for the discovery of well-preserved megafossils, although a *Glossopteris* assemblage has recently been recorded (Taylor and Smoot 1989).

Similarly, in the Neptune Range of the Pensacola Mountains, the apparently unfos-

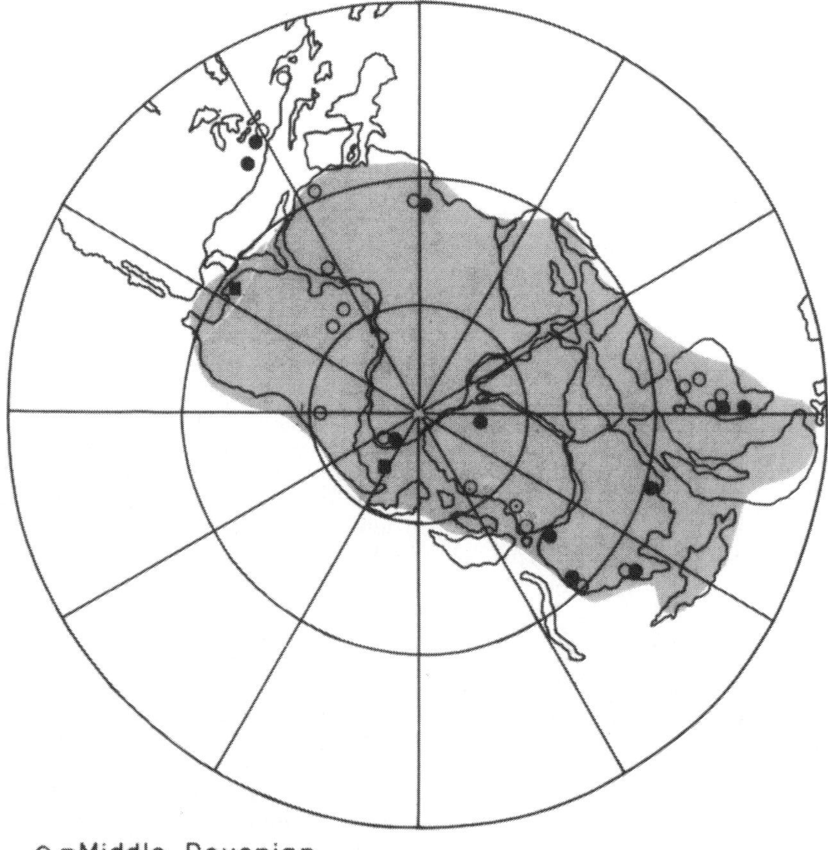

Figure 8.7. Distribution of Middle/Late Devonian megafossil assemblages on a base map produced by Denham and Scotese (1987) for the Late Devonian. Northern hemisphere occurrences not exhaustive.

siliferous Neptune Group occurs below the Dover Sandstone (considered the partial lithological equivalent of the Devonian Taylor Group in the Beacon Heights area) and may extend without break into the Late Cambrian. In this case, there are discontinuities at both top and bottom (Schmidt and Ford 1969, Webby et al. 1981). The uppermost unit of the Neptune Group may be Devonian because it is similar to rocks of Devonian age in the lower part of the Beacon Supergroup in Victoria Land. Unfortunately, the pre-Neptune Group unconformity represents an unknown length of time, so the age of the intervening sediments is conjectural.

Again, from field relations, it is possible that the unfossiliferous Cocks Formation in the Royal Society Range in South Victoria Land may be partly Early Ordovician or latest Cambrian (Webby et al. 1981).

In a paper describing trace fossils, including *Beaconichnus gouldii* and trilobite tracks, from the upper part of the New Mountain Sandstone of the Taylor Group in the Asgard Range, Victoria Land, Gevers and Twomey (1982) suggested that the sediments at the base of the Group, and below the Terra Cotta Siltstone containing Lower Devonian spores (Kyle 1977) could be earliest Devonian or older.

Finally, Late Cambrian/Early Ordovician fossils have recently been discovered in clasts in debris flow deposits near the top of the Robertson Bay Group, northern Victoria Land (Cooper et al. 1983, Wright et al. 1984, Wright and Brodie 1987). Thus, again, there is the possibility of existence of later Ordovician sediments in this succession.

We need more records of early land plants. In particular, we need them from high paleolatitudes. Antarctica may well hold the solution to our problems.

References

Anderson JM, Anderson HM (1985) Palaeoflora of southern Africa. AA Balkema, Rotterdam

Bär P, Riegel W (1974) Microfloras of the Palaeozoic of Ghana and their palaeofloristic relations. Sci Geol Bulletin 27:39–58

Barrett PJ, Grindley GW, Webb PN (1972) The Beacon Supergroup of East Antarctica. In Adie RJ (ed) Antarctic Geology and Geophysics, Universitetets forlaget, Oslo, pp 319–392

Boucot AJ, Gray J (1982) Geologic correlates of early land plant evolution. Third North American Paleontological Convention, Proceedings 1:61–66

Boucot AJ, Doumani GA, Johnson JG, Webers GF (1968) Devonian of Antarctica. In Oswald DH (ed) International Symposium on the Devonian System, Calgary, Vol. 2, Alberta Society of Petroleum Geologists, Canada, pp 639–648

Chaloner WG, Mensah MK, Crane MD (1974) Non-vascular land plants from the Devonian of Ghana. Palaeontology 17:925–947

Cocks LRM, Fortey RA (1988) Lower Palaeozoic facies and faunas around Gondwana. In Audley-Charles MG, Hallam A (eds) Gondwana and Tethys. Geological Society Special Publication, London, pp 183–200

Cooper RA, Jago JB, Rowell AJ, Braddock P (1983) Age and correlation of the Cambrian–Ordovician Bowers Supergroup, Northern Victoria Land. In Oliver RL, James PR, Jago JB (eds), Antarctic Earth Sciences, Cambridge University Press, pp 128–139

Craddock C (1969) Geology of the Ellsworth Mountains (Sheet 4, Ellsworth Mountains) Plate IV. In Bushnell VC, Craddock C (eds) Geologic maps of Antarctica. Antarctic Map Folio Series, Folio 12

Craddock C, Webers GF, Anderson JJ (1982) Geology of the Ellsworth Mountains—Thiel Mountains Ridge. In Craddock C (ed) Antarctic Geoscience, University of Wisconsin Press, Madison, p 849

Denham CR, Scotese CR (1987) Terra Mobilis™: a plate tectonics program for the Macintosh. Earth in Motion Technologies, PO Box 49245, Austin, Texas 78765

Doumani GA, Boardman RS, Rowell AJ, Boucot AJ, Johnson JG, McAlester AL, Saul J, Fisher DW, Miles RS (1965) Lower Devonian fauna of the Horlick Formation, Ohio Range, Antarctica. American Geophysical Union Antarctic Research Series 1:241–281

Edwards D, Benedetto J-L (1985) Two new species of herbaceous lycopods from the Devonian of Venezuela with comments on their taphonomy. Palaeontology 28:599–618

Edwards D, Fanning U (1985) Evolution and environment in the late Silurian–early Devonian: the rise of the pteridophytes. Philosophical Transactions of the Royal Society of London B309: 147–165

Fairon-Demaret M (1974) Nouveaux specimens du genre *Leclercqia* Banks HP, Bonamo PM et Grierson JD 1972 du Givétian du Queensland (Australie). Bullétin Institut royal Sciences naturelles de Belgique 50:1–4

Findlay RH, Jordan H (1984) The volcanic rocks of Mt Black Prince and Lawrence Peaks, North Victoria land, Antarctica. Geologisches Jahrbuch B60:143–151

Garratt MJ, Tims JD, Rickards RB, Chambers TC, Douglas JG (1984) The appearance of *Baragwanathia* (Lycophytina) in the Silurian. Botanical Journal of the Linnean Society 89:355–358

Gevers TW, Twomey A (1982) Trace fossils and their environment in Devonian (Silurian?) Lower Beacon Strata in the Asgard Range, Victoria Land, Antarctica. In Craddock C (ed) Antarctic Geoscience, University of Wisconsin Press, pp 639–647

Gray J (1985) The microfossil record of early land plants; advances in understanding of early terrestrialization 1970–1984. Philosophical Transactions of the Royal Society of London B309: 167–195

Gray J, Massa D, Boucot AJ (1982) Caradocian land plant microfossils from Libya. Geology 10: 197–201

Grierson JD, Banks HP (1963) Lycopods of the Devonian of New York State. Palaeontographica Americana 4:220–295

Grierson JD, Banks HP (1963) A new genus of lycopods from the Devonian of New York State. Botanical Journal of the Linnean Society 86: 81–101

Grindley GW, Mildenhall DC (1980) Geological background to a Devonian plant fossil discovery. In Cresswell AJ, Vella P (eds) Gondwana Five, Fifth International Gondwana Symposium, Wellington, NZ. AA Balkema, Rotterdam, pp 23–30

Grindley GW, Oliver PJ (1983) Post-Ross orogeny cratonization of northern Victoria Land. In Oliver RL, James PR, Jago JB (eds) Antarctic Earth Science, Cambridge University Press, Cambridge, pp 133–139

Grindley GW, Mildenhall DC, Schopf JM (1980) A

mid-Late Devonian flora from the Ruppert Coast, Marie Byrd Land, West Antarctica. Journal of the Royal Society of New Zealand 10:271–285

Harrington JH, Speden IG (1962) Section through the Beacon Sandstone at Beacon Heights West, Antarctica. New Zealand Journal of Geology and Geophysics 5:707–717

Helby RJ, McElroy CT (1968) Microfloras from the Devonian and Triassic of the Beacon Group, Antarctica. New Zealand Journal of Geology and Geophysics 12:376–382

Høeg OA (1942) The Downtonian and Devonian Flora of Spitsbergen. Norges Svalbard-Og Ishavs-Undersøkelser, Skrifter 83:1–228

Kemp EM (1972) Lower Devonian palynomorphs from the Horlick Formation, Ohio Range, Antarctica. Palaeontographica 139B: 105–124

Kyle RA (1977) Devonian palynomorphs from the basal Beacon Supergroup of South Victoria Land, Antarctica (Note). New Zealand Journal of Geology and Geophysics 20:1147–50

Lejal-Nicol A (1975) Sur une nouvelle flore à lycophytes du Devonien inférieur de la Libye. Palaeontographica B151:52–96

Long WE (1964) The stratigraphy of the Horlick Mountains. In Adie RJ (ed) Antarctic Geology, North Holland Publishing Company, Amsterdam, pp 352–363

McElroy CT, Rose G (1987) Geology of the Beacon Heights area, southern Victoria Land, Antarctica. 1:50,000:New Zealand Geological Survey Miscellaneous Series Map 15 (1 sheet) and notes, 47pp

McPherson JG (1979) Calcrete (caliche) palaeosols in fluvial redbeds of the Aztec Siltstone (Upper Devonian), southern Victoria Land, Antarctica. Sedimentary Geology 22:267–285

Menendez CA (1965a) *Archaeosigillaria conferta* (Frenguelli) nov. comb. del Devónico de la Quebrada de la Chavela, San Juan. Ameghiniana 4:67–68

Menendez CA (1965b) *Drepanophycus eximius* (Frenguelli) nov. comb. del Devónico de la Quebrada de la Chavela, San Juan. Ameghiniana 4: 139–140

Plumstead EP (1962) Fossil floras of Antarctica. In Trans-Antarctic Expedition 1955–1958, Scientific Report 9 (Geology), pp 1–132

Plumstead EP (1964) Palaeobotany of Antarctica. In Adie RJ (ed) Antarctic Geology. Proceedings of the first SCAR symposium on Antarctic Geology, Cape Town 1963. North Holland Publishing Co., Amsterdam, pp 637–654

Rayner R (1984) New finds of *Drepanophycus spinaeformis* Göppert from the Lower Devonian of Scotland. Transactions of the Royal Society of Edinburgh. Earth Sciences 75:353–363

Richardson JB, McGregor DC (1986) Silurian and Devonian spore zones of the Old Red Sandstone Continent and adjacent regions. Geological Survey of Canada Bulletin 364:1–79

Rigby JF, Schopf JM (1969) Stratigraphic implications of Antarctic paleobotanical studies. In Amos AJ (ed) Gondwana Stratigraphy UNESCO, Paris, Earth Sciences 2:91–106

Schmidt DL, Ford AB (1969) Geology of the Pensacola and Thiel Mountains (Sheet 5, Pensacola and Thiel Mountains) Pl.V. In Bushnell VC, Craddock C (eds) Geologic maps of Antarctica. Antarctic Map Folio Series, Folio 12

Schopf JM (1968) Studies in Antarctic paleobotany. Antarctic Journal of the United States 3:176–177

Schweitzer H-J (1965) Über *Bergeria mimerensis* und *Protolepidodendropsis pulchra* aus dem Devon Westspitzbergens. Palaeontographica B115: 117–138

Schweitzer H-J (1980) Über *Drepanophycus spinaeformis* Göppert. Bonner Paläobotanische Mitteilungen 7:1–29

Seward AC, Walton J (1923) On fossil plants from the Falkland Islands. Quarterly Journal of the Geological Society of London 79:313–333

Taylor TN, Smoot EL (in press) Permian plants from the Ellsworth Mountains, West Antarctica. In Craddock C, Splettstoesser J, Webers G (eds), Geology and Paleontology of the Ellsworth Mountains, Memoir of the Geological Society of America.

Webby BD, Vandenberg AHM, Cooper RA, Banks MR, Burrett CF, Henderson RA, Clarkson PD, Hughes CP, Laurie J, Stait B, Thomson MRA, Webers GF (1981) The Ordovician system in Australia, New Zealand and Antarctica. Correlation Chart and Explanatory Notes. Publication of the International Union of Geological Sciences 6:1–64

Webers GF, Spörli KB (1983) Palaeontological and stratigraphic investigations in the Ellsworth Mountains, West Antarctica. In Oliver RL, James PR, Jago JB (eds) Antarctic Earth Sciences, Cambridge Univ. Press, Cambridge, pp 261–264

Wright TO, Brodie C (1987) The Handler Formation, a new unit in the Robertson Bay Group, northern Victoria Land, Antarctica. In McKenzie GD (ed) Gondwana Six: Structure, Tectonics, and Geophysics, Amer. Geophysical Union, Washington, D.C. Geophys. Monogr. 40:25–29

Wright TO, Ross RJ, Repetski JE (1984) Newly discovered youngest Cambrian or oldest Ordovician fossils from the Robertson Bay terrane (formerly Precambrian), northern Victoria Land, Antarctica. Geology 12:301–305

9—Plant Distribution in Gondwana During the Late Paleozoic

Sergio Archangelsky

Global plant distribution during the Late Paleozoic (Carboniferous and Permian) was paleogeographically controlled. The latest reconstructions of sea–land distribution during the Late Paleozoic coincide in shaping a huge continental block, called Pangaea, that stretched from a northern hemisphere cold–temperate zone across the paleoequator to a southern hemisphere cold–temperate belt (Ziegler et al. 1981, Tarling 1985, Johnson and Tarling 1985, Scotese 1986). This new approach has partly changed concepts about large, paleophytogeographical units (realms, provinces, phytochoria) that were based on taxonomical lists in order to suggest consistent plant groupings in definite areas of the globe (Halle 1937, Gothan and Weyland 1954, Wagner 1962, Vakhrameiev et al. 1970, Chaloner and Meyen 1973, Chaloner and Lacey 1973). The new paleogeographical data also provide an exceptional tool for paleobotanists to explain some anomalies in plant distribution that were, if at all, marginally considered.

The development of phytopaleoecological studies placed fossil plants within a new perspective, revealing past communities by means of field techniques useful for discriminating trends in the preservation of plant debris in relation to precise paleoenvironments (Scott 1979, Spicer 1981, Iwaniw 1985). Local physical phenomena produced specific paleoenvironments (river valleys, deltas, lakes, etc.), while oceanic and atmospheric currents were responsible for global paleoclimatic patterns that also controlled the distribution of organisms (Raymond et al. 1985) though on a different scale. However, knowledge about fossil plants has improved during the last few years in several aspects, including the reconstruction of new whole plants, the structure and ultrastructure of key organs (fructifications, pollen, cuticles), and paleobiology (fungal activity, plant–animal interactions), (Stubblefield and Rothwell 1981, Taylor and Scott 1983, Smoot and Taylor 1986a, Stubblefield and Taylor 1988).

In paleobotanical literature, Late Paleozoic southern continents are called Gondwana. New paleogeographical reconstructions have shown that during the Carboniferous this traditional land mass joined a northern continent to form Pangaea. The distribution of vegetation in Pangaea varied according to paleolatitudes, and so different patterns emerged following its mobility through time. The basic information needed to understand the evolution of neopaleozoic southern floras is still scanty. However, important discoveries of new materials and the refinement of observation techniques have produced changes and additions in the information about fossil plants. In this regard, recent paleobotanical findings in Late Paleozoic strata from Antarctica add fresh information that fills a large gap in our knowledge about the distribution of organisms in strongly stressed conditions.

Taxonomy of Late Paleozoic Floras in Gondwana

The paucity of plants with preserved anatomy in Gondwana (mostly impression–compression fossils) is an obstacle for making sound comparisons with materials from the equatorial belt. This has led to confusions because generic names used in other regions for permineralized fossils were introduced for impressions or molds in Gondwanian assemblages—lycophytes are a good example. The name *Lepidodendron* was widely used for Gondwanian arborescent individuals that resembled, only superficially, their northern counterparts. Careful studies revealed that they were different taxa: *Brasilodendron,* Chaloner et al. 1979; *Bumbudendron,* Archangelsky et al. 1981 (Fig. 9.3); *Cyclodendron,* Rayner 1985 (Fig. 9.5); and *Azaniodendron,* Rayner 1986. These satellite taxa (Thomas and Meyen 1984, Thomas and Brack-Hanes 1984) are a proof that arborescent lycophytes also developed in Gondwana, where they sometimes formed compact stands in swampy environments (Cúneo and Andreis 1983); their rooting system, not of the *Stigmaria* type, resembles cormose lycophytes (Pigg and Taylor 1985). They were heterosporous but apparently had no differentiated cones: sporophylls were borne on the stems together with vegetative leaves (*Bumbudendron*). Associated dispersed megaspores have little or no ornamentation, while microspores found in the same strata are classified in known cosmopolitan genera. However, similarities based upon miospore types are in contrast with the differences that are found between northern and southern taxa, based on several organs. This accounts for some explanation other than natural affinity at a generic level, perhaps parallel evolution (Meyen 1979). Two permineralized lycophytes, namely, *Eligodendron,* Archangelsky and de la Sota 1966 (Fig. 9.4), and *Lycopodiopsis,* Kräusel 1961, have been compared with similar northern taxa in their general anatomical organization: a central pith and a thin vascular cylinder surrounded by a cortical system composed of several layers (including aerenchyma), though they lack parichnos and ligule. Gondwana lycophytes played perhaps a similar ecological role in plant communities as several northern genera, but they were adapted to more rigid climatic conditions and were surrounded by a less diversified vegetation.

The name *Lepidodendropsis* is another case of generalized use (Fig. 9.2). This genus was proposed to characterize a worldwide type of flora during the Early Carboniferous (Jongmans 1952). However, this plant is known only from external morphology of stems with leaf cushions that look similar in specimens from different parts of the world. It is indeed a poorly defined taxon and as such, not very reliable for global correlations.

Sphenophyllum is probably one of the most exciting cases among Gondwana sphenophytes. *Sphenophyllum speciosum* Feistmantel (Fig. 9.7) is a characteristic species in Gondwana: it has leaves of different sizes in a whorl (and was referred sometimes to a different genus, *Trizygia*). Other species resembling northern forms were described in Africa (Walton 1929, Huard-Moine 1965), South America (Read 1941, Archangelsky 1960, Rösler and Rohn 1983), Australia (Rigby 1966), and India (Singh et al. 1986). Recently, impressions of fertile specimens bearing cones of a *Bowmanites* type have been reported from South America by Archangelsky, Cúneo and Wagner (1982 Meeting Proj. 211, IGCP, Montevideo) (Fig. 9.6). *Sphenophyllum* is not a common component in Gondwana floras, and it is exclusively Permian. Two questions arise about this material: (1) Was the anatomy of the southern species similar to the equatorial ones? (2) Why is *Sphenophyllum* absent in Gondwana Carboniferous floras? Cautious answers could be (1) typical whorls with 6 leaves may *reasonably* correspond to an internal triarch protostele structure and (2) the genus diversified in the equatorial belt floras during most of the Carboniferous and migrated southward when climatic conditions became favorable and geographical barriers disappeared.

Permineralized fern stems are common in Permian strata, especially of western Gondwana. Some are related to the Osmundales (family Guaireaceae, Herbst 1981), while others are included in the Psaroniaceae (Herbst 1986, 1987). Impressions of fertile fronds are common, and they are referred either to en-

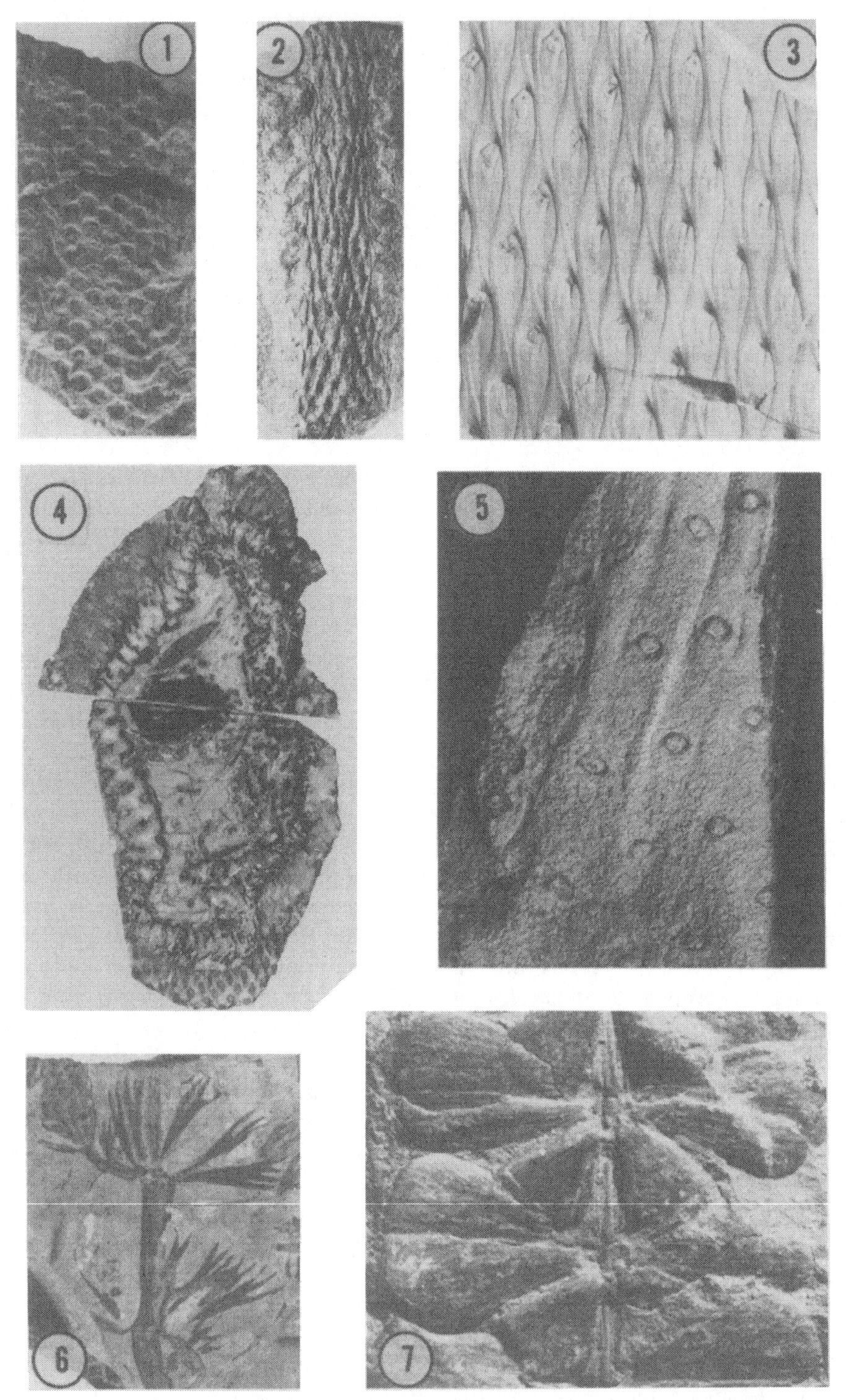

Figure 9.1. Archaeosigillaria conferta (Frenguelli) Menéndez, holotype, Early Carboniferous, Argentina, ×2.

Figure 9.2. Lepidodendropsis sekondiensis Mensah & Chaloner, Early Carboniferous, Argentina, ×1.

Figure 9.3. Bumbudendron nitidum Archangelsky, Azcuy & Wagner, holotype (replica), Late Carboniferous, Argentina, ×3.

demic taxa (*Dizeugotheca,* Archangelsky and de la Sota 1960) (Fig. 9.8 and 9.9) or cosmopolitan genera (*Asterotheca, Corynepteris*) (Fig. 9.12). A Permian age poses the same question about their absence in Carboniferous strata: they probably migrated from northern areas.

Progymnosperms were probably represented in the Carboniferous of Gondwana (Archangelsky 1984), while putative pteridosperms (Artabe et al. 1987) were different from northern materials. This group is important for Carboniferous biostratigraphy in Gondwana because several sterile fronds seem to be endemic: *Nothorhacopteris* (Fig. 9.11) *Bergiopteris, Fedekurtzia,* and *Botrychiopsis* (Fig. 9.13). Lack of anatomical data impedes a better evaluation of the real systematic status of these genera. However, the absence of typical *Neuropteris* or *Alethopteris* leaves suggests that some northern pteridosperm families were not represented in Gondwana. Yet, much has to be done with several other cases in which phylogenetic links may have existed, as with some *Sphenopteris* fronds (Fig. 9.10), *Callipteris, Eremopteris,* and *Triphyllopteris*. For the moment, we only suspect that this southern stock of fronds could have given rise to the corystosperms, a group that dominated the Triassic floras of Gondwana.

Glossopterids (Figs. 9.14–9.18) are the most widespread gymnosperms in Gondwana, and judging from the variety of fructifications, they were probably quite diversified. Essentially Permian, the glossopterids are poorly known in their anatomy, and phylogenetic links are still being debated (Schopf 1976, Gould and Delevoryas 1977, Banerjee 1984, Meyen 1987). Leaf remains are often exclusive in plant beds: this dominance probably gave rise to the concept of a *Glossopteris* flora, the genus being the essential characteristic of the Permian in Gondwana. Similar leaves have been described in other paleofloristic regions, as in the Permian of Angara (Zimina 1967), but their natural botanical affinity is still questioned (Meyen 1979). Another similar leaf is *Lesleya,* found in Carboniferous upland floras of the United States (Leary and Pfefferkorn 1977). Could they have been the forerunners of the group? Late Permian glossopterid leaves also occur in marginal areas of Gondwana, near the tropical belt of Pangaea (Turkey, Archangelsky and Wagner 1983). Were they late travelers seeking northern lands? Glossopterids had enough ecological plasticity to spread over large areas; they lived in periglacial environments, where they dominated, to subtropical regions, where they mixed with other plants as subordinate taxa. As a whole, they seem to have preferred cold temperate, seasonal climates. The fate of this group was clearly related to Late Paleozoic glaciation.

Leaves referred to as ginkgophytes are known in the Carboniferous and Permian of Gondwana (*Ginkgophyllum, Ginkgoites, Rhipidopsis*), but fertile structures have not been reported so far. These leaves may be tentatively related to the Trichopityales and the Dicranophyllales on the basis of some cuticular similarities with materials from Angara (Meyen and Smoller 1986). Fertile impressions found recently in the Permian of Patagonia suggest that some leaves may have been related to a group (order?) of primitive ginkgoalean (Archangelsky and Cúneo, in preparation).

Cordaitales were present in the Carboniferous and Permian of Gondwana, but the case is not definitely proved because there are no fertile specimens to support a precise classification. Leaves (*Noeggerathiopsis* or *Cordaites*) (Figs. 9.22 and 9.23), isolated seeds (*Cordaicarpus*), and permineralized wood (*Dadoxylon*) may be referred to this order.

The oldest conifer remains in Gondwana are

Figure 9.4. Eligodendron branisae Archangelsky & de la Sota, holotype, Early Permian, Bolivia, ×1.
Figure 9.5. Cyclodendron leslii (Seward) Kräusel, Early Permian, Vereeniging, Transvaal, S. Africa, ×3.
Figure 9.6. Sphenophyllum sp. Fertile specimen. Early Permian, Argentina, ×1.
Figure 9.7. Sphenophyllum speciosum Feistmantel, Permian, Argentina, ×2.

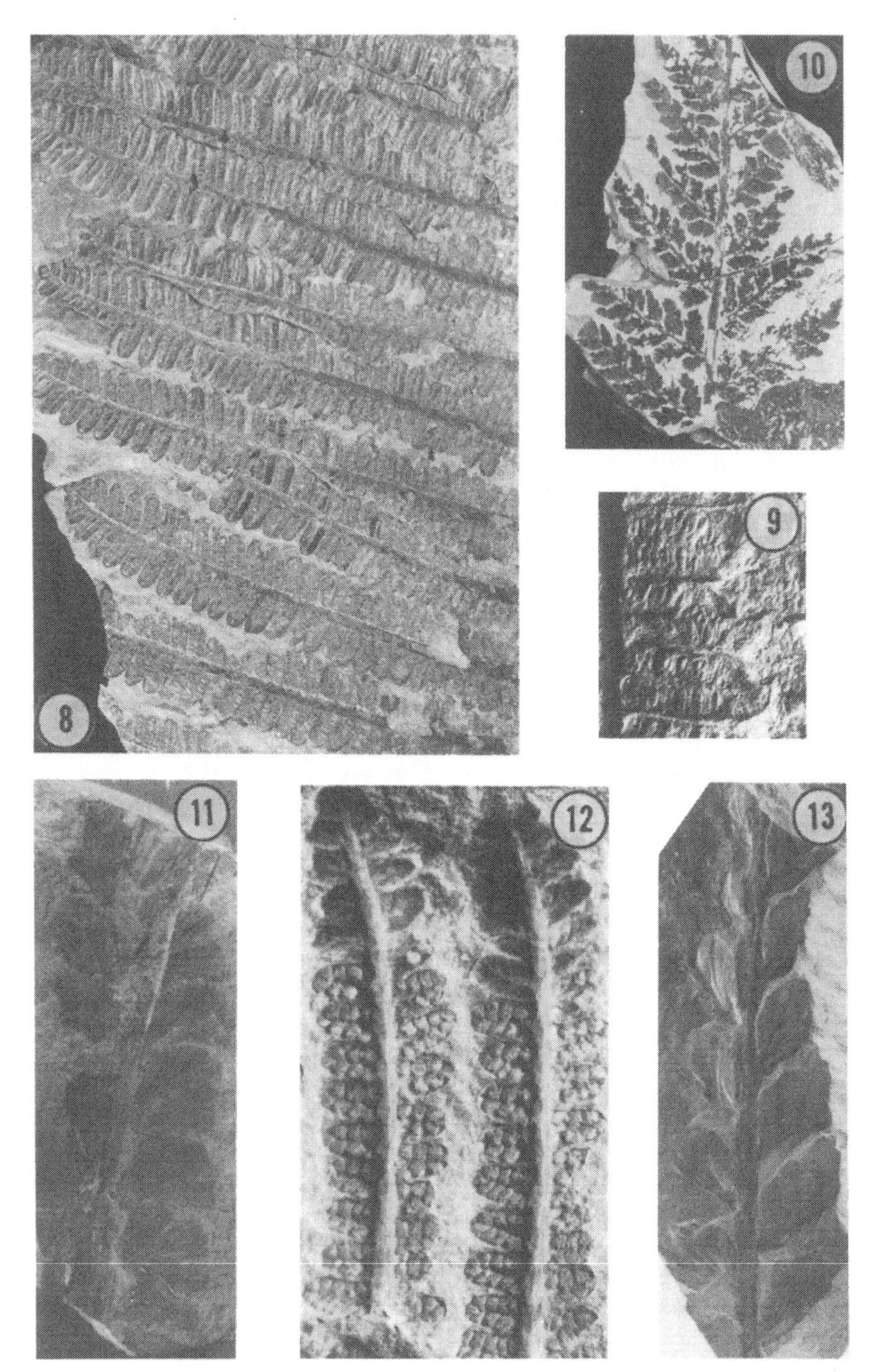

Figure 9.8,9.9. Dizeugotheca waltonii Archangelsky & de la Sota, holotype, Early Permian, Argentina. Fig. 9.8, general aspect, ×1; Fig. 9.9, detail showing transversely elongated fertile bodies, ×4.
Figure 9.10. Sphenopteris lobifolia Morris, Permian, NSW, Australia, ×1.
Figure 9.11. Nothorhacopteris argentinica (Geinitz) Archangelsky, holotype, Late Carboniferous, Argentina, ×1.
Figure 9.12. Asterotheca piatnitzkyi Frenguelli, Permian, Argentina, ×2.
Figure 9.13. Botrychiopsis weissiana Kurtz, Late Carboniferous, Argentina, ×0.5.

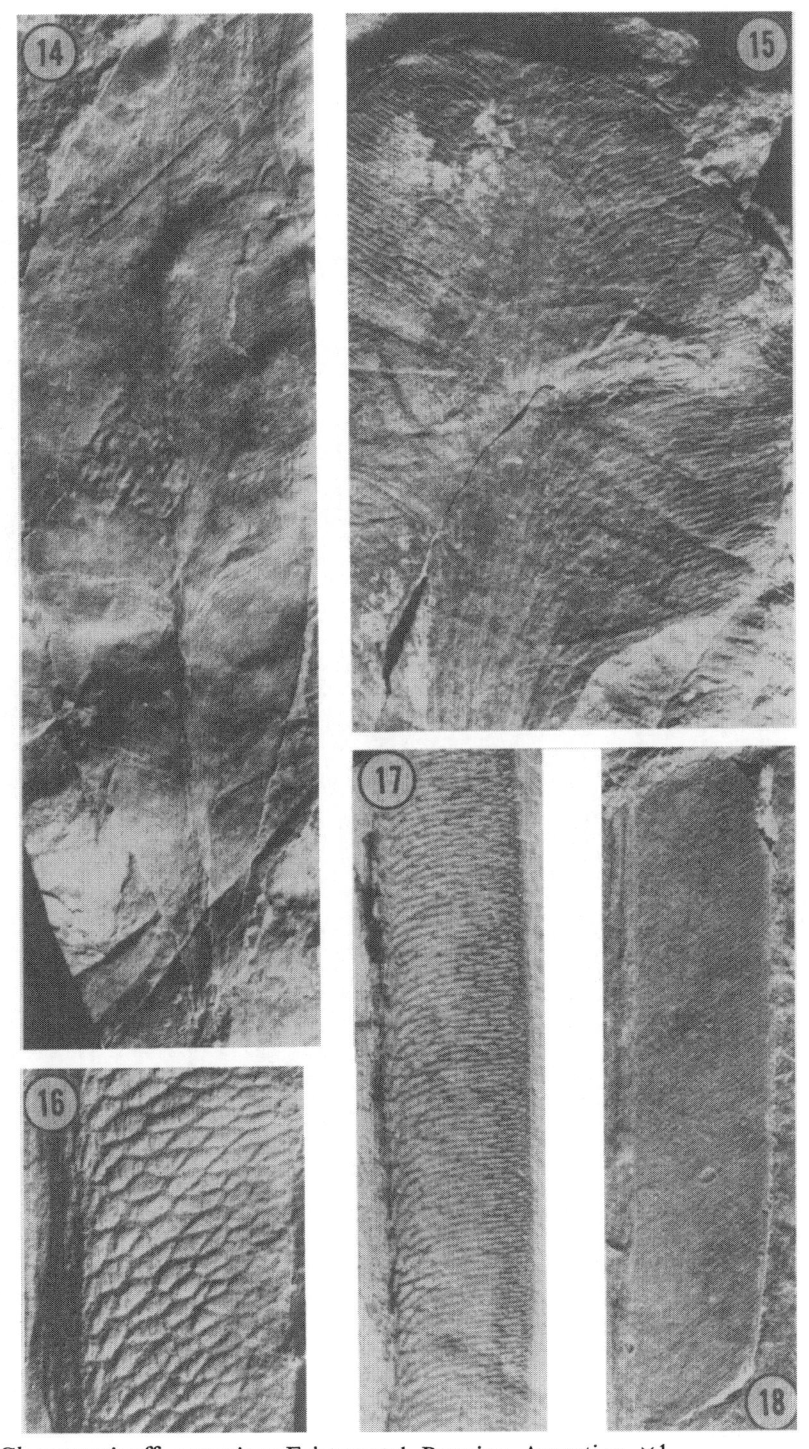

Figure 9.14. Glossopteris aff. *conspicua* Feistmantel, Permian, Argentina, ×1.
Figure 9.15. Gangamopteris castellanosii Archangelsky, holotype, Permian, Argentina, ×1.
Figure 9.16,9.17,9.18. Glossopteris spp. Three species from the same locality showing variation of the secondary venation; Fig. 9.16 with short and wide meshes, ×2; Fig. 9.17 with short and narrow meshes, ×1.5; Fig. 9.18 with few anastomoses, ×2. Permian, Argentina.

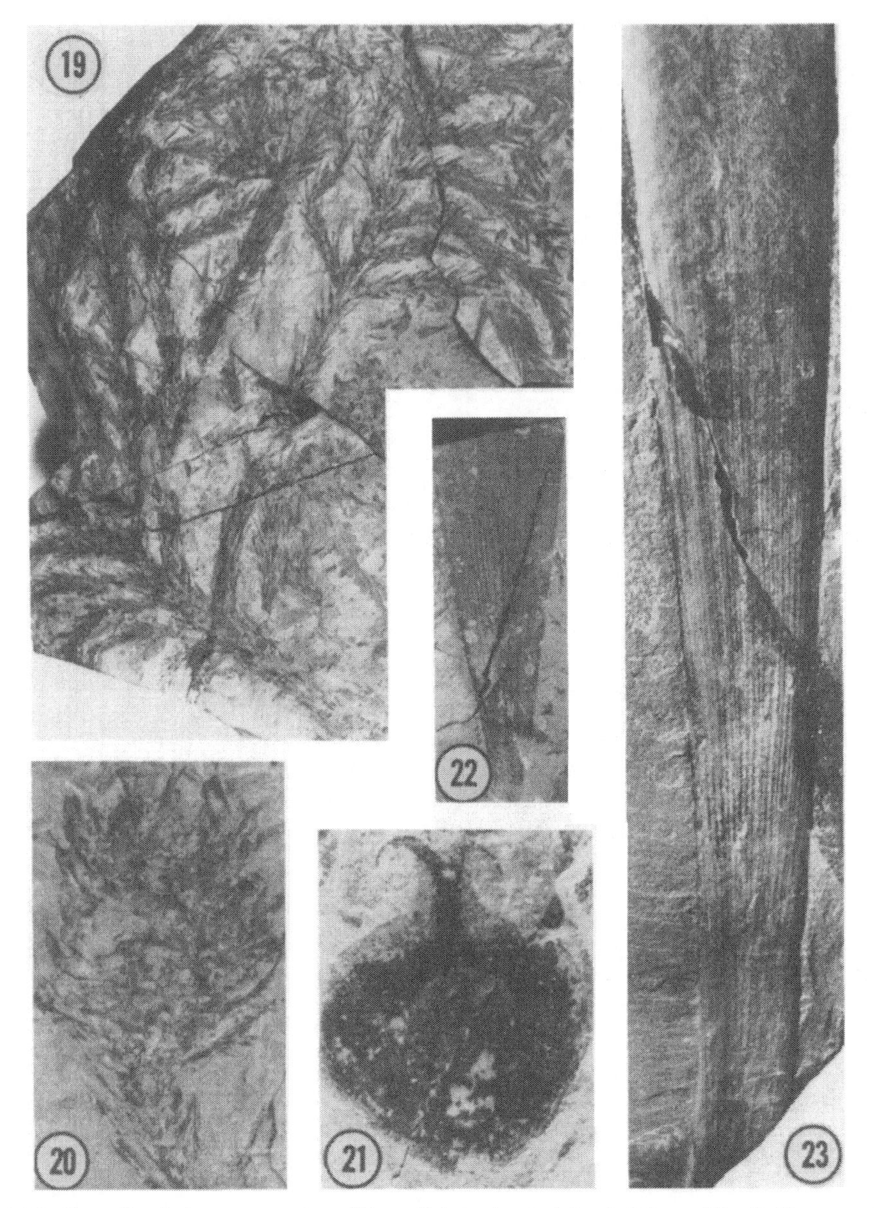

Figure 9.19,9.20. Ferugliocladus patagonicus (Feruglio) Archangelsky & Cúneo. Fig. 9.19, general aspect, ×1; Fig. 9.20, female cone, ×2. Early Permian, Argentina.
Figure 9.21. Ferugliocladus riojanus Archangelsky & Cúneo, holotype showing a seed with bifid apex and central nucellar body. Early Permian, Argentina, ×5.
Figure 9.22. Noeggerathiopsis hislopi (Bunbury) Feistmantel, holotype, early Permian, India, ×1.
Figure 9.23. Cordaites sp. Late Carboniferous, Argentina, ×1.

of Late Carboniferous age (Pant 1982, Archangelsky and Cúneo 1987). They are still rare components in Permian plant communities. Studies on compression–impression materials have shown that conifers differentiated during the Early Permian into several families. However, putative genera may well belong to other gymnosperms (*Buriadia, Walkomiella*), depending on how we define a primitive conifer. Other taxa (*Ferugliocladus,*

Ugartecladus) may be regarded as conifers in a more strict definition; they possess true male and female compact cones that have some structural differences from the northern Lebachiaceae (Figs. 9.19–9.21). Their sudden appearance in the Late Carboniferous (together with fern fronds and sphenophylls) strongly suggests that they have migrated from equatorial upland areas where their ancestors are found in slightly older strata. This southern Permian stock probably gave rise to the Podocarpaceae and Araucariaceae, two families of Gondwanic alliance (Archangelsky 1985).

Paleozoic plant assemblages are found in many localities of Antarctica and are all Permian. Glossopterids dominate as impressions (Seward 1914, Plumstead 1962, Townrow 1967). Several *Glossopteris* and *Gangamopteris* species and fructifications (*Cistella, Arberia, Arberiella*) have been described. Permineralized wood has also been found in many sites and several genera are known (Maheshwari 1972). Perhaps the most important unit yet discovered in Antarctica is the Buckley Formation that includes permineralized peat and was first reported by Schopf (1970). It has yielded "exquisitely preserved materials" that opened a new line of research now being actively developed. Bryophytes, fungi, seeds with embryos, and the anatomy of *Glossopteris* leaves have been reported (Smoot and Taylor 1986a,b, Stubblefield and Taylor 1986, Smoot and Taylor 1985, Taylor and Taylor 1987, Pigg and Taylor 1987).

The Distribution of Gondwanian Late Paleozoic Floras: A Paleogeographical–Paleoclimatical Approach

Long before the successful advent of a new global paleogeography based on plate tectonics—a theory that accepts continental drift—paleobotanists wrote about a southern supercontinent, or Gondwana. Following Wegener and Du Toit, this land mass consisted of present-day India, Australia, Antarctica, southern Africa, and South America; all these components were linked by the presence of the leaf *Glossopteris* in Permian strata. Late Paleozoic glaciation widely spread over a large area in Gondwana and the *Glossopteris* flora probably developed in cold to temperate climates. In the past, most paleobotanists tended to isolate Gondwana from other coeval continents as an independent land. However, some Permian localities in Africa yielded, along with *Glossopteris,* taxa of northern alliance, such as *Sphenophyllum, Asterotheca, Pecopteris,* and *Annularia* (Walton 1929, Teixeira 1946, Huard-Moine 1965, LeRoux 1976, Anderson and Anderson 1985). Hopping and Wagner (in Visser and Hermes 1962) also reported *Glossopteris* in association with *Sphenophyllum, Pecopteris,* and *Taeniopteris* from New Guinea. In meridional South America, northern taxa mixed together with *Glossopteris* were reported in 1908 by D. White and later by Read (1941), Frenguelli (1953), and Archangelsky (1957). Wagner (1962) reported *Glossopteris* in Turkey mixed with Cathaysian and Euramerican taxa. At that time, with this limited data base and a still fixed global paleogeography, a concept of mixed floras was proposed during the First International Gondwana Symposium (Archangelsky and Arrondo 1969). It was suggested that some northern taxa found in Gondwana were naturally related to ancestors that migrated from the tropical belt when climatic conditions improved. During that same decade, mobilistic global paleogeography developed on the basis of sound geophysical data. Since then, models of sea–land distribution for different geological periods are constantly improving. For the Westphalian, recent reconstructions (Johnson and Tarling 1985, Scotese 1986) show the formation of a land mass that included the traditional Gondwana, northern Africa–northern South America, Euramerica, and part of Asia. This Pangaea persisted until the Mesozoic and stretched from low latitudes of the Northern Hemisphere to the South Pole. This translatitudinal land strip was surrounded by sea to the west and east and probably had climatic belts that regulated the distribution of organisms on land. It is now widely accepted that the distribution of plant associations during the Late

Paleozoic was climatically controlled (Chaloner and Lacey 1973, Lemoigne 1977, Rigby 1979, Vozenin-Serra 1980, Raymond et al. 1985).

In order to understand the distribution of fossil plants in Gondwana during the Late Paleozoic, I shall consider four time segments, analyzing paleogeographic dispositions and general floristic composition for each one.

Early Carboniferous (Fig. 9.24)

During Tournaisian–Viséan times, Gondwana was an isolated block placed south of latitude 30°S. Only Australia extended northward, closer to the equator. Palynological records (Kemp et al. 1977, Playford 1985) and plant megafossils (Morris 1975) suggest a rather varied flora unknown in other Gondwanian areas. In a belt between latitude 15–55°S, fossil floras were recovered in Australia, India, northeast Africa, and central to southern South America. Lycophytes such as *Archaeosigillaria* (Fig. 9.1) and *Lepidodendropsis* (Fig. 9.2) dominate, although fronds are also common components (*Diplothmema, Rhodeopteridium, Sphenopteridium,* and *Triphyllopteris;* Archangelsky 1984, Triparthi and Singh 1985). Recently, Lejal-Nicol (1987) reported *Archaeosigillaria* and *Lepidodendropsis* from the Tournaisian of Egypt and northern Sudan, with a more diversified assemblage in younger strata (*Eskdalia, Lepidodendron,* lycophytes with infrafoliar bladders, and fronds of *Eremopteris, Rhacopteris, Nothorhacopteris,* and *Triphyllopteris*). I should mention here that we are still debating the precise age of several African and South American floras assigned either to the Late Mississippian or Early Pennsylvanian (Late Viséan to Namurian). It seems that during this interval a glaciation affected southern Gondwana, playing a role in plant distribution. The most severe climatic conditions are found in the central (inner) part of the continent; however, vegetation developed in marginal (near the sea) areas, even at high latitudes. Ocean (and perhaps atmospheric) currents controlled this distribution. The floristic associations found in northeast Africa, the Himalayas, and Perú may have been slightly older than those found in other Gondwanic areas with the same composition. These marginal floras may be considered as centers of origin and dispersal of some taxa that in slightly later times expanded and conquered the remaining inland part of Gondwana.

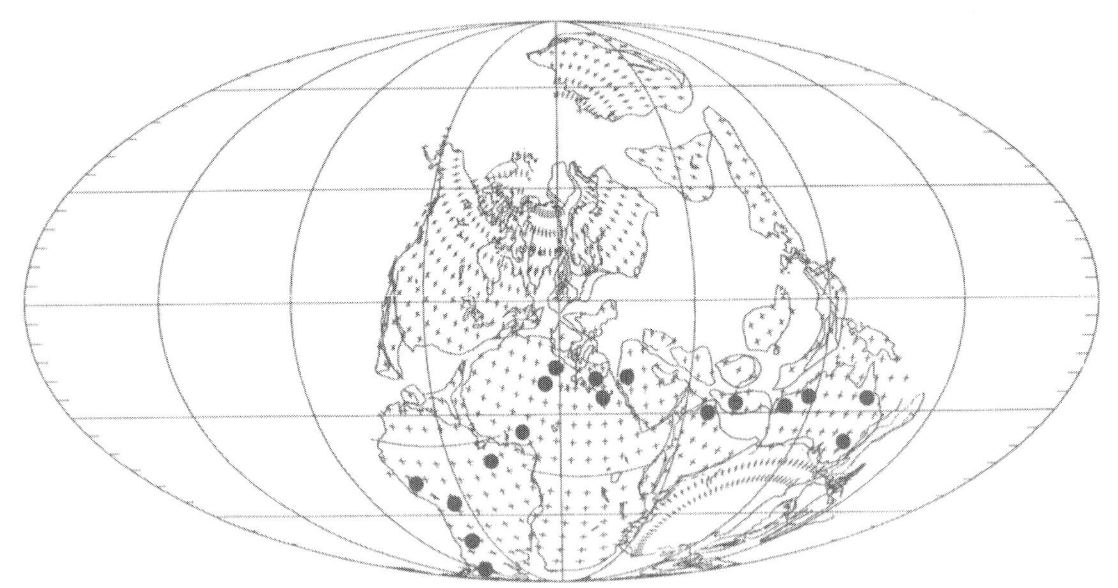

Figure 9.24. Distribution of main plant fossil sites in the Southern Hemisphere during the Early Carboniferous. (Paleogeography adapted from Scotese 1986.)

The composition of Tournaisian–Namurian floras of Gondwana should be revised in relation to coeval assemblages found in the northern components of Pangaea. If the botanical affinity is confirmed, then northern Africa could well have been a main center of radiation. If, on the other hand, there was geographical isolation, we must assume a differentiated evolution of primitive stocks, with little (if any) connecting possibilities for genetic flow. In that case, isolation would explain the fast diversification of tropical floras during the middle and upper Carboniferous, and a slow, retarded evolution of the cool southern floras.

Archaeosigillaria and *Lepidodendropsis* are also found at higher latitudes in Early Viséan strata of the Argentinian Paganzo and Tepuel–Genoa basins. At the moment, it is premature to suggest a differentiation into botanical provinces of floras that inhabited Gondwana (*s.l.*) during the Mississippian.

Middle Carboniferous (Fig. 9.25)

During the Namurian and Westphalian, Gondwana rotated clockwise so that its eastern region moved toward the pole, while the western component approached the equator and fused with Euramerica to form a vast Pangaea (Fig. 9.25). The pole was surrounded by a large land mass and a persistent glaciation acted as a huge freezing center constantly spreading cool air over the southern part of the continent. There are practically no fossil plant records beyond 60°S and most Gondwana Westphalian occurrences are situated between 30–60°S. In South America, northeast Africa, the Himalayas, and Australia, a poor but uniform flora developed. It was characterized by fronds (*Nothorhacopteris, Botrychiopsis, Triphyllopteris,* etc.), lycophytes (*Bumbudendron, Malanzania, Lepidodendropsis*), *Cordaites,* and *Ginkgophyllum* leaves. These assemblages, found in both eastern and western Gondwana, are completely different from coeval floras of tropical areas. In Venezuela, a Westphalian flora, including *Neuropteris ovata, Lobatopteris vestita,* and *Annularia,* was recorded by Odreman-Rivas and Wagner (1979), while Lejal-Nicol (1987) mentioned *Lepidophloios, Lepidodendron,* and *Rhodea* from Namurian strata in northern Africa. Therefore, Pangaea showed a latitudinal differentiation of plant associations: the richest were located in tropical areas, and the poorest near

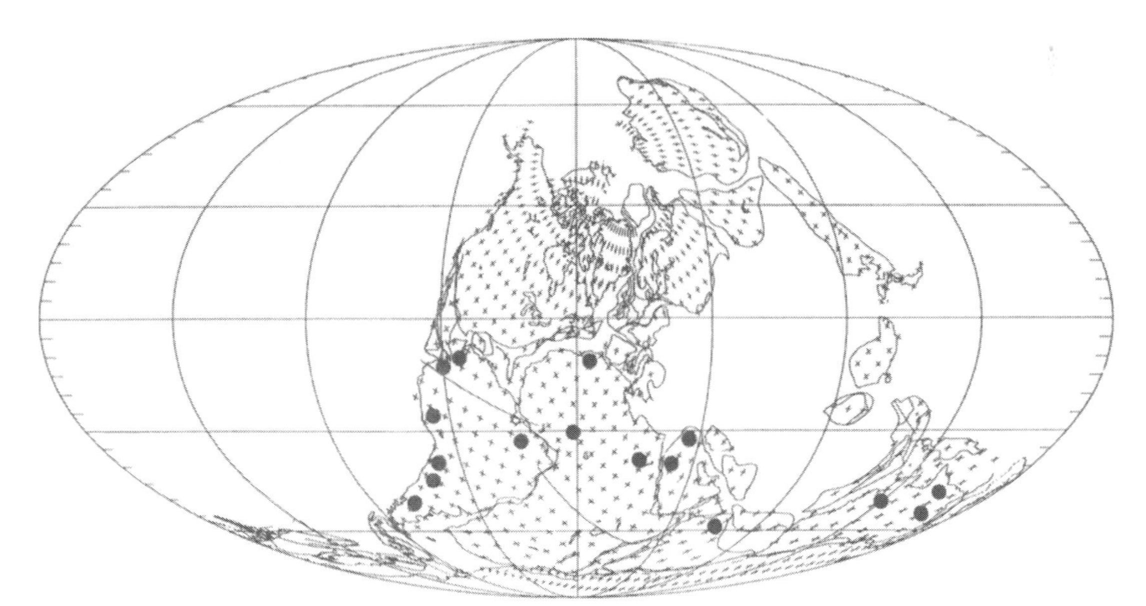

Figure 9.25. Distribution of main plant fossil sites in the Southern Hemisphere during the Late Carboniferous. (Paleogeography adapted from Scotese 1986.)

the South Pole. Gondwanian floras were situated in the 30–60°S belt, and their uniformity does not suggest a strong provincialism. Paleogeographical maps, on the other hand, show for this belt (and further south) a land mass stretching east to west. Therefore, two latitudinally expanded segments can be recognized within this continent: eastern and western Gondwana.

Late Carboniferous and Early Permian (Fig. 9.26)

The same clockwise movement of Pangaea continued in the Stephanian–Artinskian interval, during which there was a dramatic floristic change in Gondwana marked by (1) the advent and diversification of the Glossopteridales and other taxa, (2) the extinction of older Carboniferous taxa, and (3) the sudden appearance of several tropical taxa (though not in all areas; Fig. 9.26). A latitudinal control of floral distribution in the Early Permian becomes more pronounced than in the previous time segment. Floras found in northern Africa and South America and in Southwest Asia have a Euramerican alliance; they are situated in tropical latitudes (0–30°S) (Archangelsky 1984, Léjal-Nicol 1986, 1987, El Wartiti et al. 1986). A mixed *Glossopteris* flora is located in a temperate belt (from ca. 30° to 60°S). These are the richest Gondwanian assemblages in which taxa of panequatorial and subantarctic alliance are mixed forming characteristic assemblages (Archangelsky 1984). However, some floras found inside this belt lack northern taxa, though they may be located close to mixed ones. They resemble assemblages found further south (60–90°S), in which glossopterids dominate (the so-called pure *Glossopteris* floras). Tentatively, this can be explained by a paleoecological–paleoclimatological control considering that they were situated inland within the western block of Gondwana or near shore in the eastern block, where cold sea currents affected climate locally (altitude may have also played a role).

Some Antarctic floras (Whichaway Nunataks, Theron Mts.) have been referred to the early Permian (Plumstead 1962). They are dominated by glossopterids; however, the specific variety and the presence of some taxa

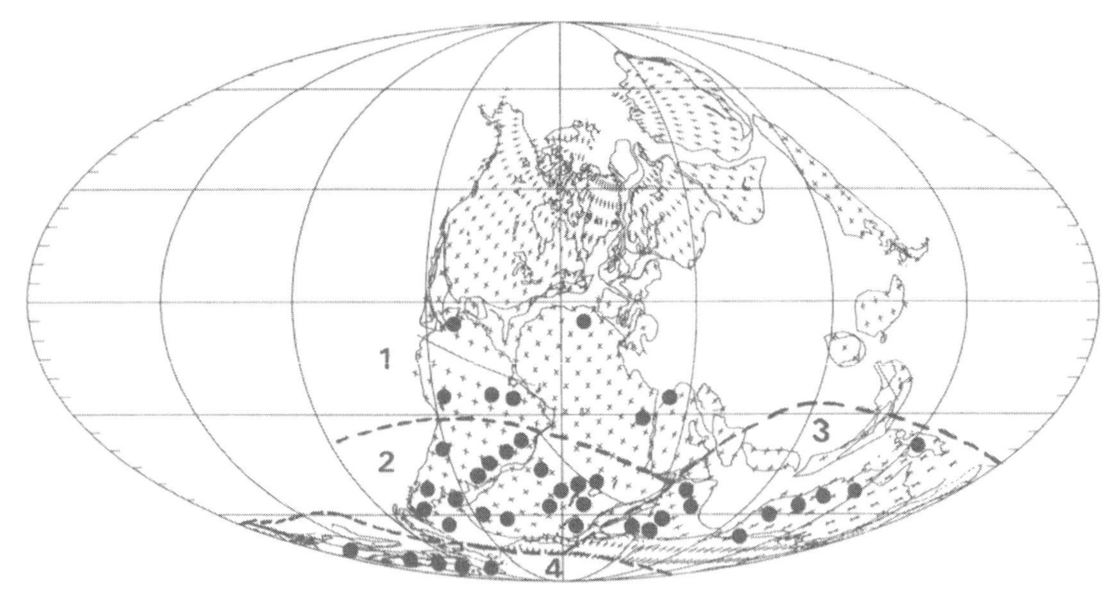

Figure 9.26. Distribution of main plant fossil sites in the Southern Hemisphere during the Early Permian. Four paleofloristic provinces are recognized: (1) paleoequatorial, (2) western Gondwana (Nothoafroamerican), (3) eastern Gondwana, and (4) Paleoantarctic. (Paleogeography adapted from Scotese 1986.)

found in younger strata in other Gondwanian areas suggest that biostratigraphy needs to be controlled.

Fossil sites found near the Late Paleozoic South Pole have that position in different paleogeographical models: this fact is difficult to explain. I believe that it is still a question to be solved either with parameters we currently use (which may need adjustment) or with new ones.

Patagonia is another special case because in all paleogeographical reconstructions it is placed near latitude 60°S. However, the richest mixed assemblages of all Gondwana are found there, including plants that could have existed only in temperate to warm conditions (Cúneo 1987). Some authors strongly suggest that Patagonia was part of a different plate, located west of South American Gondwana (Martínez 1980, Ramos 1984).

Botanical provincialism in Gondwana (including present-day Africa and South America) during the Early Permian may be outlined following climatic belts and paleogeography. Four units are recognized:

1. an equatorial (or tropical) province that related to Euramerica and extended over northern Africa, northern South America, and southeast Asia;
2. a western (Nothoafroamerican) province that extended over the rest of South America and Africa;
3. an eastern province that occupied present-day Australia (and neighboring areas) and possibly India and southern China (Li and Yao 1981, Zhang and He 1985); and
4. a southern (Paleoantarctic) province that stretched over present-day Antarctica (and eventually peripheral areas of India, South Africa and South America).

Late Paleozoic paleogeography is consistent with the presence of ecotonal areas between 1 and 2, 2 and 4, and 3 and 4. Also, no serious objection exists for accepting migrational routes from 1 to 2 and from 2 to 3. Dispersal strategies that favored isosporous pteridophytes have already been outlined in the literature (Archangelsky and Arrondo 1969, Chaloner and Lacey 1973).

Late Permian

The paleogeography of Gondwana did not change substantially, except for a slight shift northward. This allowed some floral differentiation in the eastern block (Australia and India). The same mixed floristic character was maintained in the Nothoafroamerican province.

Composition of floras also varied with the extinction of all relictual Carboniferous taxa (*Nothorhacopteris, Botrychiopsis,* etc.) and the appearance of new genera (*Dizeugotheca, Walkomiella, Thinnfeldia,* etc.). Glossopterids strongly diversified in all of Gondwana; during their acme, some probably migrated northward where they are occasionally found.

Antarctica plays an important role during this epoch. Glossopterids were strongly diversified, and there is no trace of equatorial elements. This would confirm the existence of a southern Paleoantarctic botanical province, where a pure *Glossopteris* flora developed in a cold climatic régime. Life was diversified, as shown by different plant organs (trees, roots, seeds, leaves), fungal decay, and preserved delicate bryophytic tissues. Photosynthetic activity was certainly important. The extreme southern position of these floras as suggested by all paleographical maps still needs some satisfactory explanation to make plants function under such severe conditions.

Conclusions

Geographically, during the Early Carboniferous, Gondwana occupied part of the Southern Hemisphere (including the South Pole). This huge land mass moved toward the equator, joining northern lands at the end of the Carboniferous to form a vast supercontinent, or Pangaea. The constant mobility produced vegetational changes, not only in areas that surrounded the wandering pole but also in others located closer to a temperate climatic belt. In the equatorial belt, the development of lineages in different plant groups (ferns, sphenophytes, lycophytes) can be followed in several localities where the Carboniferous and Early Permian are in continuous succession;

changes are gradual and vegetational spectra were influenced only by local ecological conditions. More important changes were possibly related to global climatic events. In Gondwana, Carboniferous and Early Permian strata show drastic changes of floral assemblages through successive plant horizons found in the same section. This pattern defines Gondwana as a Late Paleozoic floristic unit characterized by continuous migrational flows reflecting a continental (plate) mobility and consequent paleoclimatic changes. The vicinity to other floras from more stable areas, a matter of debate not long ago, explains the presence of marginal mixed floras that developed in ecotones, that is, in regions where species from different communities intermingled by finding propitious edaphic and climatic conditions.

At present, Carboniferous strata are unknown in Antarctica: this is probably related to the glaciation or merely to the fact that strata of this age still wait to be discovered. The Permian history of Antarctica is closely related to the Glossopteridales, which after a long glacial period began to develop in some appropriate environments. A paleobotanical province is postulated for this large area (Paleoantarctica). It is characterized by frigid conditions during the Carboniferous with poor or no vegetation. Although maintaining an extremely southern position in Permian times, a pioneering vegetation developed in Antarctica.

The concept of Gondwana has two approaches. On one hand, a fixed picture of an isolated southern land mass (with the classical India, Australia, Antarctica, southern Africa, and South America) and, on the other, a mobilistic view that suggests the formation of Pangaea during the Late Carboniferous, when the same components joined another continent located in tropical areas. This second approach gives better possibilities for explaining some peculiar distributional patterns of the vegetation, such as the mixed floras as contrasted with pure *Glossopteris* floras in the Permian. It also allows us to envisage provinciality, especially after the Carboniferous.

We are probably now walking the first steps in this new paleophytogeographical dimension. At this time, it is necessary to improve the data base with a better biological knowledge of the plant taxa under consideration. For this reason, the permineralized plants from Antarctica will hopefully bring good anatomical information that can be combined with what is now known from impression–compression fossils. We also have to learn about plants living in extremely unfavorable climatic and edaphic conditions to understand the high-latitude records of wood in communities showing taxonomic variety.

References

Anderson JM, Anderson HM (1985) Palaeoflora of Southern Africa. Prodromus of South African megafloras. Devonian to Lower Cretaceous. Balkema, Rotterdam

Archangelsky S (1957) Estudio geológico y paleontológico del Bajo de la Leona (Santa Cruz). Acta Geológica Lilloana 2:5–133

Archangelsky S (1960) Lycopsida y Sphenopsida del Paleozoico Superior de Chubut y Santa Cruz, Patagonia. Acta Geológica Lilloana 3:21–36

Archangelsky S (1984) Floras del Neopaleozoico del Gondwana y su zonación estratigráfica. Aspectos paleogeográficos conexos. Comunic. Servicio Geolog. Portugal 70:135–150

Archangelsky S (1985) Aspectos evolutivos de las coníferas gondwánicas del Paleozoico. Bull. Sect. Sciences, Paris 8:115–124

Archangelsky S, Arrondo OG (1969) The Permian taphofloras of Argentina with some considerations about the presence of "northern" elements and their possible significance. In Gondwana Stratigraphy, IUGS Symposium (Buenos Aires, October 1967), pp. 71–89

Archangelsky S, Cúneo R (1987) Ferugliocladaceae, a new conifer family from the Permian of Gondwana. Rev. Palaeobot. Palynol. 51:3–30

Archangelsky S, Sota ER de la (1960) Contribución al conocimiento de las Filices pérmicas de Patagonia Extraandina. Acta Geológica Lilloana 3:85–126

Archangelsky S, Sota ER de la (1966) Estudio anatómico de una nueva licópsida del Pérmico de Bolivia. Rev. Museo La Plata (ns) Secc. Paleontología 5:17–26

Archangelsky S, Wagner RH (1983) *Glossopteris anatolica* sp. nov. from uppermost Permian strata in south-east Turkey. Bull. Brit. Mus. (Nat. Hist.) Geology 37:81–91

Archangelsky S, Azcuy, CL, Wagner RH (1981) Three dwarf lycophytes from the Carboniferous of Argentina. Scripta Geologica 64:1–35

Artabe A, Archangelsky S, Arrondo OG (1987) Sobre una fructificación masculina asociada a frondas de *Botrychiopsis* del Carbonífero de Ciénaga del Vallecito, Provincia de San Juan, Argentina. Actas VII Simposio Argentino de Paleobotánica y Palinología (Buenos Aires, April 1987), pp. 21–24

Banerjee M (1984) Fertile organs of the *Glossopteris* flora and their possible relationship in the line of evolution. Evolutionary Botany and Biostratigraphy, AK Ghosh Commem. Vol., pp. 29–59

Chaloner WG, Lacey WS (1973) The distribution of Late Paleozoic floras. Special Papers in Palaeontol. 12:271–289

Chaloner WG, Meyen SV (1973) Carboniferous and Permian floras of the northern continents. In Hallam A (ed) Atlas of Paleobiogeography, Elsevier, Amsterdam, pp. 169–186

Chaloner WG, Leistikow KU, Hill A (1979) *Brasilodendron* gen. nov. and *B. pedroanum* (Carruthers) comb. nov., a Permian lycopod from Brazil. Rev. Palaeobot. Palynol. 28:117–136

Cúneo NR (1987) Estudio geológico y paleontológico de los afloramientos meridionales de la Formación Río Genoa, Pérmico inferior de Chubut, Argentina. PhD. Buenos Aires University

Cúneo NR, Andreis RR (1983) Estudio de un bosque de licofitas en la Formación Nueva Lubecka, Pérmico de Chubut, Argentina. Ameghiniana 20:132–140

El Wartiti M, Broutin J, Freytet P (1986) Premières découvertes paléontologiques dans les séries rouges carbonatées permiennes du Bassin de Tiddas (Maroc Central). CR Acad. Scienc. Paris 303 Sér. II (3):263–268

Frenguelli J (1953) Las Pecopterídeas del Pérmico de Chubut. Notas Mus. La Plata, Paleont. 16:287–296

Gothan W, Weyland H (1954) Lehrbuch der Paläobotanik. Akademie-Verlag, Berlin

Gould RE, Delevoryas T (1977) The biology of *Glossopteris:* evidence from petrified seed-bearing and pollen-bearing organs. Alcheringa 1:387–399

Halle TG (1937) The relation between Late Paleozoic floras of eastern and northern Asia. CR II Congr. Stratigr. Carbonifère (Heerlen 1935), pp. 237–245

Herbst R (1981) *Guairea milleri* nov. gen. et sp. y Guaireaceae nueva familia de las Osmundales (sensu lato) del Pérmico superior del Paraguay. Ameghiniana 18:35–50

Herbst R (1986) Studies on Psaroniaceae. I. The Family Psaroniaceae (Marattiales) and a redescription of *Tietea singularis* Solms-Laubach from the Permian of Brazil. Actas IV Congr. Argentino Paleontol. Bioestratigr. (Mendoza, November 1986), Vol. 1:163–176

Herbst R (1987) Studies on Psaroniaceae. II. *Tuvichapteris solmsii* nov. gen. et sp. from the Permian of Paraguay and Uruguay. Mem. IV Congr. Latinoameric. Paleontolog. (Santa Cruz de la Sierra, Bolivia, August 1987), Vol. 1:267–282

Huard-Moine D (1965) Contribution à l'étude de la flore dite "à *Glossopteris*" du Bassin de Wankie (Rhodésie du Sud). II. Les Sphénopsides. III. Conclusions générales. Ann. Univ. A.R.E.A.S. Reims 3:68–86

Iwaniw E (1985) Floral palaeoecology of debris flow dominated valley-fill deposits in the lower Cantabrian of NE Leon, NW Spain. An. Facultad Ciências Univers. Porto 64(Suppl.), pp. 283–357

Johnson GAL, Tarling DH (1985) Continental convergence and closing seas during the Carboniferous. CR X Congr. Intern. Strat. Géol. Carbonif. (Madrid, September 1983), Vol. 4:163–168

Jongmans WJ (1952) Some problems on Carboniferous stratigraphy. C.R. III Congr. Intern. Strat. Géol. Carbonifère (Heerlen 1951), Vol. 1:295–306

Kemp EM, Balme BE, Helby RJ, Kyle AA, Playford G, Proce PL (1977) Carboniferous and Permian palynostratigraphy in Australia and Antarctica: a review. BMR Journ. Austral. Geolog. Geophys. 2:177–208

Kräusel R (1961) "Lycopodiopsis derbyi" Renault einige andere "Lycopodiales" aus dem Gondwana-Schichten. Palaeontographica B109: 62–92

Lacey W (1975) Some problems of "mixed" floras in the Permian of Gondwanaland. Gondwana Geology (3rd. Gondwana Symposium, Canberra 1973), pp. 125–134

Leary RL, Pfefferkorn HW (1977) An early Pennsylvanian flora with *Megalopteris* and Noeggerathiales from West-Central Illinois. Illin. State Geolog. Survey, Circular 500

Lejal-Nicol A (1986) Découverte d'une flora à *Callipteris* dans la région de Suez (Egypte). CXI Congr. Nat. Soc. savantes, (Poitiers, 1986), Sciences, fasc. 2:9–22

Lejal-Nicol A (1987) Flores nouvelles du paléozoique et du mesozoique d'Egypte et du Soudan septentrional. Berliner Geowissen. Abh.A 75 (1):151–248

Lemoigne Y (1977) Sur l'individualité de la province paléofloristique gondwanienne. Ann. Soc. Géol. Nord 97:383–404

Li Xing-xue, Yao Zhao-qi (1981) Discovery of Cathaysia flora in the Qinghai-Xizang Plateau with special reference to its Permian phytogeographical provinces. Geol. and Ecolog. Studies of Qinghai-Xizang Plateau, Science Press, Beijing, vol. 1:145–148

Maheshwari HK (1972) Permian wood from Antarctica and revision of some Lower Gondwana wood taxa. Palaeontographica B138:1–43

Martínez C (1980) Structure et évolution de la

chaine hercynienne et la chaine andine dans le Nord de la Cordillère des Andes de Bolivie. Travaux et Documents de l'O.R.S.T.O.M. 119:1–352
Meyen SV (1979) Relation of Angara and Gondwana floras: a century of controversies. IV Intern. Gondwana Sympos. (Calcutta 1977), Vol. 2:45–50
Meyen SV (1987) Fundamentals of Palaeobotany. Chapman and Hall, London–New York
Meyen SV, Smoller HG (1986) The genus *Mostotchkia* Chachlov (Upper Palaeozoic of Angaraland) and its bearing on the characteristics of the Order Dicranophyllales (Pinopsida). Rev. Palaeobot. Palynol. 47:205–223
Morris N (1975) The *Rhacopteris* Flora in New South Wales. In Campbell KSW (ed) Papers III Gondwana Geology, Canberra 1973, Sect. 2, Gondwana Flora, pp. 99–108
Odreman-Rivas RO, Wagner RH (1979) Precisiones sobre algunas floras carboníferas y pérmicas de los Andes venezolanos. Bol. Geológico (Caracas, Venezuela) 13–25:77–79
Pant DD (1982) The Lower Gondwana gymnosperms and their relationships. Rev. Palaeobot. Palynol. 37:55–70
Pigg KB, Taylor TN (1985) *Cormophyton* gen. nov., a cormose lycopod from the middle Pennsylvanian Mazon Creek Flora. Rev. Palaeobot. Palynol. 44:165–181
Pigg KB, Taylor TN (1987) Anatomically preserved *Glossopteris* from Antarctica. Actas VII Simposio Argentino Paleobot. Palinolog. (Buenos Aires 1987), pp. 177–180
Playford G (1985) Palynology of the Australian Lower Carboniferous: a review. C.R.X Cong. Intern. Strat. Géolog. Carbonifère (Madrid 1983), Vol. 4:247–265
Plumstead EP (1962) Fossil floras of Antarctica. Trans-Antarctic expedition 1955–1958, Scientific Rep. 9, Geology, pp. 1–132
Ramos V (1984) Patagonia: un continente a la deriva? Actas IX Cong. Argentino Geolog. (S.C. de Bariloche 1984), Vol. 2:311–328
Rayner RJ (1985) The Permian lycopod *Cyclodendron leslii* from South Africa. Palaeontology 28:111–120
Rayner RJ (1986) *Azaniodendron* a new genus of lycopod from South Africa. Rev. Palaeobot. Palynol. 47:129–143
Raymond A, Parker WC, Parrish JT (1985) Phytogeography and paleoclimate of the early Carboniferous. In Tiffney BH (ed) Geological factors and the evolution of plants. Yale University Press, pp. 169–222
Read CB (1941) Plantas fósseis do Neo-paléozoico do Paraná e Santa Catarina. Minist. Agricult., Depart. Nacional Producc. Miner. Div. Geolog. e Mineralog., Monografía 12:1–96
Rigby J (1966) The Lower Gondwana floras of Perth and Collie basins, Western Australia. Palaeontographica B118:113–152
Rigby J (1979) Aspects concerning the identification and distribution of Late Palaeozoic plants in Gondwanaland. Geophytology 9:28–38
Rigby JF, Schopf JM (1969) Stratigraphic implications of Antarctic paleobotanical studies. In: (Ed. A.J. Amos), UNESCO, Paris, 2 (Earth Sciences) Gondwana Stratigraphy (1st Internat. Symposium on Gondwana, Buenos Aires 1967), pp. 91–106
Rösler O, Rohn R (1983) *Sphenopyhyllum paranaense* n. sp. da Formaçao Rio do Rasto (Permiano superior) de Dorizon, Estado do Parana. Bol. IG USP, Instituto de Geociencias Univerisdad de São Paulo 15:97–104
Roux SF Le (1976) On some "northern" elements in the Lower Gondwana flora of Vereeniging, Transvaal. Paleontolog. Africana 19:59–65
Schopf JM (1970) Petrified peat from a Permian coal bed in Antarctica. Science 169:274–277
Schopf JM (1976) Morphologic interpretation of fertile structures in glossopterid gymnosperms. Rev. Palaeobotan. Palynolog. 21:25–64
Schopf JM, Askin AA (1980) Permian and Triassic floral biostratigraphic zones of Southern Land Masses. In Dilcher DL, Taylor TN (eds) Biostratigraphy of fossil plants, Dowden, Hutchinson & Roso pp. 119–152
Scotese CR (1986) Phanerozoic reconstructions: a new look at the assembly of Asia. Univ. Texas Inst. Geophys., Technical Rep. 66.
Scott AC (1979) The ecology of coal measure floras from northern Britain. Proceed. Geological Assoc. 90:97–116
Seward AC (1914) Antarctic fossil plants. British Antarctic ("Terra Nova") Expedition, 1910, Nat. Hist. Report Geology 1(1):1–49
Singh UK, Srivastava AK, Maheshwari HK (1986) Sphenopsids from the Barakar Formation of the Hura Tract, Rajmahal Hills, Bihar. The Palaeobotanist 35:236–241
Smoot EL, Taylor TN (1985) Paleobotany: recent developments and future research directions. Palaeogeogr. Palaeoclimatol. Palaeoecol. 50: 149–162
Smoot EL, Taylor TN (1986a) Evidence for simple polyembryony in Permian seeds from Antarctica. Amer. Journ. Bot. 73:1079–1071
Smoot EL, Taylor TN (1986b) Structurally preserved fossil plants from Antarctica: II. A Permian moss from Transantarctic Mountains. Amer. Journ. Bot. 73:1683–1691
Spicer AA (1981) The sorting and deposition of allochtonous plant material in a modern environment at Silwood Lake, Silwood Park Berkshire, England. U.S. Geolog. Surv. Prof. Paper 1143:1–77
Stubblefield SP, Rothwell GW (1981) Embryogeny

and reproductive biology of *Bothrodendrostrobus mundus* (Lycopsida). Amer. Journ. Bot. 68:625–634

Stubblefield SP, Taylor TN (1986) Wood decay in silicified gymnosperms from Antarctica. Bot. Gazette 147:116–125

Stubblefield SP, Taylor TN (1988) Recent advances in paleomycology. Tansley review n 12. New Phytolog. 108:3–25

Tarling DH (1985) Carboniferous reconstructions based on palaeomagnetism. CR X Congr. Intern. Stratigr. Géol. Carbonifère (Madrid, September 1983) 4:153–162

Taylor TN, Scott AC (1983) Interactions of plants and animals during the Carboniferous. Bioscience 33:488–493

Taylor TN, Taylor EL (1987) Structurally preserved fossil plants from Antarctica. III. Permian seeds. Amer. Journ. Bot. 74:904–913

Teixeira C (1946) Sur la flore fossile du Karroo de Zambésie (Mozambique). CR Soc. Géolog. France 16:252–254

Thomas BA, Brack-Hanes SD (1984) A new approach to family groupings in the Lycophytes. Taxon 33:247–255

Thomas BA, Meyen SV (1984) A system of form-genera for the Upper Palaeozoic lepidophyte stems represented by compression–impression material. Rev. Palaeobot. Palynol. 41:273–281

Townrow JA (1967) Fossil plants from Allan and Carapace Nunataks, and from the Upper Mill and Shackleton Glaciers, Antarctica. New Zealand Journ. Geol. Geophys. 10:456–473

Triparthi C, Singh G (1985) Carboniferous flora of India and its contemporaneity in the world. CR X Congr. Intern. Strat. Géol. Carbonifère (Madrid, September 1983) 4:295–306

Vakhrameiev VA, Dobruskina IA, Zaklinskaia ED, Meyen SV (1970) Paleozoic and Mesozoic floras of Eurasia and phytogeography of this time. Ed. Nauka, Moscow, pp. 1–426

Visser WA, Hermes JJ (1962) Geological results of the exploration for oil in Netherlands New Guinea. Verhandl. Konin. Nederl. Geol. Mijnbouk, genootschap (Geol. Ser.) 20:53–54, 62–68

Vozenin-Serra C (1980) Sur les rapports entre provinces floristiques au Paléozoïque Supérieur. CV Congr. Nat. Soc. Savantes (Caén 1980) Sciences 1:81–98

Wagner RH (1962) On a mixed Cathaysia and Gondwana flora from SE Anatolia (Turkey). CR IV Congr. Stratigr. Carbonifère (Heerlen 1958) 3:745–752

Walton J (1929) The fossil flora of the Karroo System in the Wankie district, Southern Rhodesia. Bull. Geol. Surv. S. Rhodesia 15:62–75

White D (1908) Report on the fossil flora of the coal measures of Brazil. Relat. Final Com. Estud. Minas Carvão de Pedra, Rio de Janeiro, 3:337–617

Zhang S, He Y (1985) Late Palaeozoic palaeophytogeographic provinces in China and their relationships with plate tectonics. Palaeontol. Cathayana 2:77–86

Ziegler AM, Bambach RK, Parrish JT, Barrett SF, Gierlowski EH, Paker WC, Raymond A, Sepkoski Jr JJ (1981) Paleozoic biogeography and climatology. In Niklas K (ed) Paleobotany, Paleoecology and Evolution, vol 2, Praeger, New York, pp. 231–266

Zimina VG (1967) *Glossopteris* and *Gangamopteris* from the Permian of Primorie. Pal. Zhurnal 2:113–121 (in Russian)

10—Gondwana Floras of India and Antarctica—A Survey and Appraisal*

M.N. Bose, Edith L. Taylor, and Thomas N. Taylor

During the last three decades, there has been a steady flow of papers on the paleobotany and stratigraphy of India and Antarctica. This chapter will consider the various megafossil assemblages from these regions, which range in age from Lower Permian to Lower Cretaceous. Previously, the oldest rocks in the Gondwana Sequence of India were thought to be Late Carboniferous, but these are now believed to represent the Lower Permian. Although the sequence of formations within the Permian is fairly well established (Table 10.2), the ages and sequential positions of the various younger formations [above the Panchet Formation (lowermost Triassic), Table 10.1] are still debated. In some cases, the Jurassic–Lower Cretaceous formations rest directly on metamorphic rocks or gneisses believed to be of Archaean origin. Nowhere have these formations been observed superimposed in sequential order. The Upper Gondwana in India has been classified, and the various formations have been correlated as shown in Table 10.1.

Although the ages of some of the formations in this sequence can be questioned, the megafossil assemblages in the formations of known Lower Cretaceous age are definitely distinct from those whose ages are in doubt. Accordingly, we will discuss the various assemblages under the subheadings Permian, Triassic, Rhaeto–Liassic, Jurassic–Lower Cretaceous, and Lower Cretaceous. Detailed work on these floras, which formed the foundation of Indian Gondwana paleobotany, was done by Oldham and Morris (1863) and Feistmantel (1876–1886).

Unfortunately for Antarctic paleobotany, there are no publications that could be compared with those of Feistmantel in India (1876–1886). Antarctic paleobotanical literature is scattered and many of the plants collected have merely been recorded in association with geological research on different basins. A detailed description of Gondwana basins of Antarctica has been given by Elliot (1975). Several genera and species are noted, but descriptions or figures are not provided. Although other papers have reported the occurrence of plants, these also lack descriptive information on the taxa (e.g., Allen 1962; Grindley 1963; Adie 1964; Grindley and Warren 1964; Craddock et al. 1965; Skinner and Ricker 1968; Hjelle and Winsnes 1972; Ballance 1977). Studies dealing with the general paleobotany of Antarctica include Seward (1914), Barghoorn (1961), Plumstead (1962, 1964, and 1965), and Rigby and Schopf (1967). Other papers will be considered in association with the floras from specific formations.

Permian of India (Lower Gondwana)

In peninsular India, Permian formations with plant remains include Talchir, Karharbari,

* Contribution No. 668 of the Byrd Polar Research Center (Ohio State University).

Table 10.1. Correlation of the Upper Gondwana in India.[a]

	Lower Cretaceous		Middle–Upper Jurassic	
Gondwana division	Jabalpur Series		Rajmahal Series	
	Umia	Jabalpur	Kota	Rajmahal
Damodar Valley				Rajmahal
Rajmahal				Rajmahal
Son-Mahanadi Valley		Bansa Beds		Athgarh
Satpura		Jabalpur	Chaugan	
Pranhita-Godavari Valley		Chikiala	Kota	
East Coast				
Godavari		Tirupati	Raghavapuram	Golapilli
Guntur		Pavalur	Vemavaram	
Madras		Satyavedu	Sriperumbudur	
Ramnad			Sivaganga	
Kachchh	Umia			

[a] Adapted from Lexique Stratigraphique International, 1956.

Barakar, Barren Measures, Raniganj, and Kamthi (in ascending order, see Table 10.2). Most of these formations are present in the Son, Damodar, Mahanadi, Satpura (Pench–Kanhan–Tawa), Wardha, and Godavari Valley basins (Fig. 10.1). Among these, the Damodar basin has the best representation of all of the formations.

Talchir Formation

This formation lies at the base of the Gondwana Sequence in India and rests directly on Precambrian basement. It is overlain by the Karharbari Formation, but the contact between these two formations in different basins can be either conformable or unconformable.

Table 10.2. Permian formations of India and representative floral elements.

Epoch	Formations	Taxa present[a]
Upper Permian	Kamthi	
	Handappa	D—*Glossopteris* T—*Senia, Glossotheca, Lidgettonia, Denkania, Partha, Utkalia* R—pteridophytes
	Kamthi	D—*Glossopteris* T—*G. musaefolia, G. stricta, G. leptoneura* R—pteridophytes
	Raniganj	D—*Glossopteris, Vertebraria* T—*Plumsteadiostrobus, Venustostrobus, Jambadostrobus, Senotheca, Kendostrobus, Palaeovittaria, Pteronilssonia, Searsolia* C—articulates, pteridophytes
	Barren Measures	D—*Glossopteris* R—*Cyclodendron, Neomariopteris, Rhabdotaenia*
Lower Permian	Barakar	D—*Glossopteris, Vertebraria* T—*Barakaria, Walkomiella* R—*Gangamopteris, Noeggerathiopsis*
	Karharbari	D—*Gangamopteris, Noeggerathiopsis* T—*Botrychiopsis, Rubidgea, Euryphyllum, Arberia, Ottokaria, Palmatophyllites, Buriadia* R—*Schizoneura, Glossopteris*
	Talchir	
	Rikba	D—*Gangamopteris cyclopteroides* C—*Samaropsis* R—*Glossopteris talchirensis*
	Needle shales	R—*Gangamopteris, Noeggerathiopsis* (stunted)

[a] C, common; D, dominant; R, rare; T, characteristic. See text for further explanation.

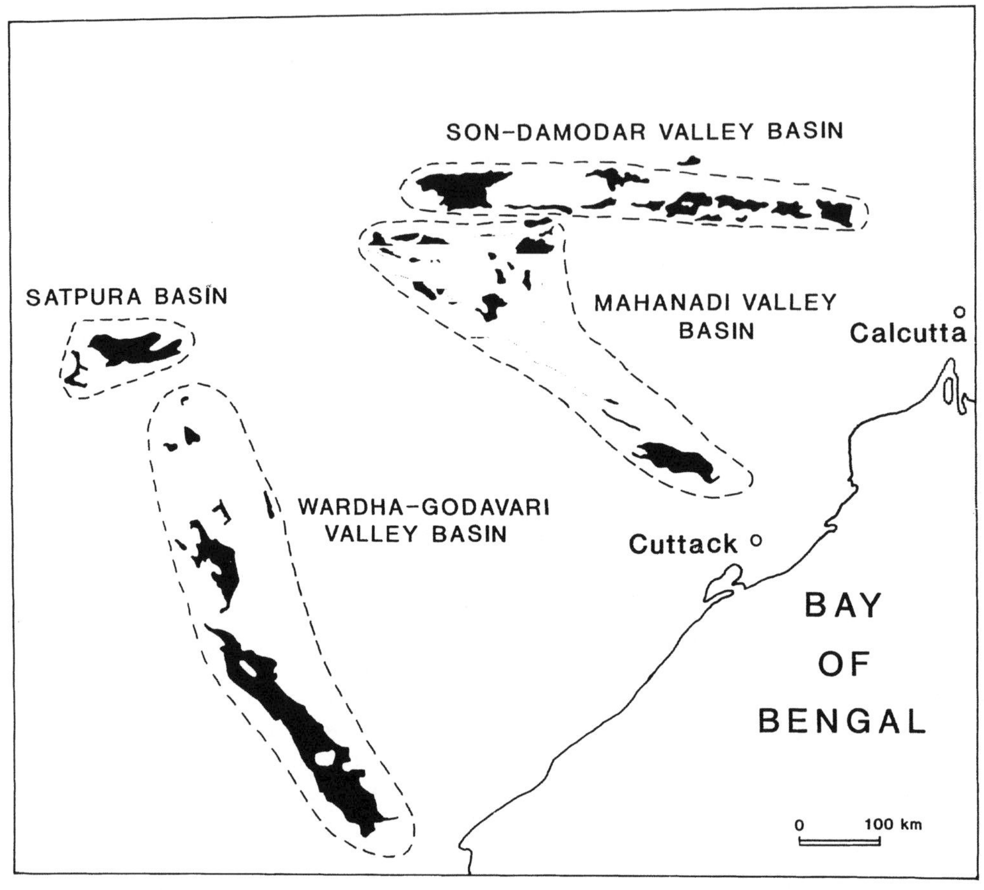

Figure 10.1. Permian basins of peninsular India (from Pareek 1969).

The Talchir comprises a varied assemblage of lithofacies, beginning with tillites and boulder beds and followed by needle shales, siltstones, and fine, soft sandstones. The earliest record of fossil plants is from about 3 m above the boulder beds. Some stunted forms of *Gangamopteris cyclopteroides, Noeggerathiopsis hislopii,* and *Noeggerathiopsis* sp. were reported by Surange and Lele (1955) from the needle shales of the Giridih Coalfield, but these were poorly preserved. The uppermost part of the Talchir Formation is dominated by *Gangamopteris cyclopteroides,* which occurs at Rikba (North Karanpura Coalfield). Also present are numerous seeds of the *Samaropsis*-type and rare *Glossopteris* leaves (*G. indica* and *G. communis*). Chandra and Surange (1979) have recently instituted a new species, *Glossopteris talchirensis,* for some of the specimens of *G. communis* initially described by Feistmantel (1879b, 1881, 1886). According to these authors, *G. communis,* which has three lectotypes (see Singh et al. 1982), does not occur in the Talchir. An interesting feature of *Gangamopteris cyclopteroides,* which occurs at Rikba, is that the margins of many of the leaves are folded. This was also observed by Høeg and Bose (1960) in some of their specimens of *G. cyclopteroides* from the Schistes Noirs at Walikale, Zaire.

Plant fossils of similar age to those in the Rikba flora have also been described by Surange and Lele (1956) from the South Rewa Basin. In addition to the known Talchir species, new taxa, such as *Cordaicarpus furcata, Samaropsis goraiensis, Paranocladus*(?) *indica,* and *Arberia umbellata,* were reported. *Paranocladus*(?) *indica* is based only on impressions, so its identity with the original specimens of *Paranocladus* Florin is not certain.

Karharbari Formation

The type area for this formation is the Giridih Coalfield. The flora has been described by Maithy (1966, 1969), Bharadwaj (1974), and Surange (1975). Unlike the Talchir, the Karharbari Formation contains a highly diverse flora. Important genera include *Phyllotheca* Brongniart, *Schizoneura* Schimper & Mougeot, *Botrychiopsis* Kurtz, *Gangamopteris* McCoy, *Glossopteris* Brongniart, *Rubidgea* Tate, *Noeggerathiopsis* Feistmantel, *Euryphyllum* Feistmantel, *Buriadia* Seward & Sahni, *Arberia* White, *Ottokaria* Zeiller, *Cordaicarpus*[1], *Samaropsis* Goeppert, and *Vertebraria* Royle. Of these taxa, *Gangamopteris* has the largest number of species (approximately 17), followed by *Noeggerathiopsis* (11). It appears that both *Gangamopteris* and *Noeggerathiopsis* achieved their maximum diversity during Karharbari time. After this they started to decline, and *Glossopteris* became the dominant element, reaching its height during the Kamthi Formation (Late Permian—Table 10.2).

The genus *Gangamopteris* has been discussed in detail by Pant and Singh (1968). They also described four new species of *Gangamopteris,* one of which, *G. hispida,* is from the Karharbari Formation. The exact horizons of the remaining species are not known. Pant and Nautiyal (1967b, 1984) have provided detailed descriptions of *Buriadia heterophylla* and *Ottokaria zeilleri. B. heterophylla* has subsequently been transferred to *B. sewardii* (Maithy 1973). Pant and Nautiyal (1984) assigned *O. zeilleri* to the leaves of *Glossopteris giridihensis* and suggested that the head of *Ottokaria* is organized in a dorsiventral manner. Additional taxa from the Karharbari Formation have been reported by Maithy (1965a, 1965b, 1965c, 1968, 1970). He instituted *Palmatophyllites* (Maithy 1965b) based on some previously described specimens of *Noeggerathiopsis lacerata* and described a new species of *Dolianitia* Millan [*D. karharbarensis* (Maithy 1970)]. A genus closely related to *Palmatophyllites* has been described by Pant and Singh (1979) as *Caulophyllites indica* from the Karharbari Formation of Giridih Coalfield.

[1] The author of this taxon is not known and there is uncertainty over the use of *Cordaicarpon* versus *Cordaicarpus*. *Cordaicarpus* seems to be a junior synonym of *Cordaicarpon* Geinitz (1862).

Barakar Formation

The Barakar Formation includes the principal coal-bearing strata in India. The flora is dominated by *Glossopteris* and *Vertebraria,* with rare *Gangamopteris* and *Noeggerathiopsis* present only in those horizons whose assignment to the Barakar is uncertain. These records are mostly from the basal parts of the formation, which at places has been suggested to belong to the underlying Karharbari Formation. In the Barakar, pteridophytic remains are not as common as in the overlying Raniganj Formation, and include species of *Phyllotheca, Lelstotheca* Maheshwari, *Schizoneura, Sphenophyllum* Koenig, *Trizygia* Royle, and *Neomariopteris* Maithy. With the exception of *Phyllotheca* and *Schizoneura,* the remaining taxa represent first occurrences in the sequence. Among the species of *Phyllotheca, P. angusta* (Surange and Kulkarni 1968) is of special interest. It has leaf sheaths with crowded stomata present on only one side, and the stems lack ridges and grooves. *Lelstotheca robusta* (Maheshwari 1972a) is fairly common in the Barakar of the Rajmahal Hills (Fig. 10.1). Other pteridophytic and gymnospermous plant remains have been described from this same area by Surange and Prakash (1960), Maheshwari and Prakash (1965), Sah and Maheshwari (1969), and Singh et al. (1987). In addition, new species of *Sphenophyllum* and *Trizygia* have recently been described from the Barakar Formation in the Raniganj Coalfield (Srivastava and Rigby 1983).

Among the gymnosperms, a new interpretation for *Diphyllopteris verticillata* has been provided by Pant and Nautiyal (1987). They suggest that it represents a seedling of *Glossopteris* that includes the first two seedling leaves and dichotomously divided cotyledons with open dichotomous venation. The cotyledons show a remarkable similarity to the segmented leaves of *Barakaria dichotoma,* which is known only from the Auranga Coalfield (Seward and Sahni 1920). Other gymnosperms from the Barakar include *Pseudoctenis ballei; Rhabdotaenia danaeoides, Macrotaeniopteris feddeni,* and *Gondwanophyton indicum* (Maithy 1974b). Based on external morphology, *M. feddeni* is indistinguishable from *Rhabdotaenia fibrosa* (Pant and Verma 1963). *Gondwanophyton indicum* is believed to be related to *Psygmophyllum* Schimper and *Rhipidopsis* Schmalhausen. The only conifer from the Barakar is *Walkomiella indica* (Surange and Singh 1951), which was described from the West Bokaro Coalfield of Bihar. Surange and Singh (1953) later described the female dwarf shoot of this same genus. Dispersed seed-like bodies, resembling seeds of *W. indica,* were described as *Walkomiellospermum indicum* by Pant and Srivastava (1964) from the Talchir Coalfield; however, the exact stratigraphic level of these fossils is not known.

The Barakar Formation contains relatively few presumed glossopterid fertile organs. These include *Eretmonia karanpurensis* and *Dictyopteridium sporiferum.* Wood is fairly common in this formation and has been referred to genera such as *Barakaroxylon* Surange and Maithy, *Araucarioxylon* Kraus, *Damudoxylon* Maheshwari, and *Polysolenoxylon* Kräusel and Dolianiti.

Barren Measures Formation

This formation is devoid of workable coal and includes mostly sandstones or ironstone shales. Plant remains collected from Jainagur, South Karanpura Coalfield, and Kulti, Raniganj Coalfield, were first described by Feistmantel (1881). Generally, the Barren Measures Formation has a rather poor assemblage of plant fossils. Kar (1968) described *Cyclodendron leslii, Neomariopteris hughesii,* and three species of *Glossopteris* from the Jharia Coalfield. Earlier, Feistmantel (1881) described six species of *Glossopteris* and *Taeniopteris danaeoides* (= *Rhabdotaenia danaeoides*). However, Chandra and Surange (1979) suggested that only four species of *Glossopteris,* that is, *G. damudica, G. raniganjensis, G. indica,* and *G. stenoneura*(?), are present in the Barren Measures.

Raniganj Formation

This formation is best developed in the Raniganj Coalfield where the plant assemblages are far more diverse than all the previously described assemblages. The assemblage has no *Gangamopteris* or *Noeggerathiopsis* (see Surange 1975) and is completely dominated by species of *Glossopteris.* More than 40 species of *Glossopteris,* based either on impressions or incrustations, have been described. Chandra and Surange (1979) have revised the various species of *Glossopteris,* emphasizing gross morphology and venation patterns, whereas Pant and Gupta (1968, 1971) and Pant and Singh (1971, 1974) have delimited species on the basis of both morphological and epidermal characters. Other taxa related to *Glossopteris,* but not known from earlier formations, include *Palaeovittaria* Feistmantel (two species) and *Belemnopteris* Feistmantel (three species). A few pollen and ovulate fructifications belonging to the "Glossopteridales" are known from the Raniganj Formation. These include *Kendostrobus cylindricus, Dictyopteridium sporiferum, D. feistmantelii, Plumsteadiostrobus ellipticus, Jambadostrobus pretiosus, Venustostrobus diademus, V. indicus,* and *Senotheca murulidihensis* (Feistmantel 1880a; Banerjee 1969; Surange and Chandra 1974d; Chandra and Surange 1976; 1977a, b, c). *Kendostrobus* represents a male fertile organ, while the others are all female fructifications. With the exception of *D. sporiferum,* all are attached to *Glossopteris*-like leaves [*D. feistmantelii* attached to *G. tenuinervis, P. ellipticus* to *G. gondwanensis, J. pretiosus* to *G. contracta* (Pant and Gupta 1971); *V. diademus* to *G. pantii,* and *S. murulidihensis* to *G. mohudaensis* (Chandra and Surange 1979)].

In the Raniganj Formation, non-glossopterid gymnosperms include *Macrotaeniopteris feddeni, Rhabdotaenia danaeoides, R. fibrosa, Pseudoctenis balli, Pterophyllum burdwanense, Pteronilssonia gopalii,* and *Rhipidopsis densinervis* (Feistmantel 1877d, 1881; Seward 1917; Pant and Mehra 1963b; Pant and Verma 1963). The only conifer known from this formation is *Searsolia oppositifolia* (Pant and Bhatnagar 1975), based on shoots with opposite leaves.

Articulates and ferns are fairly common in the Raniganj Formation. Some, like *Sphenophyllum speciosum, Phyllotheca indica, Raniganjia bengalensis,* and *Schizoneura gondwanensis,* have been described in detail by Pant and Mehra (1963a), Pant and Kidwai (1968), Pant and Nautiyal (1967a), and Pant et al. (1982). Of these, *S. speciosum* has been transferred to *Trizygia* by Maheshwari (1968). Most of the ferns described earlier using northern hemisphere generic names have now been redescribed under new genera (Maithy 1974a, c, 1975, 1977; Pant and Khare 1974; Pant and Misra 1976, 1977, 1983). Some of these include *Neomariopteris, Damudopteris, Dichotomopteris, Asansolia, Damudosorus, Trithecopteris,* and *Cuticulatopteris.* Several of these were later placed in synonymy, for example, *Neomariopteris* has priority over *Damudopteris* and *Dizeugotheca phegopteroides* over *Asansolia phegopteroides* (Srivastava and Chandra 1982). These Paleozoic ferns have been classified by Pant and Misra (1977) in two families, the Damudopteridaceae (free sporangia) and the Asterothecaceae (synangia).

The Raniganj Formation is also rich in fossil woods. A large number of species have been described under the generic names *Dadoxylon* Endlicher, *Araucarioxylon,* and *Trigonomyelon* Walton. Recently, Pant and Singh (1987) have discussed the characteristics of Lower Gondwana woods in detail and their occurrence in various parts of Gondwanaland.

Kamthi Formation

This formation was first recognized in the Wardha–Godavari Valley; plant fossils were initially described by Bunbury (1861) and Feistmantel (1881). Recent additions to the flora have been provided by Chandra and Prasad (1981). The assemblage is dominated by species of *Glossopteris,* such as *G. musaefolia, G. stricta,* and *G. leptoneura.* None of these are known from older formations. Glossopterid-type fructifications are represented by *Dictyopteridium sporiferum* and pteridophytes by *Trizygia speciosa, Schizoneura gondwanensis,* and *Neomariopteris hughesii.* A large number of fossil woods have also been reported from this region (Prasad and Chandra 1981; Biradar and Bonde 1981; Prasad 1982). Some of the genera include *Australoxylon* Marguerier, *Nandorioxylon* Biradar & Bonde, *Dadoxylon, Trigonomyelon, Kaokoxylon* Kraüsel, *Taxopitys* Kräusel, *Sclerospiroxylon* Prasad, *Araucarioxylon, Prototaxylon* Kräusel & Dolianiti, *Baieroxylon* Greguss, *Parapalaeoxylon* Prasad, and *Kamthioxylon* Mahabale & Vagyani.

One of the richest plant assemblages that is believed to occur within the Kamthi Formation is from a road cut at Hinjrida Ghati, about 4.6 km north of Handappa in the Dhenkanal District of Orissa. The assemblage at this site is dominated by species of *Glossopteris* and a variety of fertile organs. However, none of the species characteristic of the type locality of the formation (e.g., *G. musaefolia, G. stricta,* and *G. leptoneura*) are present at this site. Singh and Chandra (1987) have recently described seven new species of *Glossopteris* from Handappa. The leaves show a good deal of variation in size and shape (3.1–>30 cm long). Pollen organs related to *Glossopteris* were initially described from this area by Surange and Maheshwari (1970) as *Glossotheca utkalensis, Eretmonia utkalensis,* and *E. hinjridaensis.* Ovulate fructifications from this region include *Dictyopteridium sporiferum, Scutum sahnii, S. elongatum, S. indicum, Cistella ovata, Lidgettonia micronata, Denkania indica* (perhaps a species of *Lidgettonia*), *Partha indica,* and *Utkalia dichotoma* (Surange and Chandra 1973a, b, c; 1974a, b; Chandra 1984). It is interesting to note that with the exception of *Scutum sahnii* none of the other fructifications have been found attached to *Glossopteris*-like leaves, as is the case for most of the taxa known from the Raniganj Formation. Other gymnosperms in this assemblage include *Pseudoctenis balli* and *Senia reticulata*. If *Senia* had been found in the Jurassic, it would have been placed in *Ctenis* Lindley and Hutton. Among the pteridophytes, Chandra and Rigby (1981, 1983) have reported *Cyclodendron leslii, Phyllotheca indica, Stellotheca robusta, Trizygia speciosa, Raniganjia bengalensis, R. etheridgei, Schizoneura gondwanensis, Neomariopteris* spp, and a species of *Pantopteris*.

To date, there is not a single record of conifers or wood specimens from Handappa. According to Chandra and Rigby (1981), *Glossopteris* constitutes 70% of the total specimens, 20% are different kinds of fructifications, and the remaining 10% are articulates, ferns, and cycads. Since that report, the population of *Glossopteris* at Handappa has increased from 70 to 85% (Singh and Chandra 1987). No information is reported in either paper relative to how percentages of specimens were determined.

Table 10.2 gives a summary of the distributional pattern of some of the taxa in different formations.

Permian Floras of Antarctica

Permian plant fossils were first described by Seward (1914) from the Beacon Sandstone (Mt. Buckley, Beardmore Glacier area, Fig. 10.2) of Antarctica. Later Edwards (1928) reported the occurrence of *Glossopteris indica* Schimper in the Beacon Sandstone of the Ferrar Glacier area in South Victoria Land. Since then *Glossopteris* and related genera have been reported from several localities in Antarctica (see, e.g., Elliot 1975). There are very few assemblages that have been described from specific formations. Plumstead (1962) described three distinct assemblages of plants from the Permian of the Ross and Weddell Sea areas that she thought could be compared with the assemblages from the Karharbari to Raniganj Formations of India, in addition to assemblages from other parts of Gondwana. Her site M.S. 17 in the Ross Sea area contains a very poor assemblage of *Gangamopteris* (*G. angustifolia* and *G. obovata*), so it is not possible to determine whether it is equivalent to the upper part of the Talchir Formation. From Milorgfjella, Dronning Maud Land (Fig. 10.2, western Queen Maud Land), Plumstead (1975) described *Gangamopteris obovata, G. cyclopteroides, Euryphyllum antarcticum,* cf. *Palaeovittaria, Noeggerathiopsis hislopi,* a conifer shoot, and *Walikalia*(?) *juckesii*. These were compared with an assemblage from the Needle shales (lower Talchir Formation), which characteristically contain only some stunted leaves of *Gangamopteris cyclopteroides* and *Noeggerathiopsis hislopi*. The Milorgfjella assemblage is more similar to the flora of the Rikba beds (upper Talchir Formation). Both *Euryphyllum* and *Palaeovittaria* are absent in the Talchir Formation. It is quite likely that the leaves of *Euryphyllum arcticum* represent leaves of *G. cyclopteroides* with folded margins and cf. *Palaeovittaria* could represent poorly preserved specimens of *Glossopteris decipiens*. Plumstead's specimens of *Walikalia* exhibit distinct veins, unlike *W. cahenii* Høeg and Bose, which has only fine longitudinal striations. It is clear that the specimens from Milorgfjella will need to be reexamined before assigning a definite age to them.

Lacey and Lucas (1981a) described an assemblage of Sakmarian–Artinskian age from the Theron Mountains, Coats Land, East Antarctica (Fig 10.2). The flora is dominated by *Glossopteris* species, including *G. indica, G. communis, G. browniana, G. angustifolia, G.* cf. *conspicua,* and *G. stricta*. Of these, *G. indica* and *G.communis* are common, whereas the others appear less frequently. In addition to *Glossopteris,* Lacey and Lucas (1981a) reported cf. *Gangamopteris angustifolia,* cf. *Samaropsis* sp., and *Vertebraria indica*. This flora has been compared with those from the Theron Mountains and Whichaway Nunataks described by Plumstead (1962), the Prince Charles Mountains of East Antarctica (White 1973), and the Karharbari Formation of India. However, the Whichaway Nunataks contain

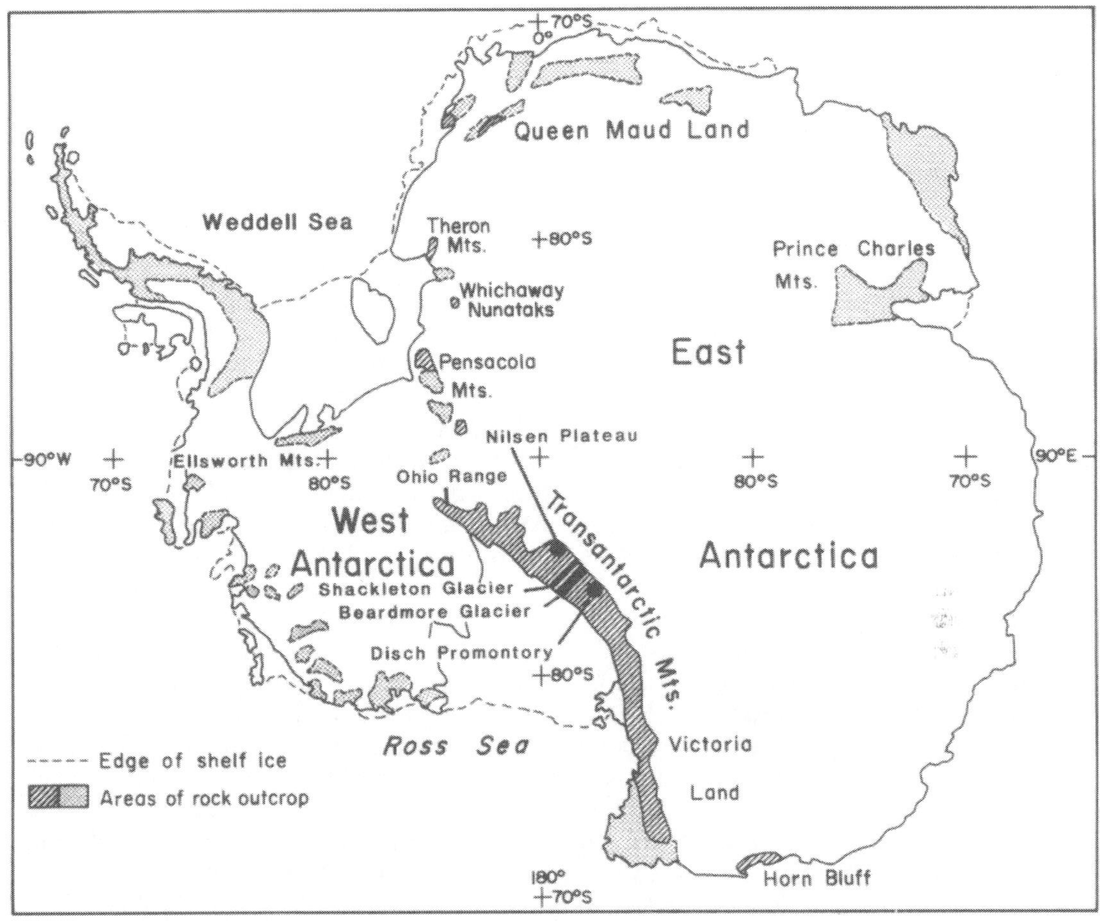

Figure 10.2. Map of Antarctica showing general location of plant collections.

more than one assemblage and the assemblage from the Theron Mountains seems to be closer to those in which *Glossopteris* is dominant. These can range in age up to the Upper Permian. The precise age of the flora from Prince Charles Mountains is not known, but the assemblage is dominated by species of *Glossopteris*. In addition to *Glossopteris*, a large number of fossil woods have been figured, as well as a doubtful specimen of *Dictyopteridium sporiferum* (White 1973). White considered this specimen to represent a male fertile frond of *Glossopteris*. In contrast to these somewhat depauperate floras, the flora of the Karharbari Formation is fairly rich and is dominated by *Gangamopteris* and *Noeggerathiopsis*. The Theron Mountains flora is quite distinct from the Karharbari assemblage and appears to be younger than Sakmarian–Artinskian.

Another assemblage dominated by species of *Glossopteris* has been described from the Mt. Glossopteris Formation in the Ohio Range (Transantarctic Mountains Fig. 10.2) (Cridland 1963). Its precise age is not known. The flora includes six species of *Glossopteris*, cf. *Gangamopteris*, *Arberiella*, and *Samaropsis longii*. Cridland also figured a specimen that he believed was intermediate between a vegetative *Glossopteris* leaf and a scale leaf. This specimen closely resembles the fertile leaves of *Glossotheca* and *Eretmonia* figured by Surange and Maheshwari (1970) and Surange and Chandra (1974c,d). The Mt. Glossopteris flora appears most similar to the assemblages known from the Barakar and Raniganj Formations in India. The former assemblage differs from the latter in an absence of articulates and pteridophytes (Table 10.2).

Grindley (1963) reported an assemblage dominated by *Glossopteris indica, G. decipiens, G.* cf. *browniana,* and *G.* cf. *angustifolia* from the Buckley Coal Measures of the Queen Alexandra Range, Beardmore Glacier area (central Transantarctic Mountains). He correlated these strata with coal-bearing terrestrial sandstones and shales found elsewhere in East Antarctica at about the middle of the Beacon Group. In addition, he suggested a number of correlatives outside Antarctica. Indian stratigraphic equivalents would include the Karharbari and Barakar Formations, with the dominance of *Glossopteris* suggesting a closer relationship with the Barakar.

Leaves similar to those described by Seward (1914) and Edwards (1928) as *Glossopteris indica* were described by Schopf (1962) from the central range of the Horlick Mountains. Associated with *G. indica,* he also found *Samaropsis longii.* Although named as a new species in this paper, *S. thomasii* was based on Thomas' specimens from South Africa and was not found in Antarctica. None of these taxa are characteristic of any Indian formations, so only a Permian age was assigned. Later, Schopf (1965–1968, 1970–1971 and 1973 field seasons [Schopf, field notes]) made further collections from different localities in Antarctica, but most of these were never described in detail. Of these, perhaps his most significant contribution is the discovery of petrified peat in Antarctica (Schopf 1970b; Taylor and Taylor, this volume). From the Ohio Range, two fertile organs have been figured by Schopf (1967, 1976). One of these, from Leaia Ledge on Mt. Schopf, represents a microsporophyll of *Eretmonia* (Schopf 1976, Fig. 10); the other, from Discovery Ridge, Mt. Glossopteris (Schopf 1976, Fig. 5), is described as an ovulate organ and termed an "antarcticoid capitulum." The morphology of this structure is similar to those glossopterid fructifications classified as cupulate types (Surange and Chandra 1974b), for example, the *Lidgettonia*-type. Both of these structures are poorly preserved. In addition to these fertile organs, Kyle (1976) also described a specimen of *Plumsteadia ovata* from southern Victoria Land. From Allan Nunatak, Townrow (1967b) recorded *Cistella stricta* and two species of *Glossopteris* (*G. communis* and *G.* cf. *cordata*).

From these sparse data on Permian floras in Antarctica, it is not presently possible to ascertain the precise age of the various assemblages. A floral list and stratigraphic implications of the Antarctic Permian flora were suggested by Rigby and Schopf (1967). Their data suggest that there are no Late Carboniferous floras present in Antarctica, with species generally comparable to the Middle and Upper Permian of India. From the data available at the time, articulate and fernlike foliage were extremely rare [Plumstead (1962) recorded doubtful species of *Annularia* sp. and *Phyllotheca* cf. *P. australis*]. *Gangamopteris*-dominated assemblages occur at site M.S. 17 in the Ross Sea area (south Victoria Land) (Plumstead 1962), Milorgfjella, Dronning Maud Land (Plumstead 1962), and south Victoria Land (Barrett and Kyle 1975) (Table 10.3.). Assemblages from these regions may be compared with those from the Rikba beds

Table 10.3. Permian of Antarctica.

Formations and localities[a]	Dominant taxa	Correlation with India
Buckley Formation	*Glossopteris*	?Raniganj, Kamthi Formations
Mt. Glossopteris Formation, Weller Coal Measures, Amery Formation	*Glossopteris*	?Barakar, Raniganj Formations
Mt Fleming (SVL)	*Gangamopteris cyclopteroides, Glossopteris decipiens*	?Karharbari Formation
Mt. Fleming (SVL), Milorgfjella, M.S. 17 (SVL)	*Gangamopteris cyclopteroides*	?Upper Talchir Formation (Rikba beds)

[a]SVL, southern Victoria Land.

(upper Talchir Formation) (Table 10.2). There is no assemblage in Antarctica that can be compared with the floras from the Karharbari Formation. The closest is the one that occurs at Mt. Fleming in southern Victoria Land from which *Gangamopteris* and a species of *Glossopteris* with an impersistent midrib have been reported (Barrett and Kyle 1975) (Table 10.3). The *Glossopteris* leaves may represent *G. decipiens*. The remaining plant fossils noted earlier may range in age from Middle to Upper Permian. In the majority of cases, however, the plants have not been collected or described sufficiently to make adequate age determinations and comparisons. If most of these assemblages are Middle to Late Permian in age, then they are perhaps contemporaneous with the floras of the Barakar and Raniganj Formations (see Table 10.2). It is noteworthy that *Glossotheca* and *Eretmonia*-type laminar structures have been found in the upper horizons in Antarctica by Cridland (1963) and Schopf (1976). In India, such laminar structures are restricted to the Raniganj and Kamthi Formations (Late Permian).

Fossil woods from Antarctica have been described by Seward (1914), Schopf (1962), Kraüsel (in Plumstead 1962), White (1973) and Maheshwari (1972b) (see Taylor and Taylor, this volume). In India fossil woods are known from the Barakar, Raniganj, and Kamthi formations. Among those known from Antarctica, *Araucarioxylon bengalense* and *A. ningahense* are known from the Barakar and Raniganj Formations, respectively; a species of *Megaporoxylon* is also known from the Raniganj Formation (Maheshwari 1966).

Triassic of India (Middle Gondwana)

The Triassic floral succession of peninsular India has been discussed by Bose (1974) and Roy Chowdhury et al. (1975). This sequence is well represented in the Damodar Basin, South Rewa, Satpura, Godavari, and Mahanadi valleys (Fig. 10.3). However, none of the basins includes a complete sequence upon which a floral succession can be constructed. The best-known plant assemblages are from Panchet, Nidhpuri, Parsora, and Tiki (Table 10.4).

Panchet Formation

Plant megafossils are known only from the lower Panchet Formation, which has a gradational contact with the underlying Late Permian of the Raniganj Formation. Near Asansol, the assemblage is characterized almost exclusively by *Glossopteris*, along with rare specimens of *Cyclopteris pachyrhachis*, *Lepidopteris*-like leaves, and *Podozamites* sp. cf. *P. lanceolatus*. The latter three species are not known from the underlying Raniganj Formation. With the exception of *Trizygia speciosa* and *Schizoneura gondwanensis*, none of the other Paleozoic pteridophytes are known from the Panchet. Although *Dicroidium* was previously regarded as a Triassic marker in India, it is now known that fronds similar to *Dicroidium* in the Panchet have cuticle comparable to *Lepidopteris* (Bose and Banerji 1976; Banerji and Bose 1977). Several species of *Glossopteris* have been described by Bose and Banerji (1976), Bose et al. (1977), and Banerji and Bose (1977) from the Auranga Valley, Ramkola–Tatapani, and Asansol regions, respectively. Most are less than 12 cm in length and have been assigned to Paleozoic taxa. Chandra and Surange (1979) reexamined some of the specimens of *Glossopteris* described from the Panchet and Nidhpuri formations and gave a list of characteristic species for the "Panchet Stage." With the exception of *G. indica*, none are yet known from the Panchet Formation. Chandra and Surange (1979) also referred most of the Panchet *Glossopteris* to species occurring in the Raniganj and Kamthi formations. These include G. *conspicua*, G. *intermedia*, G. *bosei*, and G. *leptoneura*. The last species is characteristic of the Kamthi Formation. Chandra and Surange (1979) also placed some specimens of *Glossopteris* previously described as *G. browniana* (Banerji et al. 1976) into *G. taeniensis*, which they describe as characteristic of the Karharbari Formation. Clearly, a reexamination of Panchet species of *Glossopteris* is needed before they can be used for stratigraphic purposes.

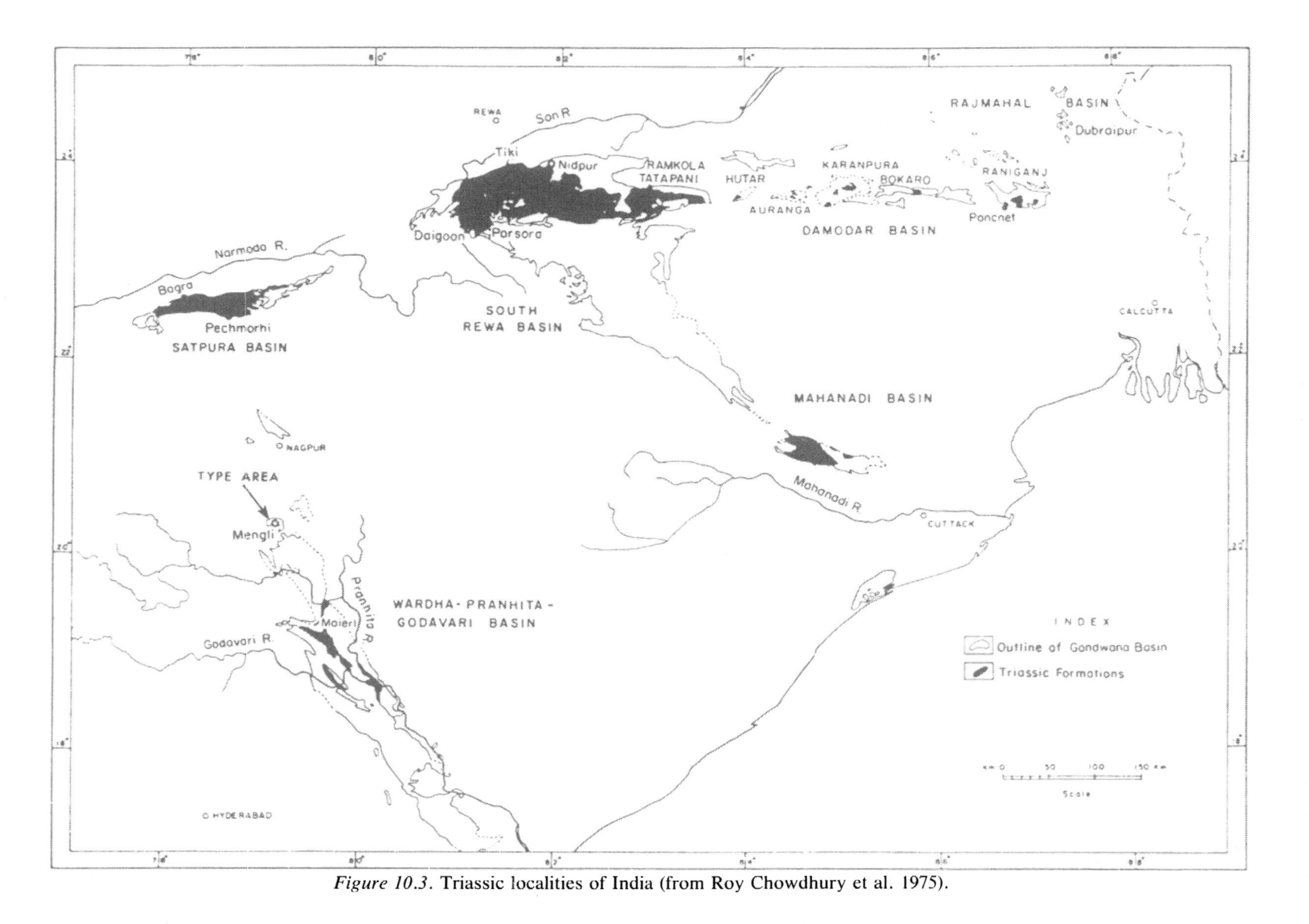

Figure 10.3. Triassic localities of India (from Roy Chowdhury et al. 1975).

Table 10.4. Triassic of India.

Formations	Taxa present[a]
Tiki	
Giar	D—*Lepidopteris strombergensis, Dicroidium giarensis*
	T—*Yabeiella, Elatocladus, Pagiophyllum*
	C—*Dicroidium hughesii*
	R—*Dicroidium zuberi*
Harai	D—*Lepidopteris madagascariensis, Dicroidium zuberi*
	T—*Xylopteris, Sphenobaiera*
	C—*Dicroidium coriaceum, Desmiophyllum*
	R—*Baiera*
Parsora	D—*Dicroidium hughesii*
	T—*Parsorophyllum, Pterophyllum*
	C—*Dicroidium odontopteroides, D. feistmantelii*
	R—*Cladophlebis, Sphenopteris*
Nidhpuri	D—*Dicroidium nidpurensis, D. papillosum*
	T—*Pteruchus, Nidistrobus, Nidia, Satsangia*
	C—*Glossopteris, Taeniopteris*
	R—*Rhabdotaenia, Lepidopteris*
Panchet	D—*Glossopteris*
	T—*?Lepidopteris, Podozamites*
	C—*Glossopteris conspicua, G. intermedia, G. bosei, G. leptoneura*
	R—*Trizygia, Schizoneura, Cyclopteris*

[a] C, common; D, dominant; R, rare; T, characteristic.

Nidhpuri Formation

One of the richest Triassic megafossil assemblages is that known from the Nidhpuri Formation in the Gopad River Valley. The flora is dominated by *Dicroidium nidpurensis* and *D. papillosum* (Bose and Srivastava 1971). Some of the species of *Glossopteris*, based on gross morphology and cuticular structure, that occur in this flora include *G. senii, G. papillosa, G. nidpurensis, G. nilssonioides*, and *G. sidhiensis*. Of these, Chandra and Surange (1979) have synonymized *G. nidpurensis* with *G. musaefolia*, a characteristic species of the Kamthi Formation, based only on gross morphology. Other gymnosperms include *Rhabdotaenia* sp., *Lepidopteris indica, Marhwaseaphyllum hastatum, Rewaphyllum nidpurensis, Gopadia coriacea, Sidhiphyllites flabellatus*, and *Glottolepis* spp. (Bose and Srivastava 1972; Srivastava 1976; Srivastava 1984a, c). Fertile organs are represented by *Pteruchus indicus, P. nidpurensis, Satsangia campanulata, Nidistrobus harrisianus*, and *Nidia ovalis* (Bose and Srivastava 1973a, b; Srivastava and Maheshwari 1973; Pant and Basu 1973; Srivastava 1974). With the exception of *P. indicus*, the cuticles of the remaining species have been reported to be similar to *Dicroidium nidpurensis*. Before any of these specimens are assigned to D. *nidpurensis*, a critical reexamination of the cuticle of all the species will be necessary. This is especially important since these reproductive structures are so different from each other: *Pteruchus* is a corystosperm, *Nidistrobus* a compact male cone, *Nidia* a female cone resembling *Zamia*, and the biological affinities of *Satsangia* are not currently known.

A few nonvascular cryptogams have been described by Pant and Basu (1978, 1981) from the Nidhpuri Formation. These include *Algacites oogonifera, Hepaticites riccardioides, H. metzgeroides, H. foliata*, and *Sphagnophyllites triassicus*. Among these, *H. riccardioides* resembles a species of *Musciphyton* described from the Ordovician of Poland, which has been suggested to represent *Carex* root remains (Kozlowski and Greguss 1959; Greguss 1961, 1962).

To date, there is no record of any pteridophyte from the Nidhpuri Formation and only one conifer shoot has been figured by Srivastava (1971) along with a specimen of *Conites* sp. A microsporangiate organ, *Lelestrobus pennatus*, believed to represent a conifer, has also been described by Srivastava (1984b).

Recently, Venkatachala (1986) has suggested that the Panchet sediments and the Nidhpuri beds of the Tiki Formation have not yielded typical *Dicroidium* specimens, but rather are dominated (in the Panchet Formation at least) by *Glossopteris*. Since unequivocal *Dicroidium* remains are found only in the Parsora Formation (see Table 10.4), the Permo–Triassic boundary should be placed at the base of the Parsora Formation. There is no evidence, however, that the Nidhpuri fossiliferous beds belong to the Tiki Formation. If so, they would then be younger than the Parsora Formation, which is definitely known to be older than the Tiki (Table 10.4). The

Nidhpuri flora differs from the Upper Permian flora of India in having an overwhelming dominance of *Dicroidium*. In addition, none of the fertile organs mentioned above from Nidhpuri have been found in Permian strata, and the pteridophytes that are common in the Upper Permian are entirely missing in the Nidhpuri flora. Based on our current knowledge, the Nidhpuri flora appears quite distinct from the assemblages known from the Raniganj, Kamthi, and Panchet Formations. Whether this assemblage from Nidhpuri should be assigned a Permian or Triassic age will require additional evaluation. However, Bharadwaj and Srivastava (1969) have compared the palynoflora from the Nidhpuri Formation with those of the Middle Triassic in Australia and have assigned an Early Triassic age to the Nidhpuri.

Parsora Formation

The flora from this formation has been synthesized by Bose (1974). It is dominated by *Dicroidium hughesii,* but also includes *D. odontopteroides* and *D. feistmantelii*. Other taxa known are *Parsorophyllum indicum, Pterophyllum sahnii,* and *Ginkgoites goiraensis* (Maheshwari and Banerji 1978). With the exception of *Noeggerathiopsis hislopi,* the Parsora Formation lacks any taxa characteristic of *Glossopteris*-dominated floras. The presence of *Noeggerathiopsis* in this flora is surprising, since there is no record of the genus in the Upper Permian of India, although fragmentary specimens of *Noeggerathiopsis* sp. have been noted in the Triassic of Nidhpuri (Srivastava, 1971). Perhaps they are ginkgoalean leaves. The Parsora Formation is extremely depauperate in pteridophytic remains, and to date there is no record of any conifers.

Tiki Formation

Recently, Pal (1984b) described plant megafossils from two different localities belonging to the Tiki Formation. The fossils were collected from the Janar River section near Harai and the Son River cutting near Giar, in the Shahdol District of central India. From the former site, *Lepidopteris madagascariensis, Dicroidium hughesii, D. zuberi, D.* sp., *Xylopteris* sp., *Sphenobaiera janarensis,* and *Baiera* sp. have been described. Of these, *L. madagascariensis* and *D. zuberi* are most common, *D. hughesii* and *D.* sp. are next in abundance, and the remaining species are rare. Pal (1984b) suggested an early Late Triassic age for the fossiliferous beds exposed near Harai.

The assemblage from Giar includes *Lepidopteris strombergensis, Dicroidium giarensis, D. coriaceum, D. zuberi, D.* sp. cf. *D. odontopteroides, Elatocladus denticulatus, E. raoi, Pagiophyllum bosei, Yabeiella indica,* and *Desmiophyllum singhii* (Pal 1984b). The Giar flora is dominated by *L. strombergensis, D. giarensis,* and *E. denticulatus*. Next in abundance are *D. coriaceum, P. bosei,* and *D. singhii,* with the remaining taxa being rare. Pal (1984b) proposed a late Late Triassic age for these fossiliferous beds. The Giar flora represents the first occurrence of *Elatocladus* and *Pagiophyllum,* which are typically found in Jurassic–Lower Cretaceous rocks in India.

Triassic Floras of Antarctica

The oldest Triassic remains were reported by Rigby and Schopf (1967) from Allan Nunatak (now Allan Hills, southern Victoria Land) (Fig. 10.2). The assemblage contains *Glossopteris indica, Taeniopteris* cf. *T. spatulata, Dicroidium odontopteroides,* and *D. feistmantelii*. Among these, *G. indica* and *Taeniopteris* occur less frequently. An age equivalent to the Panchet Formation (Lower Triassic) has been suggested for this assemblage (Rigby and Schopf 1967), although the Panchet flora does not include *D. odontopteroides* and *D. feistmantelii*. These two species are more frequent in the Parsora Formation, which may be Middle Triassic.

Plumstead (1962) described some poorly preserved plant remains from four different sites in the Ross Sea area (southern Victoria Land). They included *Schizoneura* cf. *S.* sp., *Neocalamites* cf. *N. carrerei, N.* cf. *N. foxii,*

Dicroidium odontopteroides, ?*Dicroidium*, *Nilssonia* sp., *Zamites* sp. A, *Z.* sp. B, *Z.* sp. C, *Williamsonia* sp., and other indeterminable plant fragments. Except for the *Dicroidium* species, the remaining identifications are all doubtful, based on the fragmentary nature of the specimens. Plumstead (1962) considered the fossiliferous sites to be no older than Middle Triassic and perhaps equivalent to the Parsora Formation.

Fossil plants of presumed early Upper Triassic age were described by Townrow (1967b) from the Allan Hills and Mt. Bumstead at the head of the Mill Glacier and Shackleton Glacier (Fig. 10.2). The association includes *Dicroidium odontopteroides*, *D. feistmantelii*, *D. dutoitii*, *Xylopteris elongata*, cf. *Diplasiophyllum acutum*, cf. *Johnstonia trilobita*, and *Rissikia media*. Among these, the Shackleton Glacier collection contains only one species, namely, *D. dutoitii*; *D. feistmantelii* is known only from Mt. Bumstead; and the remaining species occur at the Allan Hills and Mt. Bumstead sites. These collections do not resemble any of the Indian Triassic assemblages (see Table 10.5). At best, they may be compared with the flora of the Parsora Formation, which has a fair number of *D. odontopteroides* and *D. feistmantelii*. However, *D. hughesii* is the dominant species in this formation, and it is missing at the Antarctic sites.

Table 10.5. Triassic of Antarctica.

Localities	Taxa present[a]	Correlation with India
Livingston Island	D—*Xylopteris*, *Sagenopteris*, *Sphenobaiera* T—*Scoresbya*, *Sagenopteris*, *Elatocladus*, *Pagiophyllum* R—*Dicroidium*	? Tiki Formation
Allan Hills, Mt. Bumstead, Shackleton Glacier	D—*Dicroidium* spp. T—*Xylopteris* R—*Rissikia*	? Parsora Formation
Ross Sea Area	*Dicroidium odontopteroides*	? Parsora Formation

[a] D, dominant; R, rare; T, characteristic. See text for further explanation.

Moreover, *Xylopteris* and *Rissikia* do not occur in the Parsora.

A fairly rich assemblage of Triassic plants is known from Livingston Island in the South Shetland Islands (Fig. 10.4). Orlando (1968) first described *Asterotheca crassa*, a rachis of the Osmundaceae; *Coniopteris distans*, a fragment of dipteridacean foliage, *Thinnfeldia* sp.; and *Xylopteris* cf. *elongata*. Of these, the dipteridacean foliage, *Thinnfeldia* and *Coniopteris* seem to be of a doubtful nature. The dipteridacean lamina resembles *Scoresbya dichotoma*, described from the same locality by Banerji and Lemoigne (1987). *Thinnfeldia* sp. morphologically resembles *Dicroidium* (e.g., *D. lancifolium*), while *C. distans* may represent a fertile organ of *Dicroidium* or *Xylopteris*. Further additions to this flora were made by Lacey and Lucas (1981b). These include *Thallites* sp. A and B, *Equisetites* sp., *Neocalamites* sp., *Dicroidium* cf. *lancifolium*, *D.* cf. *spinifolia*, *Pterophyllum dentatum*, *Doratophyllum tenison-woodsii*, *Pagiophyllum* sp., *Ginkgoites* sp., and *Hexagonocaulon minutum*. In the absence of cuticle, it is perhaps appropriate to refer *Doratophyllum* to *Taeniopteris*. Banerji and Lemoigne (1987) recently made several additions to the Livingston Island flora, including *Marattiopsis* sp., *Dictyophyllum* sp., *Scoresbya dichotoma*, *Cladophlebis* sp., *Sagenopteris* sp., *Caytonia* sp., *Taeniopteris* sp., *Sphenobaiera* sp., and *Elatocladus* sp. On the basis of this assemblage, they suggested a late Upper Triassic age for the exposures at Livingston Island. In India, *Elatocladus* and *Pagiophyllum* first appear at Giar (upper Tiki Formation, Table 10.4), which is regarded as Upper Triassic.

Rhaeto–Liassic of India (?Upper Gondwana)

A few fragmentary plant remains collected from a hillock southeast of Hartala Village, Shahdol District, Madhya Pradesh, were described by Pal (1984a). The collection includes *Pagiophyllum* sp., *Brachyphyllum* sp., and *Desmiophyllum* sp. Of these, *Pagiophyllum* is

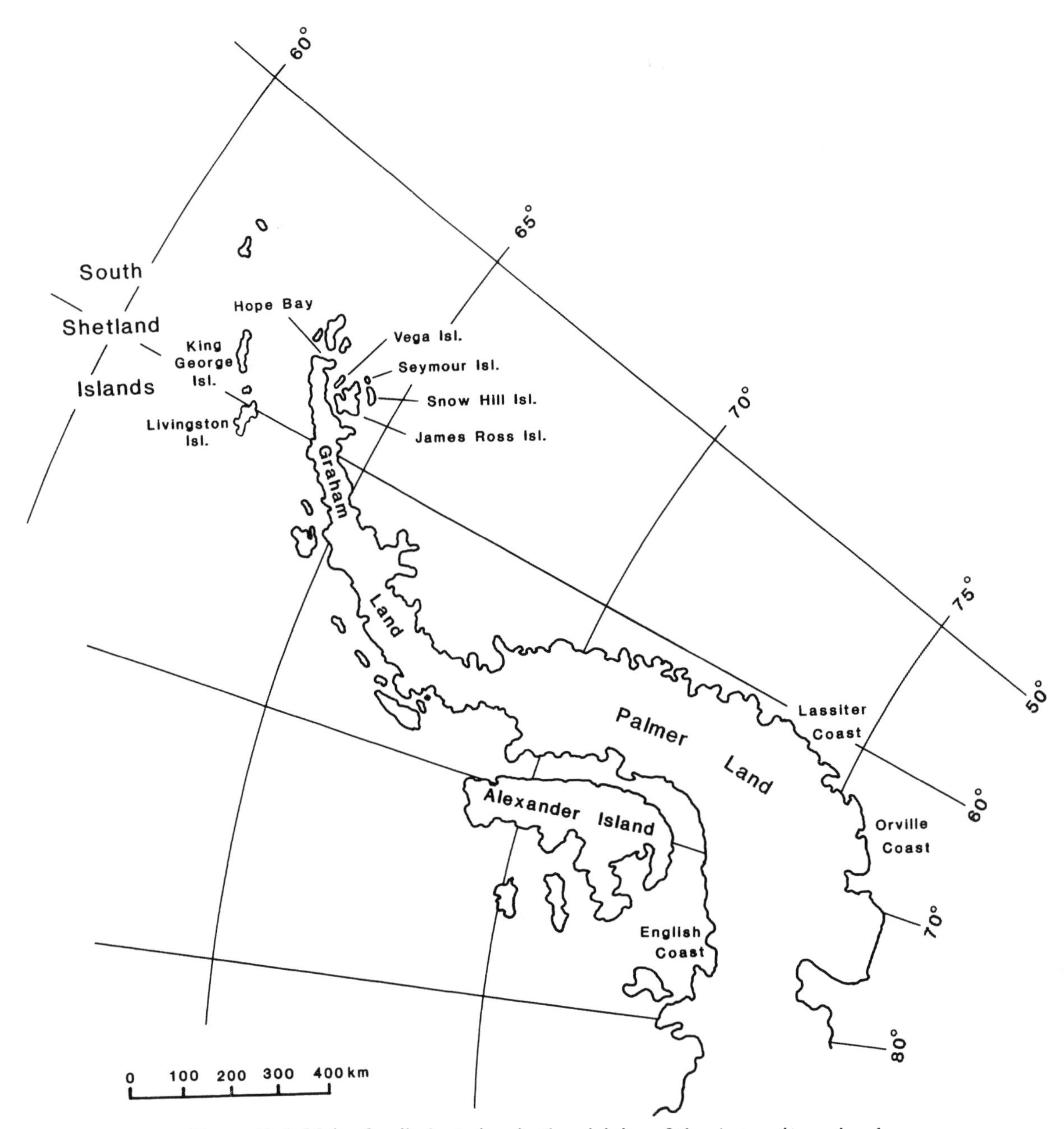

Figure 10.4. Major fossil plant sites in the vicinity of the Antarctic peninsula.

the most common element of the flora. The fossiliferous beds of Hartala Hill overlie the Upper Triassic rocks of the Tiki Formation. Because of this and the complete absence of characteristic Triassic taxa, such as *Lepidopteris* and *Dicroidium,* Pal (1984a) suggested that the Hartala Hill beds are Rhaeto–Liassic in age.

Jurassic–Lower Cretaceous (Upper Gondwana of India)

As noted earlier, there is no accepted stratigraphic sequence of the various formations within the Jurassic–Lower Cretaceous of India. Thus, it is difficult to construct a floral succession in post-Triassic sediments. How-

Table 10.6. Jurassic-Lower Cretaceous of India.

Formations (Fm) and localities	Taxa present[a]
Than Fm, Himmatnagar Fm, Gardeshwar Fm, Bansa Fm	D—*Gleichenia, Matonidium, Pagiophyllum, Brachyphyllum* T—*Gleichenia, Weichselia, Onychiopsis* R—*Ptilophyllum, Pterophyllum*
Gangapur Fm	D—*Elatocladus* T—*Gleichenia* R—*Ptilophyllum*
Jabalpur Fm	D—*Ptilophyllum* T—*Cladophlebis, Todites, Hausmannia, Sphenopteris, Allocladus, Pagiophyllum, Brachyphyllum* R—*Gleichenia*
Rajmahal Fm	
Nipania (Type 2)	D—*Nipaniophyllum* C—*Pentoxylon, Carnoconites,* Podocarpaceae R—*Ptilophyllum*
Bindaban, Onthèa, etc. (Type 1)	D—*Ptilophyllum, Pterophyllum* C—*Equisetites, Marattiopsis, Cladophlebis, Gleichenia, Taeniopteris, Elatocladus* R—*Pagiophyllum, Brachyphyllum*
East Coast	
Athgarh	T—Pteridophytes—*Gleichenia, Phlebopteris*
Golapilli	D—*Pterophyllum*
Raghavapuram	D—*Ptilophyllum, Elatocladus*
Vemavaram	C—*Dictyozamites, Pterophyllum, Otozamites*
Sriperumbudur, Sivaganga	C—*Ptilophyllum, Elatocladus, Taeniopteris*
Western India	
Kachchh District	
Bhuj Fm	D—*Ptilophyllum, Brachyphyllum, Allocladus* (Assemblage Type 2) D—*Pachypteris* T—*Linguifolium, Sagenopteris, Nilssoniopteris* (Assemblage Type 1)
Jhuran Fm	T—*Cladophlebis, Pachypteris, Pagiophyllum*
Rajasthan Province	
Sarnu Fm	C—*Phlebopteris, Otozamites* R—*Pachypteris, Pagiophyllum, Brachyphyllum*
Habur	D—*Pachypteris*
Hartala Hill	C—*Pagiophyllum* T—*Brachyphyllum, Desmiophyllum*

[a] C, common; D, dominant; R, rare; T, characteristic. See text for further explanation.

ever, it has been possible to recognize two distinct types of floras, with some local variations (see Table 10.6). The first assemblage is dominated by bennettitalean remains, with subdominant pteridophytes and conifers [e.g., Jhuran and Bhuj Formations of Kachchh, Rajmahal Formation (Type 1), East Coast Gondwanas, Table 10.6] (Fig. 10.5). In the second type of floral assemblage, bennettitalean remains are extremely rare and typical Wealden taxa, such as *Gleichenia nordenskioldii, Weichselia reticulata,* and *Onychiopsis* spp., are present. This assemblage contains more ferns, species of *Pagiophyllum,* and *Brachyphyllum,* but very few species of *Elatocladus,* and is known only from strata of undoubted Lower Cretaceous age. The floras from these time periods will be considered here under different regions (Fig. 10.5; Table 10.1) and not according to age, since in most cases the exact age is uncertain.

Western India

Plant fossils have recently been reported from two localities in Rajasthan, one occurring at Habur and the other near Sarnu village.

Habur

The fossiliferous beds, belonging to the lower part of the Pariwar Formation, are exposed about 1 km east of Habur Village. The assemblage is dominated by *Pachypteris haburensis,* followed by *Taeniopteris haburensis, Anomozamites haburensis,* and *Elatocladus con-*

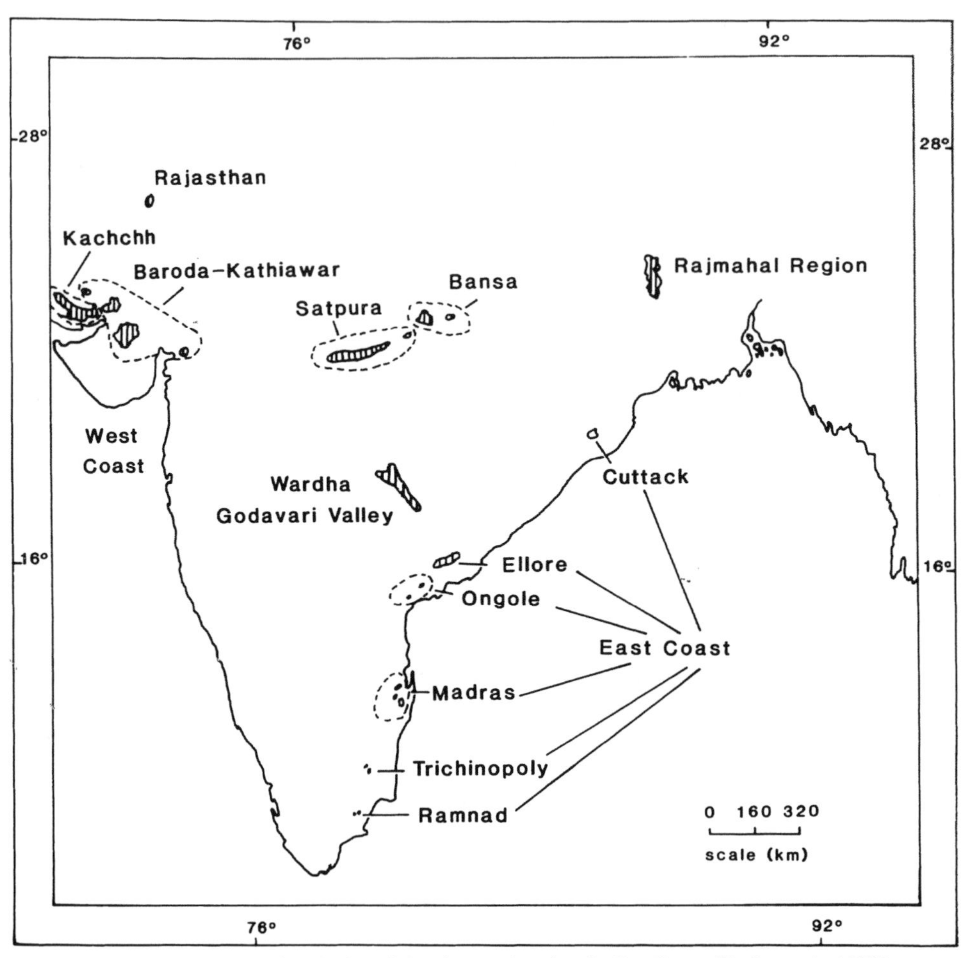

Figure 10.5. Post-Triassic localities in peninsular India (from Shah et al. 1973).

ferta (Maheshwari and Singh 1976; Bose et al. 1982a). Other taxa include *Gleichenia* sp., *Phlebopteris* sp., *Otozamites imbricatus, Ptilophyllum acutifolium,* and *Ginkgo* sp. The age of this flora has been suggested as Upper Jurassic.

Sarnu Village

Fossil plants from the Sarnu Hill Formation, Barmer Basin, were described by Baksi and Naskar (1981) and more recently by Banerji and Pal (1987). The flora is dominated by *Phlebopteris* sp. cf. *P. athgarhensis* and species of *Otozamites*. In addition to these, *Pachypteris haburensis, Pagiophyllum* sp., and *Brachyphyllum* have also been reported, although they are rare. Baksi and Naskar (1981) suggested that the Sarnu Hill Formation may be anywhere in the Upper Jurassic–Lower Cretaceous, but Banerji and Pal (1987) believe the formation is of Upper Jurassic age.

Kachchh

Biswas (1977) has provided a detailed Mesozoic stratigraphic sequence for the Kachchh region. From the Jhuran and Bhuj Formations, Bose and Banerji (1984) have described a large number of fossil plants, including more than 80 species. The Jhuran Formation was extremely poor in plant remains, with only *Cladophlebis daradensis, Cladophlebis* sp. A, *Cladophlebis* sp. C, *Pachypteris indica, P. specifica,* and *Pagiophyllum chawadensis* present. With the exception of *Cladophlebis* sp. A and C, all of these taxa are also known from the Bhuj Formation. The Bhuj Formation has two distinct types of assemblages. One is dominated by species of *Pachypteris;* also present are species of *Linguifolium, Sagenopteris, Ctenozamites, Taeniopteris, Nilssoniopteris, Pterophyllum, Otozamites, Ptilophyllum, Elatocladus, Pagiophyllum,* and *Brachyphyllum. Linguifolium, Sagenopteris,* and *Nilssoniopteris* do not occur in any other formations in India. This type of floral assemblage, which occurs at Kakadbhit and Kurbi, is reminiscent of floras outside India in the Middle Jurassic.

The other type of assemblage within the Bhuj Formation occurs at Trambau, Sukhpur, and Dharesi, where species of *Ptilophyllum, Allocladus, Brachyphyllum,* and *Araucarites* dominate. In this association, there is no *Linguifolium Sagenopteris, Ctenozamites, Nilssoniopteris,* or *Pterophyllum.* At Trambau, Bhajodi, Kakadbhit, and Dharesi, vertical and horizontal root markings are fairly common. Other interesting floral elements include *Isoetites indicus* (Bose and Roy 1964), which has been collected *in situ* at Trambau and near Nagod, and *Trambauathallites sukhpurensis* and *Marsilea*-like plants (Banerji 1987, 1989), which occur in the same section at Trambau within a single layer that extends for more than 100 m. The presence of these elements strongly suggests that the fossiliferous beds at Trambau were of freshwater origin. The evidence that the root markings at Trambau represent *in situ* rather than transported plants is the occurrence of the markings in vertical as well as horizontal positions and the presence of *Isoetites* in growth position with its roots intact. There has been some controversy over the autochthonous interpretation of these beds (see, e.g., Venkatachala 1986), which could perhaps be solved by further sedimentologic/paleoecological work on this locality. Bose and Banerji (1984) have suggested that the Bhuj Formation may range in age from Middle to Upper Jurassic, while Banerji et al. (1984) regard it as Lower Cretaceous.

Rajmahal Hills and the East Coast of India

One of the richest assemblages of fossil plants known from India comes from the Rajmahal Hills, Bihar. Floras from the various localities in the East Coast of India have often been compared with the assemblages from the Rajmahal Hills. Unlike the Rajmahal Hills, the outcrops in the East Coast are generally poor in fossils.

Rajmahal Hills

The Rajmahal has two distinct types of floras. One is dominated by cycadophytic and pteridophytic remains, while the other includes *Pentoxylon* and podocarpacean remains with extremely rare cycadophytes. The first type of assemblage is best represented at Bindaban, Sakrigalighat, and Onthea, and the second at Nipania. Fossils from these localities have been listed by Bose (1966) and Shah et al. (1973). Important components of the first floral type are *Selaginellites gracilis, Equisetites rajmahalense, Marattiopsis macrocarpa, Todites indicus, Gleichenia gleichenoides, Taeniopteris* spp., *Morrisia mcclellandi, Cycadites rajmahalensis, Anomozamites* spp., *Pterophyllum* spp., *Ptilophyllum* spp., *Dictyozamites* spp., *Weltrichia santalensis,* and *Williamsonia* spp. The second floral assemblage, occurring at Nipania, includes *Pentoxylon sahnii, Nipaniophyllum raoi, N. hobsonii, Sahnia nipaniensis, Carnoconites compactus, C. rajmahalensis, Mehtaia rajmahalensis, M. santalensis, Sitholea rajmahalensis, Indophyllum* spp., *Elatocladus sahnii, Pagiophyllum araucarioides, Brachyphyllum florini, Podostrobus rajmahalensis, Nipanioruha granthia,* and *Nipaniostrobus sahnii.* In addition to these elements, Vishnu-Mittre (1959) has described several types of pteridophytic petioles and rachises from this site. All of the species listed are based on silicified plant remains. The assemblage at Nipania is quite distinct from all the other Jurassic–Lower Cretaceous assemblages known from Bindaban, etc., in having very few cycadophytic remains. It differs from floras at Bansa and in other Lower Cretaceous formations by the dominance of *Pentoxylon* remains and conifers with podocarpacean affinities. At Bansa and in other Lower Cretaceous formations, there are more pteridophytes and araucarian remains.

In addition to the taxa mentioned, a large number of fossil woods and other petrified remains are known from the Rajmahal Hills. Some of the more well-known taxa are *Bucklandia sahnii, Williamsonia sewardiana, W. harrisiana, W. amarjolense,* and *Amarjolia dactylota* (Sahni 1932; Bose 1953, 1968; Bose et al. 1984). *Amarjolia dactylota* is the only bisexual bennettitalean fructification known from India to date. It has microsporophylls similar to those of *Weltrichia santalensis*.

East Coast

There are a number of isolated patches of post-Triassic rocks that extend along the east coast of India, from Athgarh, Cuttack District, in the north to Sivaganga, Ramnad District (Fig. 10.5). Most of these areas have their own characteristic floras, but all have been compared with the assemblages from the Rajmahal Hills. Some of the floras from this region are based on collections made from shale dumps outside well diggings.

Athgarh (Cuttack District)

The flora from this site has a fair representation of pteridophytic remains with *Phlebopteris athgarhensis* (Jain 1968) the most common. In the Rajmahal Hills, *Phlebopteris* is extremely rare with more *Gleichenia* and *Pterophyllum* present. *Gleichenia* is rare at Athgarh and *Pterophyllum* does not occur (Patra 1973).

Golapilli Sandstone (Ellore)

Pterophyllum kingianum (Feistmantel 1877b) is quite common at this site, with a moderate representation of species of *Ptilophyllum* as well. Pteridophytic remains are rare, although *Marattiopsis macrocarpa,* which is fairly common at Bindaban, Rajmahal Hills, is also present in this flora.

Raghavapuram Shales (Ellore)

The flora from the Raghavapuram shales differs from that of Golapilli in the dominance of *Ptilophyllum cutchense* and *Elatocladus plana* (Baksi 1968).

Vemavaram (Ongole)

Bennettitalean remains, represented by *Otozamites vemavaramensis, Dictyozamites feistmantelii,* and *Pterophyllum footeanum* are extremely common (Bose and Jain 1967). Next in abundance are species of *Elatocladus* and *Brachyphyllum;* pteridophytes are extremely rare.

Sriperumbudur (Madras)

This site has yielded the largest number of taxa in the East Coast region. The flora is dominated by bennettitalean remains, especially species of *Ptilophyllum* and conifers. Pteridophytes, though present, are rather rare (Shah et al. 1973).

Sivaganga (Ramnad District)

Ptilophyllum cutchense is the most abundant taxon at this site. Other species include *Taeniopteris spatulata, Elatocladus plana,* and some doubtful species of *Ginkgoites* (Gopal et al. 1957).

South Rewa and Satpura Gondwana Basins

Fossiliferous outcrops of Jurassic–Lower Cretaceous age are known from Seoni-Malwa in Hoshangabad District, Sehora, and Imjhiri villages in Narsinghpur District and Chui Hill, Jabalpur District. The localities are rich in pteridophytes (Bose 1959 and Zeba-Bano 1980), *Ptilophyllum* spp. (Bose and Kasat 1972), and conifer remains (Bose and Maheshwari 1973a, b; Maheshwari and Kumaran 1976; Sukh-Dev and Zeba-Bano 1978, 1979, 1981a). Among the pteridophytes, species of *Cladophlebis* and *Sphenopteris* are fairly common. *Hausmannia dichotoma,* which is known in India only from Kachchh and Than, has been reported from near Jatamao (Zeba-Bano 1980). From Sehora, *Onychiopsis psilotoides*, a Wealden species, has been described by Bose (1959). The richest assemblage is from Sehora and includes *Pachypteris indica, Ptilophyllum* spp., *Elatocladus, Brachyphyllum, Satpuria*, and *Araucarites. Pterophyllum* is rare, with only a single species (*P. medlicottianum*) reported by Bose and Banerji (1981) from Imjhiri. *Ctenozamites surangei* is also known from this locality (Sukh-Dev and Zeba-Bano 1981b).

Lower Cretaceous (Upper Gondwana of India)

Fossiliferous outcrops of Lower Cretaceous age are known from Gangapur; Bansa, Chandia, and Patparha (Shahdol District); and Gardeshwar (near Baroda), Himmatnagar (Gujarat), and Than and adjacent localities (Kathiawar) (Fig. 10.5).

Gangapur (Pranhita-Godavari Basin)

From the vicinity of Gangapur Village, Adilabad District, plant fossils have been described by Bose et al. (1982b). The fossiliferous beds were originally known as the Gangapur beds of the Kota Stage, but now they are grouped within the Gangapur Formation. The flora is dominated by species of *Elatocladus;* also present is the characteristic Wealden species of *Gleichenia, G. nordenskioldii*. Cycadopytes are rare, represented by *Taeniopteris* sp. cf. *T. spatulata, ?Anomozamites* sp., *Ptilophyllum cutchense*, and *Ptilophyllum* sp. cf. *P. acutifolium*. The high percentage of *Elatocladus* specimens, regarded by some to be members of the Podocarpaceae, suggests that the Gangapur assemblage is similar to the flora from Nipania, Rajmahal Hills, where podocarps are well represented. Other Cretaceous formations have more araucarian remains and include *Weichselia reticulata*. A basal Lower Cretaceous age has been suggested for the Gangapur Formation.

Bansa, Chandia, and Patparha (South Rewa Gondwana Basin)

These three localities are dominated by pteridophytes and araucarian remains. The pteridophytes are represented by three characteristic Wealden forms: *Gleichenia nordenskioldii, Onychiopsis paradoxus*, and *Weichselia reticulata*. Other pteridophytes have been described by Sukh-Dev (1970, 1971, 1972) and Pant and Srivastava (1977). Of these, *Cladophlebis medlicottiana, Gleichenia rewahensis*, and *Phlebopteris polypodioides* are fairly common, while the genus *Hausmannia* is rare, represented by only one species (*H. pachyderma;* Sukh-Dev 1972). The flora from Bansa is dominated by species of *Pagiophyllum, Brachyphyllum, Allocladus*, and *Harrisiophyllum* (Bose and Sukh-Dev 1958; Sukh-Dev and Zeba-Bano 1979; Pant et al. 1983; Srivastava et al. 1984). Other gymnosperms include *Cycadopteris* spp., *Yabeiella hirsuta*, and *Ginkgoites festmantelli*. Bennettitalean remains are extremely rare. They include only four species of *Ptilophyllum*, each of which is based on a single specimen (Bose and Sukh-Dev 1958; Sukh-Dev and Zeba-Bano 1977).

Concerning the age, the characteristic features of the megafossil assemblages from Bansa and other localities have already been discussed. Based on the palynoflora at Bansa, Maheshwari (1974) regards the Bansa Formation as Lower Cretaceous in age, which is in agreement with the megafossil data. However, Bharadwaj and Kumar (1974) have dated it as Upper Jurassic, based on the palynoflora of coals at Bansa.

Gardeshwar Formation

Fossil plants from Gardeshwar were first collected and described by Borkar and Phadke (1973, 1974). Bose et al. (1983) later provided a detailed description of the various species. The assemblage is dominated by *Pagiophyllum* and *Brachyphyllum;* bennettitaleans are absent and pteridophytes are rare. A Wealden (Lower Cretaceous) age has been assigned to this Formation.

Himmatnagar

Plant fossils preserved in sandstone were first described by Sahni (1936) from this site and included *Matonidium indicum, Weichselia reticulata, Sphenopteris* spp., and a doubtful specimen of *Thinnfeldia*. Recently, Banerji et al. (1983) have recorded *Pachypteris* sp., *Elatocladus* sp., and *Brachyphyllum* sp. from this same locality.

Than and Adjacent Localities (Kathiawar)

Megafossils from Than, Songad, and Tarnetar have been described by Roy (1968), Borkar and Chiplonkar (1973), Bose and Jana (1979), and Jana and Bose (1981). They include the following taxa: *Equisetites rajmahalense, Cladophlebis kathiawarensis, C. daradensis, Gleichenia nordenskioldii, Matonidium indicum, Dictyophyllum indicum, Hausmannia pachyderma, H. dichotoma, Pterophyllum princeps, Elatocladus longifolia, Brachyphyllum* spp., and *Araucarites* spp. Of these, *Cladophlebis kathiawarensis, C. daradensis, Matonidium indicum,* and *Brachyphyllum* species are very common. *Ptilophyllum* and other bennettitalean remains are not present, although *Pterophyllum princeps* is rare.

Jurassic-Lower Cretaceous Floras of Antarctica

Fossil plants of presumed Jurassic age are known from both East and West Antarctica (Table 10.7). In West Antarctica, plant fossils occur at Hope Bay (Graham Land; (Fig. 10.4). Stratigraphic correlation of these beds has been detailed by Adie (1964). In East Antarctica, Jurassic plants were described from Carapace Nunatak (Townrow 1967a, b). Of these localities, the flora from Hope Bay is more diverse, but the plants from the Carapace Nunatak are perhaps better preserved.

Table 10.7. Jurassic of Antarctica.

Localities	Taxa present[a]	Correlation with India
Hope Bay	C—*Scleropteris, Pachypteris, Thinnfeldia, Nilssonia, Zamites, Otozamites, Elatocladus*	?Assemblage Type-1 of Bhuj Fm
Carapace Nunatak	C—*Nothodacrium warrenii* R—*Otozamites, Masculostrobus*	?

[a] C, common; R, rare.

Hope Bay, Graham Land

Fossil plants were first described from this locality by Halle (1913), who recognized as many as 58 taxa. More recent collections made from Hope Bay and Botany Bay have been studied by Rees (1987) and Gee (1987). The flora includes one species of *Equisetites* (*E. approximatus*), 25 forms assignable to the Filicales, 19 cycadophytes, and 13 conifers. The filicaleans include *Cladophlebis* spp., *Todites williamsonii, Dictyophyllum* sp., *Coniopteris* spp., and *Sphenopteris* (8 species). Some of the *Coniopteris* and *Sphenopteris* species need critical examination. In the absence of the basal "aphlebiform" lobes, it is difficult to determine whether the specimens of *Coniopteris hymenophylloides* belong to that species. Gymnosperms include species of *Scleropteris, Pachypteris, Thinnfeldia, Nilssonia, Pseudoctenis, Zamites, Otozamites, Ptilophyllum, Elatocladus, Pagiophyllum,* and *Brachyphyllum*. In addition, *Williamsonia pusila, Cycadolepis* sp., *Araucarites cutchensis, Sphenolepidium ?oregonense, ?Conites* sp., and *Stachyopitys* cf. *annularoides* are also present. Of those described, *Nilssonia taeniopteroides* appears to differ from other species of *Nilssonia*. The lamina is attached on the upper surface on either side of the rachis, that is, the lamina is not continuous nor does it conceal the rachis from above. In addition, fronds have forked secondary veins in a few instances. In other *Nilssonia* specimens, the lamina is continuous over the rachis and secondary veins do not bifurcate. Some species of *Zamites* (*Z. antarcticus* and *Z. pachyphyllum*) are more similar to *Ptilophyllum;*

Otozamites abbreviatus appears to be a *Ptilophyllum* (Bose and Kasat, 1972), and *Williamsonia pusilla* is a doubtful form, which at best may be compared with *Bennetticarpus*.

It appears that the most common elements at Hope Bay are *Scleropteris, Pachypteris, Thinnfeldia, Nilssonia taeniopteroides, Zamites, Otozamites,* and *Elatocladus*. The flora with the most comparable composition in India is the Bhuj Type-1 (Table 10.6). However, to date, there is no record of *Scleropteris, Thinnfeldia, Nilssonia,* or *Zamites* from Kachchh. To some extent, the Hope Bay flora also resembles the floras from Sarnu, Habur, Sriperumbudur, and the Type-1 flora from the Rajmahal Hills. Halle (1913) had suggested a Middle Jurassic age for the Hope Bay flora. However, Rees (1987) indicates that sedimentological evidence favors an Early Cretaceous age for this material and Gee (1987) also describes it as Early Cretaceous.

Carapace Nunatak

Fossil plants from the Jurassic of East Antarctica were first described by Plumstead (1962) from Carapace Nunatak near the Upper Mackay Glacier (southern Victoria Land). The assemblage is rather depauperate, with only *Otozamites antarcticus, Elatocladus* cf. *E. heterophylla, Pagiophyllum* cf. *P. peregrinum,* and *Brachyphyllum* cf. *B. expansum* present. Except for *O. antarcticus,* all the remaining species described by Plumstead (1962) were placed in *Nothodacrium warrenii* by Townrow (1967a, 1967b). Townrow (1967a) also described a male cone from this locality as *Masculostrobus warrenii*. Carapace Nunatak was regarded as Lower Jurassic by Plumstead (1962); however, Townrow (1967b) suggested a probable Middle Jurassic age based on the occurrence of *Tsugaepollenites trilobatus* pollen grains. With only two genera known from this site, no stratigraphic correlation with India is possible.

Alexander Island

Thallophyte borings are known in the phosphatic cuticles and shells of decapods and brachiopods from the Lower Cretaceous of Alexander Island (Taylor 1971). From the Fossil Bluff Formation, Alexander Island, Taylor et al. (1979) have figured (with no description) *Nilssonia*(?), *Otozamites*(?), *Dictyozamites*(?), and *Ptilophyllum*(?). *Nilssonia*(?) looks more like *Taeniopteris*. From the upper part of the formation, Jefferson (1982a, 1983) has figured and given tentative names to specimens comparable to *Phlebopteris dunkeri, Hausmannia dichotoma, Cladophlebis oblonga, Gonatosorous nathorstii, Taeniopteris daintreei, Ginkgo huttoni, Bellarinea barklyi,* and *Pagiophyllum insigne*. Although some of these specimens are preserved as compressions, he notes (Jefferson 1982a) that mineral growth has destroyed the cuticle in all instances and identification is thus difficult. Based on the photographs alone, *Hausmannia dichotoma* resembles more *H. crenata,* and *Bellarinea barklyi* is comparable to some of the species of *Podozamites*. Silicified trees in growth position are known from east Titan Nunatak, Alexander Island (Jefferson 1982b, 1983). A fossil wood resembling *Circoporoxylon* has already been figured by Jefferson (1983). Earlier, Jefferson (1982b) had given a general account of fossil forests from the Lower Cretaceous of Alexander Island. The assemblage recorded by Taylor et al. (1979) is quite distinct from the one described by Jefferson (1983). The former has only cycadophytic remains, whereas the latter has no bennettitalean remains and has more pteridophytes and conifers. The age of the Fossil Bluff Formation is presumed to be Early Cretaceous. The assemblages are too poor to allow detailed comparison with any of the Jurassic–Lower Cretaceous assemblages known from India. The assemblage described by Taylor et al. (1979) seems to be closer to the assemblage Type 1 known from the Rajmahal Hills and some of the localities in the East Coast Gondwana, whereas the assemblage from the upper part of the Fossil Bluff Formation (Jefferson 1983) is more comparable to the one known from the upper part of the Jabalpur Formation (see Table 10.6). The only difference is that in the upper part of the Jabalpur Formation there are quite a few bennettitalean remains.

Concluding Remarks

The present review is incomplete, as many of the anatomically preserved plant remains are difficult to correlate with impression/compression elements. The Permian and Mesozoic floras from extrapeninsular regions of India have also not been included here.

Most of the megafloras from Antarctica are rather low in diversity when compared to those known from India. The most diverse Antarctic assemblages come from the Permian Buckley Formation, Upper Triassic of Livingston Island, and the Middle Jurassic-Lower Cretaceous of Hope Bay. The plant association at Milorgfjella, Dronning Maud Land, seems to be closest to the ones known from the Upper Talchir Formation (Rikba beds) of India. Other than this, none of the other assemblages can be compared with any of the Indian floras. Within the Permian, they may broadly be compared with the floras from Barakar to Kamthi Formations. In Antarctica, however, there is no flora currently known that resembles those from the Karharbari Formation of India.

There is no Triassic assemblage in Antarctica that compares with those from the Panchet of India. The earliest Triassic in India includes plants with *Lepidopteris*-like cuticle, whereas in Antarctica the earliest Triassic marker is *Dicroidium odontopteroides* (south Victoria Land). None of the other Antarctic Triassic assemblages are similar to the ones known from India. The assemblage from Livingston Island seems to be closest to those from the Upper Triassic of Giar (Tiki Formation). Both *Elatocladus* and *Pagiophyllum*, which are common in the Jurassic–Lower Cretaceous of India, appear for the first time in Antarctica in this flora. At Livingston Island, there is no record of *Lepidopteris*, which is common in the Tiki Formation, but the site has yielded *Sagenopteris* and *Scoresbya*, which are known from the Rhaetic of the northern hemisphere. As noted, the Hope Bay flora is most similar to the Type-1 flora from Kachchh (Bhuj Formation). No characteristic Wealden taxa, such as *Gleichenia nordenskioldii*, *Weichselia reticulata*, or *Onychiopsis psilotoides*, have been found in Antarctica.

Historically, geologists working in Antarctica have collected specimens of *Glossopteris* at a number of localities. In some instances, specific names were applied based on species described from India. These specimens need critical reexamination before they can be satisfactorily used for biostratigraphic correlation. It is important to underscore that the various Gondwana floral assemblages are not necessarily contemporaneous. Most of the sequences are of terrestrial origin, and as such, it is difficult to correlate them based on marine faunas. It is also important to note that if particular genera and species are characteristic of a certain formation in India, it does not necessarily mean that the same will hold true for formations of similar ages elsewhere in Gondwana. For example, typical Karharbari genera, such as *Botrychiopsis* and *Buriadia*, are found in the Rio Bonito Formation in the Rio Grande do Sul area of the Paraná Basin of Brazil. However, the age of this formation is believed to be equivalent to the Talchir Formation in India, which is overlain by the Karharbari Formation. Similarly, *Walkomiella*, which is confined to the Barakar Formation in India (?Middle Permian), is known from the Upper Permian in Australia.

It is surprising that conifers from both India and Antarctica are poorly represented throughout the Permian and into the Middle Triassic. From the Permian of India, there are only four identifiable species of conifers (including ?*Buriadia*), while the record from the Lower Triassic is even poorer. In the lower Upper Triassic, there are no conifers known, and it is in the late Upper Triassic that the first species of *Elatocladus* and *Pagiophyllum* appear. In Antarctica, there is a record of an unidentifiable conifer shoot from the early Permian of Milorgfjella, but the first recognizable conifer, *Rissikia media*, is found in the early Upper Triassic. As in India, species of *Elatocladus* and *Pagiophyllum* first appear in the late Upper Triassic of Antarctica. Thus, it appears that in both India and Antarctica, Jurassic–Lower Cretaceous conifers first appear in the late Upper Triassic.

In the northern hemisphere, the Upper Triassic is represented by a number of taxa belonging to the Bennettitales, while in India and Antarctica the number of these taxa are even lower than the conifers. In India, there is no record of the Bennettitales in the Upper Triassic. However, in Parsora (?Middle Triassic), there is a record of *Pterophyllum sahnii,* a form with a segmented lamina. However, it should be noted that in the absence of cuticle it is impossible to say whether this form represents a cycadean or bennettitalean frond. In Antarctica, the only Triassic bennettitalean yet known is *Pterophyllum dentatum* from Livingston Island (Lacey and Lucas 1981b). Bennettitalean remains first appear in India in floras of Jurassic–Lower Cretaceous age, most of which are regarded as probably Lower Cretaceous, whereas in Antarctica the Bennettitales are represented by a species of *Otozamites* at Carapace Nunatak (presumed Middle Jurassic age).

While it is possible to detail a stratigraphic sequence of floras in the Permian of peninsular India and to some extent in the Triassic as well, it is extremely difficult to formulate such a sequence for the Jurassic–Lower Cretaceous. As noted earlier, this is because most formations of this age are underlain by metamorphics or gneisses believed to be of Archaean age. At best, we can compare the various floral assemblages with those of similar ages outside India, without assigning any definite age. When Feistmantel (1876a) first described the fossil flora of Kachchh, he considered it to be of Jurassic age (Oolitic). We also believe the Kachchh assemblage, especially the Type-1 flora, to be more comparable to Middle Jurassic floras known elsewhere than the remaining floras from other Jurassic–Lower Cretaceous formations. The other floras that are most similar to the Kachchh assemblage are those from Sarnu and Habur. According to current practice, the Bhuj Formation of Kachchh is regarded as the youngest formation within the Gondwana Sequence of India.

There is no doubt that it is important to know the precise age of the various plant-bearing formations, both in India and Antarctica. However, it is equally clear that we have a long way to go before the age and boundary problems can be solved, even in India where the floras are relatively well known. At the present time in India, these problems cannot be solved only on the basis of megafloral and palynological evidence. An important step in more accurately defining these sequences will be the utilization of radiometric dating. Once dating of floras within individual Gondwana continents is better defined, biostratigraphic correlations between continents will become more useful.

Acknowledgments. We are grateful to Professor Svein B. Manum (Geology Department, Universitetet I Oslo) for his helpful suggestions and for kindly providing MNB with laboratory and other facilities; and to Mrs. Kristin Rangnes (Geology Institute, Oslo) and her colleagues for obtaining useful literature from different libraries. The Norwegian Academy of Science and Letters provided partial financial assistance to MNB during his stay in Oslo. This work was supported in part by National Science Foundation grants (DPP-8713685).

References

Adie RJ (1964) Stratigraphic correlation in West Antarctica. In Adie RJ (ed) Antarctic Geology, North-Holland Publishing Co, Amsterdam, pp 307–313

Allen AD (1962) Geological investigations in southern Victoria Land, Antarctica. Part 7—Formations of the Beacon Group in the Victoria Valley region. New Zealand J Geol Geophys 5:278–294

Baksi SK (1968) Fossil plants from Raghavapuram mudstone West Godavari District, A P, India. Palaeobotanist 16(3):206–215, 1967

Baksi SK, Naskar P (1981) Fossil plants from Sarnu Hill Formation, Barmer Basin, Rajasthan. Palaeobotanist 27(1):107–111, 1977

Ballance PF (1977) The Beacon supergroup in the Allan Hills, Central Victoria Land, Antarctica. New Zealand J Geol Geophys 20:1003–1016

Banerjee M (1969) *Senotheca murulidihensis,* a new glossopteridean fructification from India associated with *Glossopteris taeniopteroides* Feist. In Santapau H, Ghosh AK, Roy SK, Chanda S, Chaudhuri SK (eds) J Sen Memorial Vol, Bot Soc Bengal, Calcutta, pp 359–368

Banerji J (1987) Further contribution to the Mesozoic flora of Kutch, Gujarat. Geophytology 17(1):69–74

Banerji J (1989) Some plant remains from Bhuj Formation with remarks on the depositional environment of the bed. Palaeobotanist (in press)

Banerji J, Bose MN (1977) Some Lower Triassic plant remains from Asansol region, India. Palaeobotanist 24(3):202–210, 1975

Banerji J, Lemoigne Y (1987) Significant additions to the Upper Triassic flora of Williams Point, Livingston Island, South Shetland (Antarctic). Géobios 20:469–487

Banerji J, Pal PK (1987) Mesozoic plant remains from Sarnu, Barmer District, Rajasthan. Palaeobotanist 35(2):141–145, 1986

Banerji J, Maheshwari HK, Bose MN (1976) Some plants from the Gopad river section near Nidpur, Sidhi District, Madhya Pradesh. Palaeobotanist 23(1):59–71, 1974

Banerji J, Jana BN, Bose MN (1983) On a collection of fossil plants from Himmatnagar, Gujarat. In A K Ghosh Commem Vol—Evolutionary Botany and Biostratigraphy:463–473

Banerji J, Jana BN, Maheshwari HK (1984) The fossil floras of Kachchh. II—Mesozoic megaspores. Palaeobotanist 33:190–227

Barghoorn ES (1961) A brief review of fossil plants of Antarctica and their geologic implications. In Science in Antarctica; part 1, Life Sciences in Antarctica. Publ 839 Nat Acad Sci Nat Res Council, Washington, pp 5–9

Barrett PJ, Kyle RA (1975) The Early Permian glacial beds of South Victoria Land and the Darwin Mountains, Antarctica. In Campbell KSW (ed) Gondwana Geology (papers presented at the Third Gondwana Symposium, Canberra, Australia, 1973), Australian National University Press, pp 333–346

Bharadwaj DC (1974) Palaeobotany of Talchir and Karharbari formations and Lower Gondwana glaciation. In Surange KR, Lakhanpal RN, Bharadwaj DC (eds) Aspects and Appraisal of Indian Palaeobotany, Birbal Sahni Institute of Palaeobotany, Lucknow, pp 369–385

Bharadwaj DC, Kumar P (1974) Palynostratigraphy of coals from Machrar Nala, Bansa, M P, India. Geophytology 4:147–152

Bharadwaj DC, Srivastava SC (1969) A Triassic mioflora from India. Palaeontographica B125: 119–149

Biradar NV, Bonde SD (1981) *Nandorioxylon saksenae* gen. et sp. nov.—a new gymnospermous wood from the Kamthi Stage of Chandrapur District, Mahrashtra State, India. Geophytology 11(1):90–95

Biswas SK (1977) Mesozoic rock-stratigraphy of Kutch, Gujarat. Quart J Geol Min & Metall Soc India 49(3–4):1–51

Borkar VD, Chiplonkar GW (1973) New plant fossils from the Umias of Saurashtra. Palaeobotanist 20(3):269–279, 1971

Borkar VD, Phadke AV (1973) Some plant fossils from Gardeshwar, Gujarat. M V M Patrika 8(2):44–46

Borkar VD, Phadke AV (1974) Fossil flora from the Gardeshwar Formation—a new formation of the Upper Gondwana of India. Bull Earth Sciences 57:64

Bose MN (1953) *Bucklandia sahnii* sp. nov. from the Jurassic of the Rajmahal Hills, Bihar. Palaeobotanist 2:41–50

Bose MN (1959) The fossil flora of the Jabalpur Series—2. Filicales. Palaeobotanist 7(2):90–92

Bose MN (1966) Fossil plant remains from the Rajmahal and Jabalpur Series in the Upper Gondwana of India. In Symposium on Floristics and Stratigraphy of Gondwanaland, Birbal Sahni Institute of Palaeobotany, Lucknow, pp 143–154

Bose MN (1968) A new species of *Williamsonia* from the Rajmahal Hills, India. J Linn Soc (Bot) 61:121–127

Bose MN (1974) Triassic floras. In Surange KR, Lakhanpal RN, Bharadwaj DC (eds) Aspects and Appraisal of Indian Palaeobotany, Birbal Sahni Institute of Palaeobotany, Lucknow, pp 258–293

Bose MN, Banerji J (1976) Some fragmentary plant remains from the Lower Triassic of Auranga Valley, District Palamau. Palaeobotanist 23(2): 139–144, 1974

Bose MN, Banerji J (1981) Cycadophytic leaves from Jurassic–Lower Cretaceous rocks of India. Palaeobotanist 28–29:218–300, 1979–1980

Bose MN, Banerji J (1984) The fossil floras of Kachchh. I—Mesozoic megafossils. Palaeobotanist 33:1–189

Bose MN, Jain KP (1967) *Otozamites vemavaramensis* sp. nov. from the Upper Gondwana of the East Coast of India. Palaeobotanist 15(3): 314–315, 1966

Bose MN, Jana BN (1979) *Dictyophyllum* and *Hausmannia* from the Lower Cretaceous of Saurashtra, India. Palaeobotanist 26(2):180–184

Bose MN, Kasat ML (1972) The genus *Ptilophyllum* in India. Palaeobotanist 19(2):115–145, 1970

Bose MN, Maheshwari HK (1973a) *Brachyphyllum sehoraensis,* a new conifer from Sehora, Narsinghpur District, Madhya Pradesh. Geophytology 3(2):121–125

Bose MN, Maheshwari HK (1973b) Some detached seed-scales belonging to Araucariaceae from the Mesozoic rocks of India. Geophytology 3(2): 205–214

Bose MN, Roy SK (1964) Studies on the Upper Gondwana of Kutch—2. Isoetaceae. Palaeobotanist 12(3):226–228

Bose MN, Srivastava SC (1971) The genus *Dicroi-*

dium from the Triassic of Nidpur, Madhya Pradesh, India. Palaeobotanist 19(1):41–51, 1970
Bose MN, Srivastava SC (1972) *Lepidopteris indica* sp. nov. from the Lower Triassic of Nidpur, Madhya Pradesh, India. J Palaeont Soc India 15:64–68
Bose MN, Srivastava SC (1973a) *Nidistrobus* gen. nov., a new pollen-bearing fructification from the Lower Triassic of Gopad river valley, Nidpur. Geophytology 2(2):211–212
Bose MN, Srivastava SC (1973b) Some micro- and megastrobili from the Lower Triassic of Gopad river valley, Nidpur. Geophytology 3(1):69–80
Bose MN, Sukh-Dev (1958) A new species of *Ptilophyllum* from Bansa, South Rewa Gondwana Basin. Palaeobotanist 6(1):12–14
Bose MN, Banerji J, Maithy PK (1977) Some fossil plant remains from Ramkola–Tatapani coalfields, Madhya Pradesh. Palaeobotanist 24(2):108–117, 1975
Bose MN, Kumaran KPN, Banerji J (1982a) *Pachypteris haburensis* n. sp. and other plant fossils from the Pariwar Formation. Palaeobotanist 30(1):1–11, 1981
Bose MN, Kutty TS, Maheshwari HK (1982b) Plant fossils from the Gangapur Formation. Palaeobotanist 30(2):121–142
Bose MN, Banerji J, Jana BN (1983) Mesozoic plant remains from Gardeshwar, Gujarat. In A K Ghosh Commem Vol—Evolutionary Botany and Biostratigraphy:483–498
Bose MN, Banerji J, Pal PK (1984) *Amarjolia dactylota* (Bose) comb. nov., a bennettitalean bisexual flower from the Rajmahal Hills, India. Palaeobotanist 33(3):217–229, 1983
Bunbury CJF (1861) Notes on a collection of fossil plants from Nagpur, Central India. Quart J Soc London 17:325–346
Chandra S (1984) *Utkalia dichotoma* gen. et sp. nov.—A fossil fructification from the Kamthi Formation of Orissa, India. Palaeobotanist 31(3): 208–212
Chandra S, Prasad MNV (1981) Fossil plants from the Kamthi Formation of Maharashtra and their biostratigraphic significance. Palaeobotanist 28–29:99–121
Chandra S, Rigby JF (1981) Lycopsid, sphenopsid and cycadaceous remains from the Lower Gondwana of Handappa, Orissa. Geophytology 11(2):214–219
Chandra S, Rigby JF (1983) The Filicales of the Lower Gondwanas of Handappa, Orissa. Palaeobotanist 31(2):143–147, 1982
Chandra S, Surange KR (1976) Cuticular studies of the reproductive organs of *Glossopteris* Part I—*Dictyopteridium feistmanteli* sp. nov. attached on *Glossopteris tenuinervis*. Palaeontographica B156(4–6):87–102
Chandra S, Surange KR (1977a) Cuticular studies of the reproductive organs of *Glossopteris* Part II—*Cistella* type fructification *Plumsteadiostrobus ellipticus* gen. et sp. nov. attached on *Glossopteris taenioides* Feistmantel. Palaeobotanist 23(3):161–175, 1974
Chandra S, Surange KR (1977b) Cuticular studies of the reproductive organs of *Glossopteris* Part III—Two new female fructifications—*Jambadostrobus* and *Venustostrobus*—borne on *Glossopteris* leaves. Palaeontographica B164(4–6):127–152
Chandra C, Surange KR (1977c) Cuticular studies of the reproductive organs of *Glossopteris* Part IV—*Venustostrobus indicus* sp. nov. Palaeobotanist 24(3):149–160
Chandra S, Surange KR (1979) Revision of the Indian species of *Glossopteris*. Monograph No. 2:1–291. Birbal Sahni Institute of Palaeobotany, Lucknow, India
Congrès Géologique International—Commission de Stratigraphie (1956) Lexique stratigraphique International, Asie 3(8):1–404. Paris
Craddock C, Bastien TW, Rutford RH, Anderson JJ (1965) *Glossopteris* discovered in West Antarctica. Science 148:634–637
Cridland AA (1963) A *Glossopteris* flora from the Ohio Range, Antarctica. Amer J Bot 50:186–195
Edwards WN (1928) The occurrence of *Glossopteris* in the Beacon Sandstone of Ferrar Glacier, South Victoria Land. Geol Mag 65:323–327
Elliot DH (1975) Gondwana basins of Antarctica. In Campbell KSW (ed) Gondwana Geology (papers presented at the Third Gondwana Symposium, Canberra, Australia, 1973), Australian National University Press, pp 493–536
Feistmantel O (1876a) Fossil flora of Gondwana system: Jurassic (Oolitic) flora of Kach. Mem Geol Surv India Palaeont Indica, Ser 11, 2(2): 1–80
Feistmantel O (1876b) Notes on the age of some fossil floras in India—III, IV & V. Rec Geol Surv. India 9(3):63–114
Feistmantel O (1876c) On some fossil plants from the Damuda Series in the Raniganj Coalfield, collected by Mr. J. Wood-Mason. J Asiat Soc Beng 45:329–380
Feistmantel O (1877a) Jurassic (Liassic) flora of the Rajmahal Group in the Rajmahal Hills. Mem Geol Surv India Palaeont Indica, Ser 2, 1(2):53–162
Feistmantel O (1877b) Jurassic (Liassic) flora of the Rajmahal Group from Golapili (near Ellore), South Godavari District. Mem Geol Surv India Palaeont Indica, Ser 2, 1(3):163–190
Feistmantel O (1877c) Flora of the Jabalpur Group (Upper Gondwana), in the Son-Narbada region. Mem Geol Surv India Palaeont Indica, Ser 2, 2(2):81–105
Feistmantel O (1877d) Notes on fossil floras in India-X. On *Pterophyllum* from the Raniganj

Field and the Cycadaceae from the Damuda Series. Rec Geol Surv India 10(2):70–73
Feistmantel O (1879a) The fossil flora of the Upper Gondwana. Outliers on the Madras Coast. Mem Geol Surv India Palaeont Indica, Ser 2, 1(4):191–224
Feistmantel O (1879b) The fossil flora of the Lower Gondwana-1. The flora of the Talchir-Karharbari beds. Mem Geol Surv India Palaeont Indica, Ser 12, 3(1):1–48
Feistmantel O (1880a) The fossil flora of the Lower Gondwana-2. The flora of the Damuda and Panchet divisions. Mem Geol Surv India Palaeont Indica, Ser. 12, 3(1):1–77.
Feistmantel O (1880b) The fossil flora of the Lower Gondwana-1 (supplement). The flora of the Talchir–Karharbari beds. Mem Geol Surv India Palaeont Indica, Ser 12, 3(1):49–64
Feistmantel O (1881) The fossil flora of the Gondwana System. The flora of the Damuda–Panchet Division. Mem Geol Surv India Palaeont Indica, Ser 12, 3(3):78–149
Feistmantel O (1882) Fossil flora of Gondwana System in India-1. The fossil flora of the South Rewa Gondwana Basin. Mem Geol Surv India Palaeont Indica, Ser 12, 4(1):1–52.
Feistmantel O (1886) The fossil flora of the Gondwana System-2. The fossil flora of some of the coalfields in western Bengal. Mem Geol Surv India Palaeont Indica, Ser 12, 4(4):1–71
Gee CT (1987) Revision of the Early Cretaceous flora from Hope Bay, Antarctica. XIV International Botanical Congress. Berlin (West), Germany. Abstracts:333
Geinitz HB (1862) Dyas oder die Zechsteinformation und das Rothliegende—Band 2, Die Pflanzen der Dyas und Geologisches: Leipzig, Wilhelm Engelmann, pp 131–342
Gopal V, Jacob C, Jacob K (1957) Stratigraphy and palaeontology of the Upper Gondwana of the Ramnad District on the East Coast. Rec Geol Surv India 84(4):477–496
Greguss P (1961) Die Entdechkung von Urkormophyten aus dem Ordovizium (2). Acta Biologica N.S. 7(1–2):3–30
Greguss P (1962) Some new data on the Ordovician land plants from Poland (3) Acta Biologica N.S. 8(1–4):45–58
Grindley GW (1963) The geology of the Queen Alexandra Range, Beardmore Glacier, Ross Dependency, Antarctica; with notes on the correlation of Gondwana sequences. New Zealand J Geol Geophys 6:307–347
Grindley GW, Warren G (1964) Stratigraphic nomenclature and correlation in the western Ross Sea region. In Adie RJ (ed) Antarctic Geology, North-Holland Publishing Co, Amsterdam, pp 315–333
Halle TG (1913) The Mesozoic flora of Graham Land. Wiss Ergebn Schwed Südpolar-Exped 3(14):1–123
Hjelle A, Winsnes T (1972) The sedimentary and volcanic sequence of Vestfjella. In Adie RJ (ed) Antarctic Geology and Geophysics (SCAR Symp. on Antarctic Geol. and Solid Earth Geophys., Oslo 1970), Universitetetsforlaget, Oslo, pp 539–546
Høeg OA, Bose MN (1960) The *Glossopteris* flora of the Belgian Congo with a note on some fossil plants from the Zambesi Basin (Mozambique). Ann Mus v. Congo Belge 32:1–106
Jain KP (1968) Some plant remains from the Upper Gondwana of East Coast, India. Palaeobotanist 16(2):151–155, 1967
Jana BN, Bose MN (1981) *Hausmannia dichotoma* Dunker and *Pterophyllum princeps* Oldham & Morris from Than, Saurashtra. Geophytology 11(1):41–44
Jefferson TH (1982a) The preservation of fossil leaves in Cretaceous volcaniclastic rocks from Alexander Island, Antarctica. Geol Mag 119: 291–300
Jefferson TH (1982b) Fossil forests from the Lower Cretaceous of Alexander Island, Antarctica. Palaeontology 25:681–708
Jefferson TH (1983) Palaeoclimatic significance of some Mesozoic Antarctic fossil floras. In Oliver RL, James PR, Jago JB (eds) Antarctic Earth Sciences (4th Int. SCAR Symp. on Antarctic Earth Sci., Adelaide, 1982), Cambridge Univ. Press, Cambridge, pp 593–599
Kar RK (1968) Studies in the *Glossopteris* flora of India-36. Plant fossils from Barren Measures succession of Jharia Coalfield, Bihar, India. Palaeobotanist 16(3):243–248, 1967
Kozlowski R, Greguss P (1959) Discovery of Ordovician land plants. Acta Palaeont Polonica 4(1):1–9
Kyle RA (1976) *Plumsteadia ovata* n. sp., a glossopterid fructification from South Victoria Land, Antarctica (Note). New Zealand J Geol Geophys 17:719–721
Lacey WS, Lucas RC (1981a) A Lower Permian flora from the Theron Mountains, Coats Land. Br Antarct Surv Bull 53:153–156
Lacey WS, Lucas R (1981b) The Triassic flora of Livingston Island, South Shetland Islands. Br Antarct Surv Bull 53:157–173
Maheshwari HK (1966) Studies in the *Glossopteris* Flora of India-28. On some fossil woods from the Raniganj Stage of the Raniganj Coalfield, Bengal. Palaeobotanist 15(3):243–257
Maheshwari HK (1968) Studies in the *Glossopteris* flora of India-38. Remarks on *Trizygia speciosa* Royle with reference to the genus *Sphenophyllum* Koenig. Palaeobotanist 16(3):283–287, 1967
Maheshwari HK (1972a) *Lelstotheca:* a new name

for *Stellotheca* Surange & Prakesh. Geophytology 2(1):106
Maheshwari HK (1972b) Permian wood from Antarctica and revision of some Lower Gondwana wood taxa. Palaeontographica B138:1–43
Maheshwari HK (1974) Lower Cretaceous palynomorphs from the Bansa Formation, South Rewa Gondwana Basin, India. Palaeontographica B146:21–55
Maheshwari HK, Banerji J (1978) On a ginkgoalean leaf from the Triassic of Madhya Pradesh. Palaeobotanist 25:249–253, 1976
Maheshwari HK, Kumaran KPM (1976) Some new conifer remains from the Jabalpur Group. Palaeobotanist 23(1):30–39, 1974
Maheshwari HK, Prakash G (1965) Studies in the *Glossopteris* flora of India-21. Plant megafossils from the Lower Gondwana exposures along the Bansloi River in the Rajmahal Hills, Bihar. Palaeobotanist 13(2):115–128, 1964
Maheshwari HK, Singh MP (1976) On some plant fossils from the Pariwar Formation, Jaisalmer Basin, Rajasthan. Palaeobotanist 23(3):116–123
Maithy PK (1965a) Studies in the *Glossopteris* flora of India-17. On the genus *Rubidgea* Tate. Palaeobotanist 13(1):42–44
Maithy PK (1965b) Studies in the *Glossopteris* flora of India-18. Gymnospermic seeds and seed-bearing organs from the Karharbari beds, Giridih Coalfield, Bihar. Palaeobotanist 13:45–56
Maithy PK (1965c) Studies in the *Glossopteris* flora of India-25. Pteridophytic, ginkgoalean remains from the Karharbari beds, Giridih Coalfield, India. Palaeobotanist 13:239–243
Maithy PK (1966) Palaeobotany and stratigraphy of the Karharbari Stage with particular reference to the Giridih Coalfield. In Symposium on Floristics and Stratigraphy of Gondwanaland. Birbal Sahni Institute of Palaeobotany, Lucknow, pp 102–109
Maithy PK (1968) Studies in the *Glossopteris* flora of India-35. On two new fossil plants from the Ganjra Nalla beds, South Rewa Gondwana Basin. Palaeobotanist 16(3):219–221, 1967
Maithy PK (1969) The contemporaneity of the Karharbari flora in Gondwanaland. In Santapau H, Ghosh AK, Roy SK, Chanda S, Chaudhuri SK (eds) J Sen Memorial Volume, Bot Soc Bengal, Calcutta, pp 279–298
Maithy PK (1970) Studies in the *Glossopteris* flora of India-39. On some new plant fossils from the Karharbari beds, Giridih Coalfield, India. Palaeobotanist 18(2):167–172
Maithy PK (1973) *Buriadia sewardii* Sahni: The correct name of *Buriadia heterophylla* Feistmantel. Geophytology 3(1):111–112
Maithy PK (1974a) A revision of Lower Gondwana *Sphenopteris* from India. Palaeobotanist 21(1): 70–80, 1972
Maithy PK (1974b) Studies in the *Glossopteris* flora of India—*Gondwanophyton* gen. nov. with a revision of allied plant fossils from the Lower Gondwana of India. Palaeobotanist 21(3):298–304, 1972
Maithy PK (1974c) *Dichotomopteris*, a new type of fern frond from the Lower Gondwana of India. Palaeobotanist 21(3):365–367, 1972
Maithy PK (1975) Some contributions to the knowledge of Indian Lower Gondwana ferns. Palaeobotanist 22(1):29–38, 1973
Maithy PK (1977) Three new fern fronds from the *Glossopteris* flora of India. Palaeobotanist 24(2): 96–101
Oldham T, Morris J (1863) Fossil flora of the Rajmahal Hills. Mem Geol Surv India Palaeont Indica, Ser 2, 1(1):1–52
Orlando HA (1968) A new Triassic flora from Livingston Island, South Shetland Islands. Br Antarct Surv Bull 16:1–13
Pal PK (1984a) Fragmentary plant remains from the Hartala Hill, South Rewa Gondwana Basin, India. Palaeobotanist 32(2):126–129, 1983
Pal PK (1984b) Triassic plant megafossils from the Tiki Formation, South Rewa Gondwana Basin, India. Palaeobotanist 32(3):253–309, 1983
Pant DD, Basu N (1973) *Pteruchus indicus* sp. nov. from the Triassic of Nidpur, India. Palaeontographica B144:11–24
Pant DD, Basu N (1978) On two structurally preserved bryophytes from the Triassic of Nidpur, India. Palaeobotanist 25:340–352
Pant DD, Basu N (1981) Further contributions on the non-vascular cryptogams from the Middle Gondwana (Triassic) beds of Nidpuri, India. Part II. Palaeobotanist 28–29:188–200, 1979–1980
Pant DD, Bhatnagar S (1975) A new kind of foliage shoot *Searsolia oppositifolia* gen. et sp. nov. from Lower Gondwanas of Raniganj Coalfield, India. Palaeontographica B152:191–199
Pant DD, Gupta KL (1968) Cuticular structure of some Indian Lower Gondwana species of *Glossopteris* Brongniart. Part I. Palaeontographica B124:45–81
Pant DD, Gupta KL (1971) Cuticular structure of some Indian Lower Gondwana species of *Glossopteris* Brongniart. Part II. Palaeontographica B132:130–152
Pant DD, Khare PK (1974) *Damudopteris*—a new genus of ferns from the Lower Gondwanas of the Raniganj Coalfield, India. Proc. R. Soc. Lond B186:121–135
Pant DD, Kidwai P (1968) On the structure of stems and leaves of *Phyllotheca indica* Bunbury and its affinities. Palaeontographica B121:102–121
Pant DD, Mehra B (1963a) On the epidermal structure of *Sphenophyllum speciosum* (Royle) Zeiller. Palaeontographica B112:51–57
Pant DD, Mehra B (1963b) On a cycadophyte leaf, *Pteronilssonia gopalii* gen. et sp. nov., from the

Lower Gondwanas of India. Palaeontographica B113:126–134
Pant DD, Misra L (1976) Compression of a new type of pteridophyll, *Asansolia* gen. nov. from the Lower Gondwana of the Raniganj Coalfield, India. Palaeontographica B155:129–139
Pant DD, Misra L (1977) On two genera of pteridophylls, *Damudosorus* gen. nov. and *Trithecopteris* gen. nov. from the Lower Gondwana of the Raniganj Coalfield. Palaeontographica B164:76–86
Pant DD, Misra L (1983) *Cuticulatopteris* gen. nov. and some other pteridophylls from the Raniganj Coalfield, India (Lower Gondwana). Palaeontographica B185:27–37
Pant DD, Nautiyal DD (1967a) On the structure of *Raniganjia bengalensis* (Feistmantel) Rigby with discussion of its affinities. Palaeontographica B121:52–64
Pant DD, Nautiyal DD (1967b) On the structure of *Buriadia heterophylla* (Feistmantel) Seward & Sahni and its fructification. Phil Trans Roy Soc London 252B(774):27–48
Pant DD, Nautiyal DD (1984) On the morphology and structure of *Ottokaria zeilleri* sp. nov.—a female fructification of *Glossopteris*. Palaeontographica B172:127–152
Pant DD, Nautiyal DD (1987) *Diphyllopteris verticillata* Srivastava, the probable seedling of *Glossopteris* from the Palaeozoic of India. Rev. Palaeobot Palynol 51:31–36
Pant DD, Singh KB (1968) On the genus *Gangamopteris* McCoy. Palaeontographica B124:83–99
Pant DD, Singh KB (1971) Cuticular structure of some Indian Lower Gondwana species of *Glossopteris* Brongniart. Part 3. Palaeontographica B135:1–40
Pant DD, Singh RS (1974) On the stem attachment of *Glossopteris* and *Gangamopteris* leaves.Part II—Structural features. Palaeontographica B147: 42–73
Pant DD, Singh S (1979) *Caulophyllites indica* gen. et sp. nov. from the Giridih Coalfield, India. Palaeontographica B169:107–115
Pant DD, Singh VK (1987) Xylotomy of some woods from Raniganj Formation (Permian), Raniganj Coalfield, India. Palaeontographica B203:1–82
Pant DD, Srivastava GK (1964) On *Walkomiellospermum indicum* gen. et sp. nov., seed-like bodies and alete megaspores from Talchir Coalfield, India. Proc Nat Inst Sci India 29B(6):575–584, 1963
Pant DD, Srivastava GK (1977) On the structure of *Gleichenia rewahensis* Feistmantel and allied fossils from the Jabalpur Series, India. Palaeontographica B163:152–161
Pant DD, Verma BK (1963) On the structure of leaves of *Rhabdotaenia* Pant from the Raniganj Coalfield, India. Palaeontology 6:301–314
Pant DD, Misra L, Nautiyal DD (1982) On the structure of stems and leaves of *Schizoneura gondwanensis* Feistmantel. Palaeontographica B183:1–7
Pant DD, Srivastava GK, Pant R (1983) On the cuticular structure of leaves of *Desmiophyllum* type from Bansa beds, India and their assignment to genus *Harrisiophyllum* nov. Palaeontographica B185:38–55
Pareek HS (1969) Some observations on the lithology of coal-bearing beds and the nature of coal of the major Gondwana basins of India. In Amos AJ (ed) Gondwana Stratigraphy (1st IUGS Gondwana Symp., Buenos Aires, 1967), UNESCO, Paris, 2 (Earth Sciences), pp 883–903
Patra BP (1973) Notes on some Upper Gondwana plants from the Athgarh sandstones, Cuttack District, Orissa. Palaeobotanist 20(3):325–333, 1971
Plumstead EP (1962) Fossil floras of Antarctica (with an appendix on Antarctic fossil woods by Richard Kräusel). Trans-Antarctic Exped 1955–1958, Sci Rep Geol 9:1–154
Plumstead EP (1964) Palaeobotany of Antarctica. In Adie RJ (ed) Antarctic Geology (Proc. 1st Int. SCAR Symp. on Antarctic Geol., Cape Town, 1963), North-Holland Publishing Company, Amsterdam, pp 637–654
Plumstead EP (1965) Glimpses into the history and prehistory of Antarctica. Antarktiese Bulletin 9:1–7
Plumstead EP (1975) A new assemblage of plant fossils from Milorgfjella, Dronning Maud Land. Brit Antarct Surv Sci Rep 83:1–30
Prasad MNV (1982) An annotated synopsis of Indian Palaeozoic gymnospermous woods. Rev Palaeobot Palynol 38:119–156
Prasad MNV, Chandra S (1981) Two species of *Australoxylon* from the Kamthi Formation of Chandrapur district, Maharashtra. Geophytology 11(1):1–5
Rees PM (1987) A diverse Middle–Lower Cretaceous flora from the northern Antarctic Peninsula. XIV International Botanical Congress. Berlin (West), Germany. Abstracts:287
Rigby JF, Schopf JM (1967) Stratigraphic implications of Antarctic paleobotanical studies. In Amos AJ (ed) Gondwana Stratigraphy (1st IUGS Gondwana Symp., Buenos Aires, 1967), UNESCO,Paris, 2 (Earth Sciences), pp 91–106
Roy SK (1968) Pteridophytic remains from Kutch and Kathiawar, India. Palaeobotanist 16(2): 108–114, 1967
Roy Chowdhury MK, Sastry MVA, Shah SC, Singh G, Ghosh SC (1975) Triassic floral succession in the Gondwana of Peninsular India. In Campbell KSW (ed) Gondwana Geology (papers presented at the Third Gondwana Symposium, Canberra, Australia, 1973), Australian National University Press, pp 149–158

Sah SCD, Maheshwari HK (1969) Plant megafossils from the Bansloi valley, Rajmahal Hills, Bihar, India, with remarks on the position of the Dubrajpur Group. In Santapau H, Ghosh AK, Roy SK, Chanda S, Chaudhuri SK (eds) J Sen Memorial Volume, Bot Soc Bengal, Calcutta, pp 369–374
Sahni B (1932) A petrified *Williamsonia* (*W. sewardiana,* sp. nov.) from the Rajmahal Hills, India. Mem Geol Surv India Palaeont Indica N.S. 20(3):1–19
Sahni B (1936) The occurrence of *Matonidium* and *Weichselia* in India. Rec Geol Surv India 71(2):152–175
Schopf JM (1962) A preliminary report on plant remains and coal of the sedimentary section in the central range of the Horlick Mountains, Antarctica. Inst Polar Studies Ohio State Univ Report 2:1–61
Schopf JM (1965) Anatomy of the axis in *Vertebraria*. In Hadley JD (ed) Geology and Paleontology of the Antarctic, Amer. Geophys. Union, Antarctic Res. Ser. 6:217–228
Schopf JM (1966) Antarctic paleobotany and palynology. Antarctic Journal of the U.S. 1(4):135
Schopf JM (1967) Antarctic fossil plant collecting during the 1966–1967 season. Antarctic Journal of the U.S. 2(4):114–116
Schopf JM (1968) Studies in Antarctic paleobotany. Antarctic Journal of the U.S. 3(5):176–177
Schopf JM (1970a) Antarctic collections of plant fossils, 1969–1970. Antarctic Journal of the U.S. 5:89–90
Schopf JM (1970b) Petrified peat from a Permian coal bed in Antarctica. Science 169:276–277
Schopf JM (1971) Notes on plant tissue preservation and mineralization in a Permian deposit of peat from Antarctica. Amer J Sci 271:522–543
Schopf JM (1976) Morphologic interpretation of fertile structures in glossopterid gymnosperms. Rev Palaeobot Palynol 21:25–64
Seward AC (1914) Antarctic fossil plants. Brit. Antarct Terra Nova Exped (Geol) 1:1–49
Seward AC (1917) Fossil Plants, Volume III. Cambridge University Press, Cambridge
Seward AC, Sahni B (1920) Indian Gondwana plants: a revision. Mem Geol Surv India Palaeont Indica, N.S. Ser. 7(1):1–41
Shah SC, Singh G, Gururaja MN (1973) Observations of the post-Triassic Gondwana Sequence of India. Palaeobotanist 20(2):221–237
Singh G, Maithy PK, Bose MN (1982) Upper Palaeozoic flora of Kashmir Himalaya. Palaeobotanist 30(2):185–232, 1981
Singh KJ, Chandra S (1987) Some new species of *Glossopteris* from the Kamthi Formation of Handappa, Orissa. Geophytology 17(1):39–55
Singh VK, Srivastava AK, Maheshwari HK (1987) Sphenopsids from the Barakar Formation of Hura Tract, Rajmahal Hills, Bihar. Palaeobotanist 35(3):236–241, 1986
Skinner DNB, Ricker J (1968) The geology of the region between the Mawson and Priestley Glaciers, North Victoria Land, Antarctica. Part II—Upper Palaeozoic to Quaternary geology. New Zealand J Geol Geophys 11:1041–1075
Srivastava AK, Chandra S (1982) Pteridophytic remains from the Selected Searsole Colliery, Raniganj Coalfield, West Bengal, India. Geophytology 12(1):95–104
Srivastava AK, Rigby JF (1983) *Sphenophyllum, Trizygia* and *Gondwanophyton* from Barakar Formation of Raniganj Coalfield, with a revision of Lower Gondwana Sphenophyllales. Geophytology 13(1):55–62
Srivastava GK, Nautiyal DD, Pant DD (1984) Some coniferous shoots from Bansa beds of the Jabalpur Formation (Lower Cretaceous). Palaeontographica B194:131–150
Srivastava SC (1971) Some gymnospermic remains from the Triassic of Nidpur, Sidhi District, Madhya Pradesh. Palaeobotanist 18(3):280–296, 1969
Srivastava SC (1974) Pteridospermic remains from the Triassic of Nidpur, Madhya Pradesh, India. Geophytology 4(1):54–99
Srivastava SC (1976) Some macroplant fossils from the Triassic rocks of Nidpur, India. Palaeobotanist 23(1):44–48, 1974
Srivastava SC (1984a) *Sidhiphyllites:* a new ginkgophytic leaf genus from the Triassic of Nidpur, India. Palaeobotanist 32(1):20–25, 1983
Srivastava SC (1984b) *Lelestrobus:* a new microsporangiate organ from the Triassic of Nidpur, India. Palaeobotanist 32(1):86–90, 1983
Srivastava SC (1984c) New leaf compression from the Triassic of Nidpur, India. Geophytology 14(2):199–207
Srivastava SC, Maheshwari HK (1973) *Satsangia,* a new plant organ from the Triassic of Nidhpuri, Madhya Pradesh. Geophytology 3(2):222–227
Suk-Dev (1970) Some ferns from the Lower Cretaceous of Madhya Pradesh-1. Palaeobotanist 18(2):197–207, 1970
Sukh-Dev (1971) Ferns from the Cretaceous of Madhya Pradesh–2. Palaeobotanist 19(3):277–280
Sukh-Dev (1972) Ferns from the Cretaceous of Madhya Pradesh–3. Palaeobotanist 19(3):281–283, 1970
Sukh-Dev, Zeba-Bano (1977) Three species of *Ptilophyllum* from Bansa, Madhya Pradesh. Palaeobotanist 24(3):161–169, 1975
Sukh-Dev, Zeba-Bano (1978) *Araucaria indica* and two other conifers from the Jurassic–Cretaceous rocks of Madhya Pradesh, India. Palaeobotanist 25:496–508, 1976
Sukh-Dev, Zeba-Bano (1979) Observations on the genus *Allocladus,* and its representatives in the

Jabalpur Formation. Palaeontographica B169: 116–121
Sukh-Dev, Zeba-Bano (1981a) Some fossil gymnosperms from the Satpura Basin, Madhya Pradesh, India. Palaeobotanist 27(1):1–11, 1977
Sukh-Dev, Zeba-Bano (1981b) Occurrence of the genus *Ctenozamites* in the Jabalpur Formation. Palaeobotanist 28–29:324–328, 1979–1980
Surange KR (1975) Indian Lower Gondwana floras: a review. In Campbell KSW (ed) Gondwana Geology (papers presented at the Third Gondwana Symposium, Canberra, Australia, 1973), Australian National University Press, pp 135–147
Surange KR, Chandra S (1973a) *Dictyopteridium sporiferum* Feistmantel—a female cone from the Lower Gondwana of India. Palaeobotanist 20(1):127–136, 1971
Surange KR, Chandra S (1973b) *Denkania indica* gen. et sp. nov. A glossopteridian fructification from the Lower Gondwana of India. Palaeobotanist 20(2):264–268
Surange KR, Chandra S (1973c) *Partha*, a new type of female fructification from the Lower Gondwana of India. Palaeobotonist 20(3):356–360
Surange KR, Chandra S (1974a) Fructifications of Glossopteridae from India. Palaeobotanist 21(1): 1–17, 1972
Surange KR, Chandra S (1974b) *Lidgettonia mucronata* sp. nov.: a female fructification from the Lower Gondwana of India. Palaeobotanist 21(1):121–125, 1972
Surange KR, Chandra S (1974c) Further observations on *Glossotheca* Surange & Maheshwari, a male fructification of Glossopteridales. Palaeobotanist 21(2):248–254
Surange KR, Chandra S (1974d) Some male fructifications of Glossopteridales. Palaeobotanist 21(2):255–266, 1972
Surange KR, Kulkarni S (1968) On two new species of *Phyllotheca* from the South Karanpura Coalfield, Bihar. Palaeobotanist 16(1):95–100, 1967
Surange KR, Lele KM (1955) Studies in the *Glossopteris* flora of India–3. Plant fossils from the Talchir needle shales from Giridih coalfield. Palaeobotanist 4:153–157
Surange, KR, Lele KM (1956) Studies in the *Glossopteris* flora of India–6. Plant fossils from Talchir beds of South Rewa Gondwana Basin. Palaeobotanist 5(2):82–90
Surange KR, Maheshwari KH (1970) Some male and female fructifications of Glossopteridales from India. Palaeontographica B129:178–192
Surange KR, Prakash F (1960) Studies in the *Glossopteris* flora of India–12. *Stellotheca robusta* nov. comb: a new equisetaceous plant from the Lower Gondwanas of India. Palaeobotanist 9(1–2):49–52, 1960
Surange KR, Singh P (1951) *Walkomiella indica*, a new conifer from the Lower Gondwanas of India. J Indian Bot Soc 30(1–4):143–147
Surange KR, Singh P (1953) The female dwarf shoots of *Walkomiella indica*—a conifer from the Lower Gondwana of India. Palaeobotanist 2:5–8
Taylor BJ (1971) Thallophyte borings in phosphatic fossils from the Lower Cretaceous of southeast Alexander Island, Antarctica. Paleontology 14:294–302
Taylor BJ, Thomson MRA, Willey LE (1979) The geology of the Ablation Point to Keystone Cliffs area, Alexander Island. Sci Rept Br Antarct Surv 82:1–65
Townrow JA (1967a) On a conifer from the Jurassic of East Antarctica. Proc Roy Soc Tasmania 101:137–146
Townrow JA (1967b) Fossil plants from Allan and Carapace Nunataks, and from the Upper Mill and Shackleton Glaciers, Antarctica. New Zealand J Geol Geophys 10:456–473
Venkatachala BS (1986) Palaeobotany in India—Quo vadis? Geophytology 16(1):1–24
Vishnu-Mittre (1959) Studies on the fossil flora of Nipania, Rajmahal Series, India—Pteridophyta, and general observations on Nipania fossil flora. Palaeobotanist 7(1):47–66
White, ME (1973) Permian flora from the Beaver Lake area, Prince Charles Mountains, Antarctica. 2. Plant fossils. Austral Bur Min Res Geol Geophys (Bull 126), Palaeont Papers 1969:13–18
Zeba-Bano (1980) Some pteridophytic remains from the Jabalpur Formation. Palaeobotanist 26(3):237–247, 1977

11—Structurally Preserved Permian and Triassic Floras from Antarctica*

Edith L. Taylor and Thomas N. Taylor

Introduction

The Permian and Triassic floras of Gondwana are known almost exclusively from impression/compression specimens. Generally, Permian floras from the southern hemisphere demonstrate a low diversity and are often poorly preserved as impressions. By comparison, floras of Triassic age display much greater diversity and are often better preserved, some with cuticular remains. The importance of preservation in the interpretation of floral elements is perhaps best demonstrated by the controversy surrounding the organization and understanding of the reproductive organs of the Glossopteridales (e.g., Pant 1987; Surange and Chandra 1975; Gould and Delevoryas 1977; Taylor 1987). The reconstruction of reproductive organs in this group is based almost completely on impression remains. For example, *Dictyopteridium* has been interpreted by some as representing a three-dimensional cone bearing helically arranged ovules (Surange and Chandra 1975; Rigby 1978); others believe it should more accurately be depicted as a flattened, somewhat fleshy structure with bilateral symmetry (e.g., Pant 1987). With the discovery of anatomically preserved glossopterid remains from the Bowen Basin of Australia, Gould and Delevoryas (1977) showed that at least one of the reproductive organs borne by the glossopterids consisted of a partially inrolled leaf bearing ovules on one surface. On the basis of this evidence, these authors classified the glossopterids with the seed ferns. Although much remains to be learned about the structure and organization of glossopterid reproductive organs known from impression/compression fossils, the silicified material from Bowen Basin graphically demonstrates the value of three-dimensionally preserved plant fossils in interpreting complex morphologies. Our extensive knowledge of Upper Carboniferous coal ball floras, including the reproductive biology and development of many of the plants, is based on the excellent anatomical preservation of these floral elements.

In Antarctica, as elsewhere in Gondwana, Permian and Triassic floras are generally represented by compression/impression remains (e.g., Plumstead 1962; Lacey and Lucas 1981). The floras in some areas have been reported to be very poorly preserved (e.g., Schopf 1962), and in many cases, this may be due to the proximity of extensive volcanism. However, cuticle has been reported from a Triassic site in the central Transantarctic Mountains (Taylor and Taylor 1988). No doubt due to the proximity of volcanic sediments, silicified wood is fairly widespread and there are several sites that have yielded permineralized peat and other examples of anatomically preserved specimens.

---

* Contribution No. 669 of the Byrd Polar Research Center (Ohio State University).

Permineralized Wood

It is difficult to determine the earliest reports of silicified wood from Antarctica, although its presence has been noted in many geological contributions. Variously preserved fragments occur at scattered localities both on the continent (Fig. 11.1) and in the vicinity of the Antarctic peninsula (Fig. 11.2). Wood fragments are known that range in age from Permian to Tertiary.

Permian Wood

Petrified wood of Permian age has been noted at many sites in Antarctica. These extend throughout the Transantarctic Mountains, including northern Victoria Land (Maheshwari 1972), southern Victoria Land (Seward 1914; Kräusel 1962; Matz et al. 1972; Maheshwari 1972), Queen Maud Mountains (Doumani and Minshew 1965; Maheshwari 1972), Queen Alexandra Range (Grindley 1963; Grindley et al. 1964; Barrett 1972; Jefferson and Taylor 1983; Barrett et al. 1986), and the Horlick Mountains (Schopf 1962; Doumani and Long 1962; Long 1965; Maheshwari 1972) (Fig. 11.1). In East Antarctica, petrified wood has also been found in association with the compression/impression floras of western Queen Maud Land (Plumstead 1975) and the Prince Charles Mountains (White 1973). In West Antarctica, Permian wood was reported in the Falkland Islands (=Las Islas Malvinas)

Figure 11.1. Map of Antarctica showing principal plant localities. The Ohio Range is part of the Horlick Mountains.

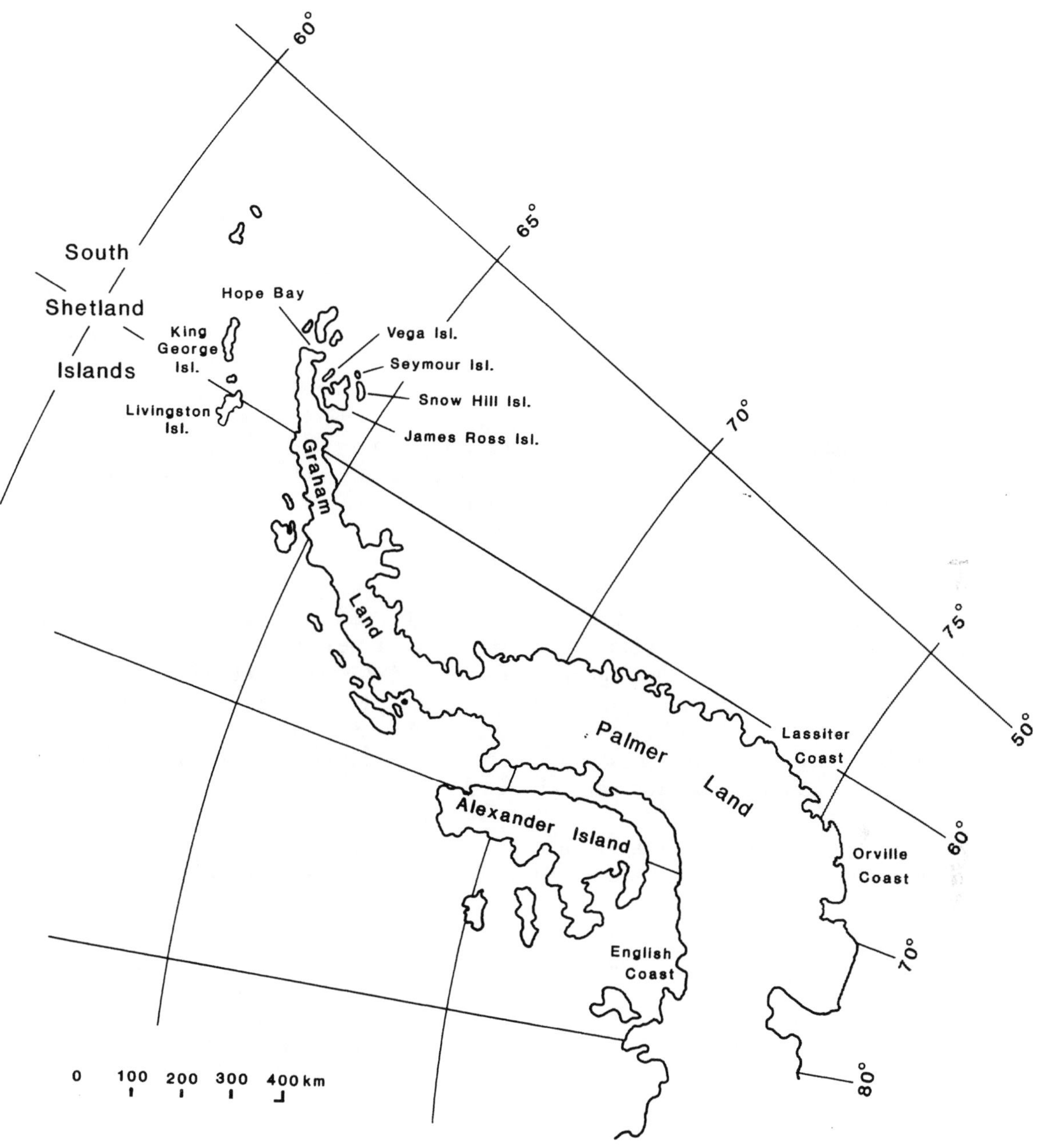

Figure 11.2. Map of Antarctic peninsula and surrounding islands showing principal plant localities.

flora by Halle (1911) and Seward and Walton (1923). Many of these contributions, however, consider the geology of a particular region and as such only casually note the presence of wood and other plant fossils (e.g., Grindley et al. 1964; Doumani and Minshew 1965; Barrett 1972; Matz et al. 1972; Barrett et al. 1986, and others). Most of the wood has never been described in detail.

One of the earliest studies of Permian wood from Antarctica was that of Seward (1914). He described a poorly preserved specimen that came from an erratic boulder collected along the Priestley Glacier (northern Victoria Land)

(Fig. 11.1) on Scott's "Terra Nova" expedition in 1910. Since the pith and primary xylem were not preserved, the wood could not be allied with any existing Paleozoic wood genus. Seward therefore instituted a new genus for this material, *Antarcticoxylon*, which has subsequently been used as a repository for poorly preserved Antarctic wood that lacks important diagnostic features.

Edwards (1928) later reexamined some of the material collected on the National Antarctic Expedition of 1901–1904 (originally described by Arber 1907) from the Ferrar Glacier region (southern Victoria Land, Fig. 11.1) and placed these woody fragments into another form genus, *Dadoxylon*. In 1962, Kräusel described petrified wood that had been collected on the Trans-Antarctic Expedition of 1955–1958 (compression/impression flora from this expedition detailed by Plumstead 1962). Only two specimens were sufficiently preserved so that their identity could be determined, and both were found in coal beds on Allan Nunatak (southern Victoria Land). Pith and primary xylem were preserved in *Taeniopitys scottii*, and Kräusel noted the presence of centripetal wood in the pith region. *Dadoxylon allanii* consisted entirely of crushed secondary xylem.

The most comprehensive study of Permian Antarctic woods is the work of Maheshwari (1972). This material came from a variety of sites throughout the Transantarctic Mountains (see Long 1965 for an initial description of some specimens). Maheshwari described 10 taxa, of which 3 represented new species and 4 new combinations. These included 3 species of *Araucarioxylon*, 2 from Mt. Schopf (Ohio Range, Horlick Mountains, Fig. 11.1), and 1 from southern Victoria Land [*A. allanii* (Kräusel) comb. nov.]; 1 species of *Dadoxylon* [*D. weaverense* from Mt. Weaver in the Queen Maud Mts. (central Transantarctic Mountains)]; *Protophyllocladoxylon dolianitii*, *Damudoxylon*, *Megaporoxylon antarcticum*, *M. canalosum*, and *Polysolenoxylon kraeuselii* from the Ohio Range (Horlick Mts.); and ?*Antarcticoxylon* sp. from the Mt. Nansen Massif (northern Victoria Land). Schopf had earlier described *Antarcticoxylon* cf. *A. priestleyi*, also from the Horlick Mountains (Schopf 1962). Some of Maheshwari's material was silicified and some preserved in limonite, but all exhibited distinct growth rings that ranged from 1.0 mm wide (in *Damudoxylon*) to 9.0 mm in *Polysolenoxylon*.

In addition to Maheshwari's detailed study, White (1973) and Plumstead (1975) also described woody fragments from the Prince Charles Mountains and western Queen Maud Land, respectively. White was able to distinguish growth rings and pitting in her 2 specimens, but did not give the material a generic designation. Plumstead's specimens were very poorly preserved and were classified on external features only.

More recently, Jefferson and Taylor (1983) examined wood from the central Transantarctic Mountains as a means of obtaining paleoclimate data (also see the following section, *Mesozoic Wood*). They found growth rings up to 8.0 mm wide in *Araucarioxylon*. Mean sensitivity, a measurement of water availability (Fritts 1976) was also calculated for these plants. Based on these and additional sedimentological data, a warm, ice-free summer climate was suggested for this area during the Permian.

In the peninsula area, Halle (1911) and Seward and Walton (1923) described wood fragments from the Falkland Islands. Halle named two new taxa, based on specimens from Darwin Harbor, East Falkland; *Dadoxylon lafoniense* and *D.* cf. *angustum*. *D.* cf. *angustum* consisted only of secondary xylem, but *D. lafoniense* included a well-preserved pith and primary body. Seward and Walton (1923) later delineated another species of *Dadoxylon* from Choiseul Sound, East Falkland, as *D. bakeri*. These authors described *D. bakeri* as closely related to both *D. angustum* and *D. lafoniense*. All three taxa exhibited marked growth rings.

Mesozoic Wood

Triassic woods have been noted in Beacon rocks of the Transantarctic Mountains from Victoria Land (Skinner and Ricker 1968; Matz et al. 1972) to the Queen Alexandra Range

(Beardmore Glacier area, Fig. 11.1) (Barrett 1972; Jefferson and Taylor 1983; Barrett et al. 1986). In situ stumps have been noted in the vicinity of Fremouw Peak (Beardmore Glacier area, Barrett et al. 1986). These woods are typically silicified and exhibit well-developed growth rings (e.g., Jefferson and Taylor 1983; Barrett et al. 1986; Taylor and McCallister 1989); however, none have previously been described in detail. The woody axes present in the silicified peat from Fremouw Peak (see later section, *Triassic Peat*) are currently being studied for general anatomy and systematic affinities (Meyer-Berthaud and Taylor 1989) and for growth ring parameters as a measure of climatic sensitivity (Taylor and McCallister 1989).

Due no doubt to volcanism during the Jurassic, fossils of this age are rare in the Transantarctic Mountains. However, Jefferson et al. (1983) noted the presence of tree trunks that had been engulfed by lava flows in northern Victoria Land. This wood, which compared to *Protocupressinoxylon,* was described as containing "possible growth rings," although the authors noted that some cell rows crossed the ring boundaries. Recent reexamination of some of these woods shows extensive crushing of the secondary xylem, giving the appearance of false growth rings. No true rings could be distinguished and growth appears for the most part to have been continuous.

The majority of Mesozoic woods from Antarctica are known from the peninsula area (e.g., Laudon et al. 1983, 1985; Gee 1984) (Fig. 11.2), its surrounding islands (Lacey and Lucas 1981; Jefferson 1982, 1983, 1987; Francis 1986; Fig. 11.2), and the subantarctic islands north and east of the peninsula (Gordon 1930; Jefferson and MacDonald 1981). As was the case with Permian wood localities, some of these reports represent geologic works in which plant fossils are only mentioned (e.g., Laudon et al. 1983, 1985; Matz et al. 1972). To date, more structural and especially paleoenvironmental work has been completed on Mesozoic rather than on Paleozoic woods, most notably the contributions of Jefferson on Late Cretaceous material from Alexander Island (Fig. 11.2) (Jefferson 1982, 1983, 1987). He studied in situ fossil forests on Alexander Island as paleoclimate indicators. The wood was described as similar to *Circoporoxylon* Kräusel although no formal description was given. The trees exhibited distinct growth rings with a gradual transition from early to late wood. The size and structure of growth rings and the spacing of in situ stumps was used to reconstruct the Late Cretaceous forest. Based on this reconstruction and paleolatitude information, the problem of maintaining plant growth at such high latitudes was discussed (see also Creber and Chaloner 1984). More recently, Francis (1986) examined Cretaceous woods from James Ross, Vega, and Livingston islands for paleoclimate inferences based on growth-ring data, including ring width and mean sensitivity. However, none of these specimens represented in situ remains, and some specimens were recovered from marine sediments, so the possibility of driftwood cannot be ruled out.

Tertiary Wood

Rocks of Tertiary age occur on many of the peninsular islands as well as some subantarctic islands, and a number of these localities have yielded angiosperm-dominated leaf floras (e.g., Dusén 1908; Zastawniak et al. 1985). Permineralized wood is known from Seymour Island (Sharman and Newton 1894, 1898; Gothan 1908), Snow Hill Island (Gothan 1908), King George Island (Cortemiglia et al. 1981; Lucas and Lacey 1981; Torres 1984, Torres et al. 1984; Torres and Lemoigne 1988), Livingston Island (Lemoigne and Torres 1988), and Kerguelen Island in the Indian Ocean (Edwards 1921). Well-preserved wood specimens, including large trunks, were originally detailed by Halle (1912) from West Point Island in the Falkland Islands (Las Islas Malvinas). Based on their excellent preservation, he suggested a Quaternary age for the material. Based on sedimentologic relationships, however, Birnie and Roberts (1986) have recently described this deposit as Tertiary. Wood found there has been assigned to *Nothofagus*.

A number of the Tertiary wood floras listed above consist entirely of coniferous wood. However, it should be noted that none of the Antarctic floras, of any age, have been extensively sampled and adequately investigated. Lucas and Lacey (1981) described a flora of 11 specimens of probable early Tertiary age from King George Island. Only 3 of these were well-preserved enough for identification. These included *Dadoxylon pseudoparenchymatosum, D. kellerense*, and *D.* sp. Edwards (1921) described 2 coniferous woods from Kerguelen Island (50°S, 70°E) as *Cupressinoxylon antarcticum* and *D. kerguelense*.

Other wood floras include both angiosperms and gymnosperms. Cortemiglia et al. (1981) described two silicified woods found in a moraine on King George Island (Fig. 11.2). One has affinities with the Fagaceae with characters intermediate between *Fagus* and *Nothofagus*, and the other is compared to extant *Araucaria* wood based on its structure. The flora is thought to be Upper Oligocene to Lower Miocene in age. Torres (1984; Torres et al. 1984) also described wood from King George Island, including *Cupressinoxylon, Araucarioxylon, Podocarpoxylon, Nothofagoxylon,* and other dicots. However, in their recent study of petrified woods from Chile, Nishida et al. (1988a) consider *N. antarcticus* Torres (1984) as a synonym of *N. scalariforme* Gothan (1908). More recently, Torres and Lemoigne (1988) described an assemblage of wood specimens from Admiralty Bay, King George Island. These included *Phyllocladoxylon,* 2 species of *Araucarioxylon,* and 2 new species of *Nothofagoxylon.* They compared this assemblage with the Tertiary woods from Seymour Island, as well as those known from Chile and Argentina.

The most extensive contribution on Tertiary Antarctic woods is Gothan's (1908) work on fossil wood from Seymour and Snow Hill Islands (Fig. 11.2) collected during the Swedish South Polar Expedition of 1901–1903. He delineated 3 new species of gymnospermous wood and 3 angiospermous taxa. Of the gymnosperms, 2 were comparable to podocarpaceous types (*Phyllocladoxylon antarcticum* and *Podocarpoxylon parenchymatosum*), and 1 was regarded as araucarian based on its anatomy [*Dadoxylon* (*Araucaria*) *pseudoparenchymatosum*]. The angiosperm woods were lauraceous (*Laurinoxylon uniseriatum* and *L.* sp.) and fagaceous (*Nothofagoxylon scalariforme*) (see also Nishida et al. 1988b). All were relatively well preserved and exhibited distinct growth rings. As in comparable aged leaf floras from high southern latitudes, the wood floras commonly contain podocarps, araucarians, and *Nothofagus* relatives.

Perhaps the most interesting Tertiary wood discovery in Antarctica was that made during the 1985–1986 field season by Webb and Harwood (1987). They discovered wood fragments within the Late Cenozoic Sirius Formation in the Dominion Range (Beardmore Glacier area) of the central Transantarctic Mountains. These were sectioned and found to represent a species of *Nothofagus* (Carlquist 1987). Because of patchy preservation, the wood cannot be placed in an extant species, but it is considered to be close to *N. betuloides* and *N. gunnii. Nothofagus betuloides* occurs today in southern Chile and Argentina, while *N. gunnii* is found only in alpine Tasmania. A low-diversity palynoflora had previously been found at this same locality (Askin and Markgraf 1986). Based on a diatom age of Late Pliocene, these floral fragments are assumed to represent the last presence of a flora that apparently existed during a brief warm period prior to the final climatic cooling in the early Pleistocene.

Permineralized Peat

Despite the relatively widespread occurrence of silicified wood in Antarctica, it is surprising that more silicified plant organs have not been reported. It was not until 1970 that Schopf first mentioned an occurrence of permineralized peat of Permian age from the central Transantarctic Mountains (Schopf 1970a, b). During the same field season (1969–1970), Schopf also collected silicified peat of Early–Middle Triassic age from a col north of Fremouw Peak (Schopf 1970b). In 1978, he described a fern from this peat as *Osmundacaulis beardmorensis*. Schopf also discussed various aspects of mineralization and preservation for these peat occurrences (e.g., Schopf 1971; 1977). Although he realized the significance of the petrified plants (e.g., Schopf 1970a), his untimely passing prevented him from undertaking a systematic examination of the floral elements from these two sites. Extensive collections of permineralized peat from both sites were made during the 1985–1986 field season (Taylor et al. 1986a; 1986b). In the following sections, we present the floras from these two localities as they are presently understood and the implications that these plants have on our understanding of certain groups.

Permian Peat

Silicified peat of Permian age occurs as erratics at the eastern end of Skaar Ridge in the central Transantarctic Mountains (84° 47′S, 163° 15′E, Buckley Island Quadrangle, Barrett and Elliot 1973). Although the peat is known to occur within the Buckley Formation, it has not yet been found in situ; only occurring as scattered boulders that could have been emplaced by previous glacial action. Based on the occurrence of mixed sandstone–peat boulders, however, it is believed that permineralization took place within a fluvial system (Taylor et al. 1986a; 1989). The Buckley Formation has been variously dated from Early to Late Permian (Barrett et al. 1986), but a recently completed study of palynomorphs from Buckley deposits in the same area suggests an age of Late Permian (equivalent to Australian Stage 5; Farabee et al. 1989).

The flora from the Skaar Ridge locality is comparable to those of similar age throughout Gondwana, consisting almost exclusively of glossopterid remains. These include leaves (Figs. 11.3, and 11.5), stems with abundant secondary growth, *Vertebraria* roots (Fig. 11.7), and a variety of reproductive organs. The most common floral element at this site is *Glossopteris*. Two different species are present, and these have been described in detail by Pigg and Taylor (1985; 1987; 1989). In addition to the vascular plant remains, one of the interesting features of this peat is the abundance of several types of fungi, including wood-rotting forms (Stubblefield and Taylor 1985; 1986; Taylor and Stubblefield 1987) (Fig. 11.7) and the occurrence of a small, well-preserved moss gametophyte (Smoot and Taylor 1986a) (Fig. 11.6). The wood-rotting fungi occur in *Antarcticoxylon*-type wood as well as in *Vertebraria* axes (Fig. 11.7) and produce decay patterns that are similar to those formed by modern white rot and white pocket rot fungi, both of which involve basidiomycetes. Associated with the decay areas in the fossil wood are septate hyphae with simple and medallion clamp connections (Stubblefield and Taylor 1986; Taylor and Stubblefield 1987). The moss remains (*Merceria augustica*) consist of unistratose leaves and small axes bearing rhizoids. The leaves bear a slightly thickened midrib region and can be up to several millimeters in width.

A number of reproductive organs also occur in the Permian peat. These include pollen sacs with striate, bisaccate grains and at least two types of dispersed ovules. One of these ovules was first noted by Schopf (1976) and compared to *Nummulospermum*, a genus that Walkom established based on impression remains from Queensland, Australia (Walkom 1921). The permineralized ovules were described in detail as *Plectilospermum elliotii* (Smoot and Taylor 1986b; Taylor and Taylor 1987a) and found to be unlike Walkom's genus both in size and in overall morphology (Walkom's ovule had a nucellar beak and distal, bicornute flange). *Plectilospermum* is a small (3×5 mm long), winged ovule with a complex integument (Fig. 11.4). A number of the ovules contain well-preserved megagametophyte tissue (e.g., Fig. 11.4), several with archegonia and probable embryos inside them (Smoot and Taylor 1986b).

In addition to the *Plectilospermum* ovules, another smaller ovule occurs in the peat from Skaar Ridge. This ovule is also winged, but is only about one-third the size of *Plectilospermum*. At the micropylar end are two oval pads of tissue consisting of cells with spiral thicken-

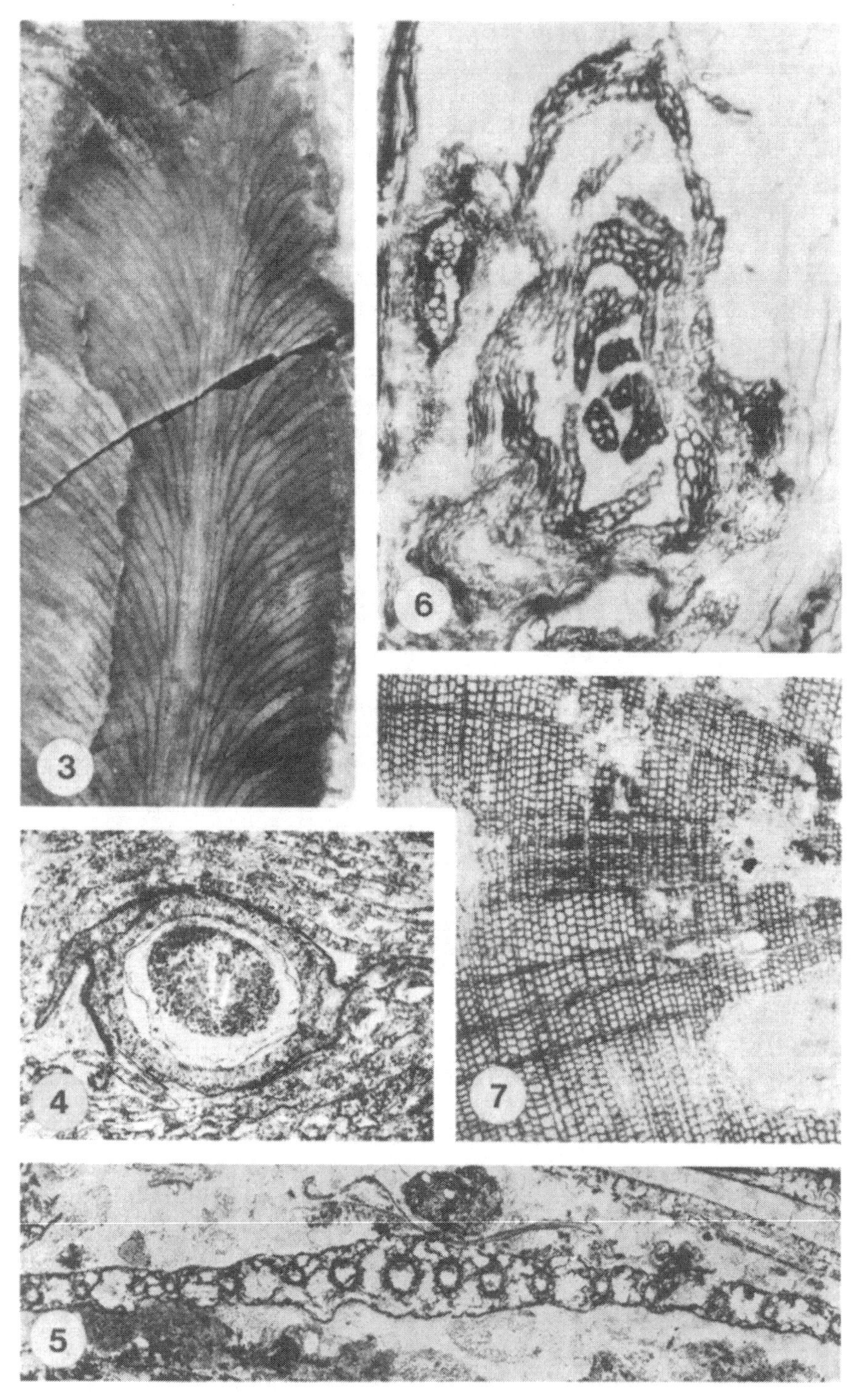

Figures 11.3–11.7. Permian plants from Skaar Ridge peat deposit, central Transantarctic Mountains.

ings on their walls. The pollen chamber of this ovule often contains bisaccate grains. These ovules also occur attached to a delicate, laminar megasporophyll that measures about 1 mm thick and 5–6 mm in width (Taylor 1987). Three to four vascular bundles are present in the megasporophyll, and the anatomy of this structure is similar to one of the two types of *Glossopteris* present at the same locality. Since the vascular bundles are preserved, it is possible to determine with certainty that the ovules are attached on the adaxial surface of the structure. Only a fragment of the megasporophyll is preserved, so the full size of this structure is not certain. Since the discovery of this megasporophyll, *Plectilospermum* ovules have also been found attached to a similar megasporophyll with *Glossopteris*-like anatomy.

Triassic Peat

A similar deposit of permineralized peat occurs in a col north of Fremouw Peak, in the Beardmore Glacier area (84° 18′S, 164° 20′E) (Smoot et al. 1985). This peat is found at the base of the upper part of the Fremouw Formation (Beacon Supergroup) and is considered to be Early–Middle Triassic in age, based on the occurrence of the vertebrate *Cynognathus* in the same formation (Hammer 1989). Unlike the Permian deposit, the Triassic peat occurs in situ in a sandy stream bed. The peat blocks are underlain by a carbonaceous shale (paleosol) that contains rootlets, and it has been suggested (Taylor et al. 1986b; 1989) that the peat represented part of a forested island that was undercut during a flood.

The flora of the Triassic peat exhibits much greater diversity than its Permian counterpart (Smoot et al. 1987). Like compression floras of comparable age, foliage attributed to *Dicroidium* appears to dominate (Fig. 11.13). However, there are two types of plant associations present at this locality. One type is dominated by *Dicroidium* foliage (Pigg and Taylor 1989) and contains a number of small dispersed ovules, sphenophyte stems, and gymnospermous wood. In the second association are found the cycad remains, *Antarcticycas* (Taylor and Smoot 1985; Smoot et al. 1985) and a variety of small fern axes (Millay 1987; Millay et al. 1987b; Millay and Taylor 1989) (e.g., Fig. 11.11).

The first plants described from this peat included stems and petiole traces of an osmundaceous fern, *Osmundacaulis beardmorensis* (Schopf 1978). Smoot et al. (1985) described cycad stems and roots from this site as *Antarcticycas schopfii*. The anatomy of *Antarcticycas* is remarkably similar to modern cycads, including the presence of so-called girdling traces in the cortex. Millay et al. (1987a) later detailed the occurrence of phi thickenings in roots of *Antarcticycas*. This site has also yielded five small filicalean axes with petioles attached (Millay 1987, Millay et al. 1987b; Millay and Taylor 1989) (e.g., Fig. 11.11) and a sphenophyte stem that exhibits equisetalean anatomy (Osborn and Taylor 1989).

One of the unusual aspects of the Triassic peat is the presence of numerous fossil fungi, including typical wood-rotting forms (Stubblefield and Taylor 1985, 1986), a clamp-bearing fungus (Osborn et al. 1989), endogonaceous types (Stubblefield et al. 1987a; Taylor and White 1989) (Fig. 11.8), and the first remains

Figure 11.3. *Glossopteris* leaf exposed on the surface of silicified peat block. × 2.5.

Figure 11.4. Transverse section of the seed *Plectilospermum elliotii* showing two wings and central megagametophyte tissue. ×13.

Figure 11.5. Transverse section of *Glossopteris* leaf with numerous vascular bundles, each surrounded by a bundle sheath. ×15.

Figure 11.6. Transverse section of the moss *Merceria augustica* near the apex showing position of leaves. ×160.

Figure 11.7. Transverse section of segment of *Vertebraria* axis. Note ring boundaries and degraded areas caused by fungal attack. ×30.

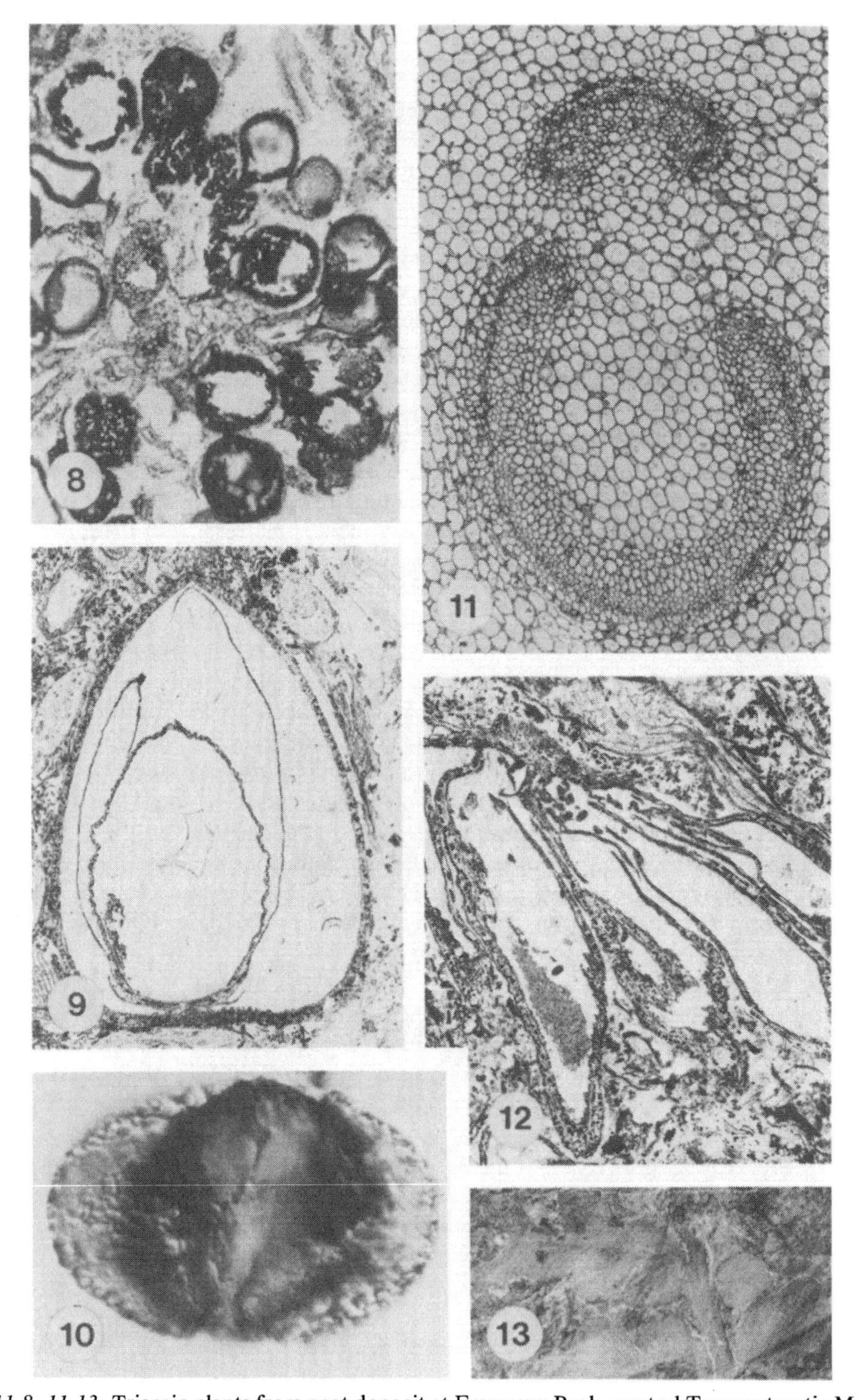

Figure 11.8–11.13. Triassic plants from peat deposit at Fremouw Peak, central Transantarctic Mountains.

of a trichomycete (White and Taylor 1989). Perhaps one of the most exciting aspects of the Triassic fungi is the presence of arbuscles, the physiological exchange structures of endomycorrhizae, in roots of *Antarcticycas* (Stubblefield et al. 1987b, c). These specimens represent the only known examples of arbuscles from a vesicular–arbuscular mycorrhizal fossil fungus. White and Taylor (1988) have also described an interesting fungus from the Triassic peat that shows a number of primitive characters. *Endochaetophora* could not be assigned to any extant group, but because it shares features with several fungal groups, it may represent an example of a group that was ancestral to the living Ascomycotina.

A number of reproductive organs are also present in this peat, including several types of isolated seeds. One of these is a small, radially symmetrical seed that has recently been described as *Ignotospermum monilii* (Perovich and Taylor 1989). *Ignotospermum* has a complex integument and a characteristic bilayered nucellus (Fig. 11.9). Also present is a *Pteruchus*-like pollen organ (Fig. 11.12) that contains bisaccate pollen of the *Alisporites* type (DeVore and Taylor 1988) (Fig. 11.10). But, perhaps the most interesting reproductive organ found in the Triassic peat to date is a cupulate organ that contains from 2–5 ovules (Taylor and Taylor 1987b). Each ovule is triangular in transverse section, with the integument thickened in the corners. In transverse section, each cupule measures about 1.5 × 2.5 cm and is leaf-like in its anatomy. At present, the affinities of this organ are unknown.

Conclusions

Antarctica offers tremendous potential for further work on structurally preserved plants. Few of the localities that have yielded permineralized wood have been collected adequately, nor have most of the wood samples been studied systematically. Although it is important to be able to place wood specimens within particular genera, Antarctic permineralized wood is even more important as a source of tree ring data useful for reconstructing paleoclimate records for high-latitude floras.

The Permian peat from Skaar Ridge and the Triassic locality of Fremouw Peak are unique deposits that have the potential to provide information comparable to that known from Carboniferous coal ball floras. Since the plants are anatomically preserved, they are useful in reconstructing complex three-dimensional structures, understanding reproductive biology and development, and providing a picture of a specialized flora. In addition, there is some indication that other petrified floras may yet be found in Antarctica. For example, a Triassic age compression flora has been described by several authors from Livingston Island (Fig. 11.1) (e.g., Lacey and Lucas 1981; Lemoigne 1987; Banerji and Lemoigne 1987). However, Lacey and Lucas also noted the presence of several partially petrified specimens in this flora, including fern and seed fern rachides and an unusual axis (*Hexagonocaulon*) of uncertain affinities. From Carapace Nunatak, Townrow (1967) described specimens of the Jurassic conifer, *Nothodacrium*, that occurred in cherty bands. The plants were

Figure 11.8. *Sclerocystis*-like fungus showing stalked chlamydospores. ×400.

Figure 11.9. Longitudinal section of *Ignotospermum* ovule. ×16.

Figure 11.10. *Alisporites* pollen grain extracted from peat matrix. ×800.

Figure 11.11. Portion of fern axis with leaf trace. Note excellent cellular preservation, including phloem. ×60.

Figure 11.12. Several elongate pollen sacs, comparable to *Pteruchus*. ×20.

Figure 11.13. Portion of a *Dicroidium* leaf exposed on the surface of a peat block. ×1.6.

not permineralized, although the internal tissues were sometimes replaced by silica. Schopf (1965) described a limonite petrifaction of *Vertebraria* from the Ohio Range. A permineralized peat deposit has also been reported from the Allan Hills (Gabites 1985). A deposit similar to the one from Skaar Ridge was found by a field party during the 1987–1988 field season in the vicinity of Starshot Glacier (158° 8′E, 82° 59′S) (M.N. Rees, personal communication). These scattered reports suggest that systematic collecting in certain areas of Antarctica will reveal additional sites containing structurally preserved plants. The information that these floras can provide is critical, not only in expanding our knowledge of Paleozoic and Mesozoic plants in Gondwana, but also in using fossil floras to more accurately circumscribe the climate of the continent through geologic time.

Acknowledgments. We would like to thank Dr. Margaret N. Rees, University of Nevada, Las Vegas for information about a fossil peat deposit near Starshot Glacier. Thanks to past and present members of the Byrd Polar Research Center and members of U.S. Navy squadron VXE-6 for assistance in the field. Some of the research that is cited in this review was supported by the National Science Foundation (DPP-8611884 and DPP-8716070).

References

Arber EAN (1907) Report on the plant-remains from the Beacon sandstone. In Ferrar HT Report on the field-geology of the region explored during the "Discovery" Antarctic expedition, 1901–4. Brit. Natl. Antarctic Expd. Rept. 1901–1904, Nat. Hist. 1:48

Askin RA, Markgraf V (1986) Palynomorphs from the Sirius Formation, Dominion Range, Antarctica. Antarctic Jour. of the U.S. 21(5):34–35

Banerji J, Lemoigne Y (1987) Significant additions to the Upper Triassic flora of Williams Point, Livingston Island, South Shetlands (Antarctica). Géobios 20:469–487

Barrett PJ (1972) Stratigraphy and petrology of the mainly fluviatile Permian and Triassic part of the Beacon Supergroup, Beardmore Glacier Area. In Adie RJ (ed) Antarctic Geology and Geophysics (SCAR Symp. on Antarctic Geol. and Solid Earth Geophys., Oslo 1970). Universitetetsforlaget, Oslo, pp 365–372

Barrett PJ, Elliot DH (1973) Reconnaissance geologic map of the Buckley Island Quadrangle, Transantarctic Mountains, Antarctica. U.S. Geol. Surv. Map A-3

Barrett PJ, Elliot DH, Lindsay JF (1986) The Beacon Supergroup (Devonian–Triassic) and Ferrar Group (Jurassic) in the Beardmore Glacier Area, Antarctica. In Geology of the Central Transantarctic Mountains, Ant. Res. Ser., Amer. Geophys. Union, 36(14):339–428

Birnie JF, Roberts DE (1986) Evidence of Tertiary forest in the Falkland Islands (Islas Malvinas). Palaeogeog., Palaeoclimatol., Palaeoecol. 55: 45–53

Carlquist S (1987) Pliocene *Nothofagus* wood from the Transantarctic Mountains. Aliso 11:571–583

Cortemiglia GC, Gastaldo P, Terranova R, (1981) Studio di piante fossili trovate nella King George Island delle Isole Shetland del Sur (Antartide). Atti Soc. Ital. Sci. Nat. Museo Civ. Stor. Nat. Milano 122(1–2):37–61

Creber GT, Chaloner WG (1984) Influence of environmental factors on the wood structure of living and fossil trees. Bot. Rev. 50:357–448

DeVore ML, Taylor TN (1988) Mesozoic seed plants: A pollen organ from the Triassic of Antarctica. Amer. J. Bot. 75(6, part 2):106

Doumani GA, Long WE (1962) The ancient life of the Antarctic. Sci. Amer. 207(3):168–184

Doumani GA, Minshew VH (1965) General geology of the Mount Weaver area, Queen Maud Mountains, Antarctica. In Geology and Paleontology of the Antarctic, Antarctic Res. Ser., Amer. Geophys. Union, 6:127–139

Dusén P (1908) Über die tertiäre Flora der Seymour-Insel. Wissenschaftliche Ergebnisse der Schwedischen Südpolar-Expedition, 1901–03, 3(3):1–27

Edwards WN (1921) Fossil coniferous wood from Kerguelen Island. Ann. Bot. 35:609–617

Edwards WN (1928) The occurrence of *Glossopteris* in the Beacon Sandstone of Ferrar Glacier, South Victoria Land. Geol. Mag. 65:323–327

Farabee MJ, Taylor EL, Taylor TN (1989) Correlation of Permian and Triassic palynomorphs from the central Transantarctic Mountains, Antarctica. Proc. VII Int. Palynol. Congr., Brisbane (in press)

Francis JE (1986) Growth rings in Cretaceous and Tertiary wood from Antarctica and their palaeoclimatic implications. Palaeontology 29:665–684

Fritts HC (1976) Tree-rings and climate, Academic Press, New York.

Gabites HI (1985) Triassic paleoecology of the Lashly Formation, Transantarctic Mountains, Antarctica. Unpubl. M.S. Thesis, Victoria Univ. of Wellington, New Zealand

Gee CT (1984) Preliminary studies of a fossil flora

from the Orville Coast, eastern Ellsworth Land, Antarctic Peninsula. Antarctic Jour. of the U.S. 19(5):36–37

Gordon WT (1930) A note on *Dadoxylon* (*Araucarioxylon*) from the Bay of Isles. In Smith WC (ed) Report on the Geological Collections Made During the Voyage of the "Quest" on the Shackleton–Rowett Expedition to the South Atlantic and Weddell Sea in 1921–1922. British Museum (Natural History), London, pp 24–27

Gothan W (1908) Die fossilen Hölzer von der Seymour und Snow Hill Insel. Wissenschaftliche Ergebnisse der Schwedischen Südpolar-Expedition 1901–1903, 3(8):1–33

Gould RE, Delevoryas T (1977) The biology of *Glossopteris:* evidence from petrified seed-bearing and pollen-bearing organs. Alcheringa 1:387–399

Grindley GW (1963) The geology of the Queen Alexandra Range, Beardmore Glacier, Ross Dependency, Antarctica; with notes on the correlation of Gondwana sequences. New Zealand Jour. Geol. Geophys. 6:307–347

Grindley GW, McGregor VR, Walcott RI (1964) Outline of the geology of the Nimrod–Beardmore–Axel Heiberg Glaciers region, Ross Dependency. In Adie RJ (ed) Antarctic Geology (Proc. 1st Int. SCAR Symp. on Antarctic Geol., Cape Town, 1963). North-Holland Publ. Co., Amsterdam, pp 206–219

Halle TG (1911) On the geological structure and history of the Falkland Islands. Bull. Geol. Inst. Univ. Uppsala 11:115–129

Halle TG (1912) The forest-bed of West Point Island. Bull. Geol. Inst. Univ. Uppsala 11:206–218

Hammer WJ (1989) Triassic terrestrial vertebrate faunas of Antarctica. In Taylor TN and Taylor EL (eds) Antarctic Paleobiology and its Role in the Reconstruction of Gondwana. Springer-Verlag, New York, *this volume,* pp. 42–50

Jefferson TH (1982) Fossil forests from the Lower Cretaceous of Alexander Island, Antarctica. Palaeontology 25:681–708

Jefferson TH (1983) Palaeoclimatic significance of some Mesozoic Antarctic fossil floras. In Oliver RL, James PR, Jago JB (eds) Antarctic Earth Sciences (4th Int. SCAR Symp. on Antarctic Earth Sci., Adelaide, 1982). Cambridge Univ. Press, Cambridge, pp 593–599

Jefferson TH (1987) The preservation of conifer wood: examples from the Lower Cretaceous of Antarctica. Palaeontology 30:233–249

Jefferson TH, MacDonald DIM (1981) Fossil wood from South Georgia. British Antarctic Survey Bull. 54:57–64

Jefferson TH, Taylor TN (1983) Permian and Triassic woods from the Transantarctic Mountains: Paleoenvironmental indicators. Antarctic Jour. of the U.S. 18(5):55–57

Jefferson TH, Siders MA, Haban MA (1983) Jurassic trees engulfed by lavas of the Kirkpatrick Basalt Group, northern Victoria Land. Antarctic Jour. of the U.S. 18(5):14–16

Kräusel R (1962) Antarctic fossil wood. In Trans-Antarctic Expedition, 1955–1958, Sci. Rept. 9 (Geology), pp 133–154

Lacey WS, Lucas RC (1981) The Triassic flora of Livingston Island, South Shetland Islands. Brit. Antarctic Survey Bull. 53:157–173

Laudon TS, Thomson MRA, Williams PL, Milliken KL, Rowley PD, Boyles JM (1983) The Jurassic Latady Formation, southern Antarctic peninsula. In Oliver RL, James PR, Jago JB (eds) Antarctic Earth Sciences (4th Int. SCAR Symp. on Antarctic Earth Sci., Adelaide, 1982). Cambridge Univ. Press, Cambridge, pp 308–314

Laudon TS, Lidke DJ, Delevoryas T, Gee CT (1985) Sedimentary rocks of the English Coast, eastern Ellsworth Land, Antarctica. Antarctic Jour. of the U.S. 20(5):38–40

Lemoigne Y (1987) Confirmation de l'existence d'une flore triasique dans l'île Livingston des Shetland du sud (Ouest Antarctique). C.R. Acad. Sci. Paris, Sér. II 304:543–546

Lemoigne Y, Torres T (1988) Paléoxylologie de l'Antarctide: *Sahnioxylon antarcticum* n. sp. et interprétation de la double zonation des cernes des bois secondaires du genre de structure (parataxon) *Sahnioxylon* Bose et Sah, 1954. C.R. Acad. Sci. Paris, Série II, 306:939–945

Long WE (1965) Stratigraphy of the Ohio Range, Antarctica. In Geology and Paleontology of the Antarctic, Antarctic Res. Ser., Amer. Geophys. Union, 6:71–116

Lucas RC, Lacey WS, (1981) A permineralized wood flora of probable early Tertiary age from King George Island, South Shetland Islands. Brit. Antarctic Survey Bull. 53:147–151

Maheshwari HK (1972) Permian wood from Antarctica and revision of some Lower Gondwana wood taxa. Palaeontographica 138B:1–43

Matz DB, Pinet PR, Hayes MO (1972) Stratigraphy and petrology of the Beacon Supergroup, Southern Victoria Land. In Adie RJ (ed) Antarctic Geology and Geophysics (SCAR Symp. on Antarctic Geol. and Solid Earth Geophys., Oslo 1970). Universitetetsforlaget, Oslo, pp 353–358

Meyer-Berthaud B, Taylor TN (1989) The structure and affinities of woody stems from the Triassic of Antarctica. Amer. J. Bot. 76(6, part 2):170–171

Millay MA (1987) Triassic fern flora from Antarctica. Actas VII Simposio Argentino de Paleobotanica y Palinologia:173–176

Millay MA, Taylor TN (1989) New fern stems from the Triassic of Antarctica. Rev. Palaeobot. Palynol. (in press)

Millay MA, Taylor TN, Taylor EL (1987a) Phi thickenings in fossil seed plants from Antarctica. IAWA Bull. 8:191–201

Millay MA, Taylor TN, Taylor EL (1987b) Studies of Antarctic fossil plants: An association of ferns from the Triassic of Fremouw Peak. Antarctic Jour. of the U.S. 22(5):31–32

Nishida M, Nishida N, Rancusi M (1988a) Notes on the petrified plants from Chile (1). Jour. Jap. Bot. 63:39–48

Nishida M, Nishida N, Nasa T (1988b) Anatomy and affinities of the petrified plants from the Tertiary of Chile V. Bot. Mag. 101:293–309

Osborn JM, Taylor TN (1989) Structurally preserved sphenophytes from the Triassic of Antarctica: Vegetative remains of *Spaciinodum*. Amer. J. Bot. (in press)

Osborn JM, Taylor TN, White JF (1989) *Palaeofibulus*, gen. nov., a clamp-bearing fungus from the Triassic of Antarctica. Mycologia (in press)

Pant DD (1987) Reproductive biology of the glossopterids and their affinities. Bull. Soc. Bot. France, 134, Actual. Bot. 1987(2):77–93

Perovich NE, Taylor EL (1989) Structurally preserved fossil plants from Antarctica. IV. Triassic ovules. Amer. J. Bot. 76:992–999

Pigg KB, Taylor TN (1985) Anatomically preserved *Glossopteris* from the Beardmore Glacier area of Antarctica. Antarctic Jour. of the U.S. 20(5): 8–10

Pigg KB, Taylor TN (1987) Anatomically preserved *Glossopteris* from Antarctica. Actas VII Simposio Argentino de Paleobot. y Palinol., Buenos Aires, 177–180

Pigg KB, Taylor TN (1989) Permineralized *Glossopteris* and *Dicroidium* from Antarctica. In Taylor TN and Taylor EL (eds) Antarctic Paleobiology and its Role in the Reconstruction of Gondwana. Springer-Verlag, New York, pp 164–172

Plumstead EP (1962) Fossil floras of Antarctica. In Trans-Antarctic Expedition, 1955–1958, Sci. Rept. 9 (Geology):1–132

Plumstead EP (1975) A new assemblage of plant fossils from Milorgfjella, Dronning Maud Land. Brit. Antarctic Survey Sci. Rept. 83:1–30

Rigby JF (1978) Permian glossopterid and other cycadopsid fructifications from Queensland. Geol. Surv. Qnslnd. Publ. 367 (Palaeontol. Paper 41):1–21

Schopf JM (1962) A preliminary report on plant remains and coal of the sedimentary section in the central range of the Horlick Mountains, Antarctica. Ohio St. Univ. Inst. Polar Studies Rept. 2:1–61

Schopf JM (1965) Anatomy of the axis in *Vertebraria*. In Hadley JB (ed) Geology and Paleontology of the Antarctic, Amer. Geophys. Union, Antarctic Res. Ser. 6:217–228

Schopf JM (1970a) Petrified peat from a Permian coal bed in Antarctica. Science 169:274–277

Schopf JM (1970b) Antarctic collections of plant fossils, 1969–1970. Antarctic Jour. of the U.S. 5(4):89

Schopf JM (1971) Notes on plant tissue preservation and mineralization in a Permian deposit of peat from Antarctica. Amer. Jour. Sci. 271:522–543

Schopf JM (1976) Morphologic interpretation of fertile structures in glossopterid gymnosperms. Rev. Palaeobot. Palynol. 21:25–64

Schopf JM (1977) Coal forming elements in permineralized peat from Mt. Augusta (Queen Alexandra Range). Antarctic Jour. of the U.S. 12(4):110–112

Schopf JM (1978) An unusual osmundaceous specimen from Antarctica. Can. J. Bot. 56:3083–3095

Seward AC (1914) Antarctic fossil plants. Brit. Mus. (Nat. Hist.) Geology 1:1–49

Seward AC, Walton J (1923) On a collection of fossil plants from the Falkland Islands. Quarterly Jour. Geol. Soc. London 79:313–333

Sharman G, Newton ET (1894) Note on some fossils from Seymour Island, in the Antarctic regions, obtained by Dr. Donald. Trans. Roy. Soc. Edinburgh 39:707–709

Sharman G, Newton ET, (1898) Notes on some additional fossils collected at Seymour Island, Graham's Land, by Dr. Donald and Captain Larsen. Proc. Roy. Soc. Edinburgh 22:58–61

Skinner DNB, Ricker J (1968) The geology of the region between the Mawson and Priestley Glaciers, North Victoria Land, Antarctica. Part II-Upper Paleozoic to Quaternary geology. New Zealand Jl. Geol. Geophys. 11:1041–1075

Smoot EL, Taylor TN (1986a) Structurally preserved fossil plants from Antarctica. II. A Permian moss from the Transantarctic Mountains. Amer. J. Bot. 73:1683–1691

Smoot EL, Taylor TN (1986b) Evidence of simple polyembryony in Permian seeds from Antarctica Amer. J. Bot. 73:1077–1079

Smoot EL, Taylor TN, Delevoryas T (1985) Structurally preserved fossil plants from Antarctica. I. *Antarcticycas*, gen. n., a Triassic cycad stem from the Beardmore Glacier area. Amer. J. Bot. 71:410–423

Smoot EL, Taylor TN, Collinson JW (1987) Lower Triassic plants from Antarctica: Diversity and paleoecology. Actas VII Simposio Argentino Paleobotanica y Palinologia, Buenos Aires:193–197

Stubblefield SP, Taylor TN (1985) Fossil fungi in antarctic wood. Antarctic Jour. of the U.S. 20(5):7–8

Stubblefield SP, Taylor TN (1986) Wood decay in silicified gymnosperms from Antarctica. Bot. Gaz. 147:116–125

Stubblefield SP, Taylor TN, Seymour RL (1987a) A possible endogonaceous fungus from the Triassic of Antarctica. Mycologia 79:905–906

Stubblefield SP, Taylor TN, Trappe JM (1987b) Fossil mycorrhizae: A case for symbiosis. Science 237:59–60

Stubblefield SP, Taylor TN, Trappe JM (1987c)

Vesicular–arbuscular mycorrhizae from the Triassic of Antarctica. Amer. J. Bot. 74:1904–1911

Surange KR, Chandra S (1975) Morphology of the gymnospermous fructifications of the *Glossopteris* flora and their relationships. Palaeontographica 149B:153–180

Taylor EL (1987) *Glossopteris* reproductive organs: an analysis of structure and morphology. XIV Int. Bot. Congr. Abstr., Berlin, 24 July–1 August 1987, p 286

Taylor EL, McCallister ER (1989) Tree-ring structure and implications for paleoclimate in the Triassic of Antarctica. Amer. J. Bot. 76(6, part 2): 175

Taylor EL, Taylor TN (1988) Late Triassic flora from Mt. Falla, Queen Alexandra Range. Antarctic Jour. of the U.S. 23(5)

Taylor EL, Taylor TN, Collinson JW, Elliot DH (1986a) Structurally preserved Permian plants from Skaar Ridge, Beardmore Glacier region. Antarctic Jour. of the U.S. 21(5):27–28

Taylor EL, Taylor TN, Collinson JW (1989) Depositional setting and paleobotany of Permian and Triassic permineralized peat from the central Transantarctic Mountains, Antarctica. Int. Jour. Coal Geol. 12:657–679

Taylor TN, Smoot EL (1985) A new Triassic cycad from the Beardmore Glacier area of Antarctica. Antarctic Jour. of the U.S. 20(5):5–7

Taylor TN, Stubblefield SP (1987) A fossil mycoflora from Antarctica. Actas VII Simposio Argentino de Paleobotanica y Palinologia, Buenos Aires 13–15 April 1987, pp 187–191

Taylor TN, Taylor EL (1987a) Structurally preserved fossil plants from Antarctica. III. Permian seeds. Amer. J. Bot. 74:904–913

Taylor TN, Taylor EL (1987b) An unusual gymnospermous reproductive organ of Triassic age. Antarctic Jour. of the U.S. 22(5):29–30

Taylor TN, White JF (1989) Fossil fungi (Endogonaceae) from the Triassic of Antarctica. Amer. J. Bot. 76:389–396

Taylor TN, Taylor EL, Collinson JW (1986b) Paleoenvironment of Lower Triassic plants from the Fremouw Formation. Antarctic Jour. of the U.S. 21(5):26–27

Torres T (1984) *Nothofagoxylon antarcticus* n. sp., madera fósil del Terciario de la isla Rey Jorge, islas Shetland del Sur, Antártica. Ser. Cien., Inst. Antartico Chileno 31:39–52

Torres T, Lemoigne Y (1988) Maderas fósiles terciarias de la Formación Caleta Arctowski, Isla Rey Jorge, Antártica. Ser. Cient. INACH 37:69–107

Torres T, Roman A, Deza A, Rivera C (1984) Anatomía, mineralogía y termoluminiscencia de madera fósil de la isla Rey Jorge, Islas Shetland del Sur. Memoria III Cong. Latinoamericano de Paleontología, Mexico 2:566–574

Townrow JA (1967) On a conifer from the Jurassic of East Antarctica. Roy. Soc. Tasmania, Papers and Proc. 101:137–148

Walkom AB (1921) On *Nummulospermum,* gen. nov., the probable megasporangium of *Glossopteris*. Quart. Jour. Geol. Soc. 77:289–296

Webb PN, Harwood DM (1987) Terrestrial flora of the Sirius Formation: its significance for Late Cenozoic glacial history. Antarctic Jour. of the U.S. 22(4):7–11

White JF, Taylor TN (1988) Triassic fungus from Antarctica with possible ascomycetous affinities. Amer. J. Bot. 75:1495–1500

White JF, Taylor TN (1989) A trichomycete-like fossil from the Triassic of Antarctica. Mycologia (in press)

White ME (1973) Permian flora from the Beaver Lake area, Prince Charles Mountains, Antarctica. 2. Plant fossils. Austral. Bur. Mineral Res. Geol. Geophys. Bull. 126, Palaeont. Papers 1969:13–18

Zastawniak E, Wrona R, Gazdzicki A, Birkenmajer K (1985) Plant remains from the top part of the Point Hennequin Group (Upper Oligocene), King George Island (South Shetland Islands, Antarctica). Studia Geol. Polonica 81:143–164

12—Permineralized *Glossopteris* and *Dicroidium* from Antarctica

Kathleen B. Pigg and Thomas N. Taylor

Introduction

Historically, megafossil floras of Permian and Triassic ages in the southern hemisphere have been dominated by two characteristic leaf forms, *Glossopteris* and *Dicroidium*. Both are known from numerous localities in South Africa (e.g., Townrow 1957; Anderson and Anderson 1983, 1985), South America (Archangelsky 1957, 1968; Petriella 1979), Australia (Gould 1975; Retallack 1977), Antarctica (e.g., Rigby and Schopf 1969), and India (Lele 1961; Chandra and Surange 1979) and have served as important index fossils in Gondwana biostratigraphic correlations (e.g., Retallack 1977; Chandra and Surange 1979; Schopf and Askin 1980, and references cited therein). In fact, leaves of these types typify the megafossil assemblages to such an extent that Permian and Triassic Gondwana floras are commonly known as the *Glossopteris* and *Dicroidium* floras, respectively (e.g., Gould 1975; Retallack 1977; Schopf and Askin 1980).

Although representatives of these leaf forms are widespread and easily recognized, botanical information, particularly about their structural features, has been known in only a limited way. Leaves of these types occur almost entirely as compression and impression fossils that provide external morphological (surface) features, but give little information on cellular structure. Although cuticular remains have been described for over 60 species of *Glossopteris* (e.g., Srivastava 1956; Pant 1958; Pant and Singh 1971), the great degree of variability and the typically poor preservational states have precluded a definitive characterization of diagnostic features. Cuticular detail is known for many species of *Dicroidium* (e.g., Townrow 1957; Archangelsky 1968; Anderson and Anderson 1983), but in this taxon, too, cuticle is limited in the amount of information it can provide about leaf structure. Perhaps more importantly, with the exception of the material described in this chapter, three-dimensionally preserved cellular structure has been reported for only one putative member of the glossopterids (Gould and Delevoryas 1977) and is otherwise unknown in *Dicroidium*. Thus, although their fossil records are extensive, *Glossopteris* and *Dicroidium* leaves have remained, in many ways, poorly known.

Moreover, information regarding the plants that bore these characteristic leaf forms also remains obscure. Although they clearly dominated the Gondwana landscape for over 80 million years, we currently have little sense of the structural and ecological diversity, as well as the evolutionary relationships of the *Glossopteris* plants (e.g. Schopf 1976; Gould and Delevoryas 1977; Pant 1977). The major environmental changes that characterized the end of the Paleozoic presumably led to the shift in floristic domination from *Glossopteris* to plants with *Dicroidium* fronds (e.g., Delevoryas 1973; Schopf and Askin 1980), but plants of this latter group, the corystosperms, are equally enigmatic.

The discovery of Permian and Triassic

permineralized peats containing three-dimensionally preserved plant organs from the Transantarctic Mountains (Schopf 1970; Taylor et al. 1986; Smoot et al. 1986) has provided the first opportunity to describe the anatomy of *Glossopteris* and *Dicroidium* leaves in detail (Pigg and Taylor 1985, 1987a, b). The description of internal anatomy provides an additional level of resolution that has not been available from the study of external morphological form and cuticular features alone and promises to greatly increase our understanding of the leaves that characterized Gondwana.

Methods and Localities

The collecting localities occur near the Beardmore Glacier in the central Transantarctic Mountains. Specimens of *Glossopteris* are found in a permineralized peat of Permian age, along with numerous *Vertebraria* axes, stems, a moss (*Merceria*, Smoot and Taylor 1986), seeds of several types, and glossopterid reproductive structures (Schopf 1976; Smoot et al. 1986; Taylor 1987). This material occurs at the Skaar Ridge locality, on the southeastern end of Skaar Ridge, south east of Mt. Augusta (84° 47′S, 163°15′E, Buckley Island Quadrangle, Barrett and Elliot 1973) at an elevation of approximately 2300 m (Smoot et al. 1986). Stratigraphically, the material occurs within the upper Buckley Formation and is considered to be early to late Permian in age.

Anatomically preserved *Dicroidium* specimens occur at a nearby Triassic locality, Fremouw Peak (84°16′S 164°21′E, Buckley Island Quadrangle, Barrett and Elliot 1973). This locality contains a floristically diverse assemblage of plants, including a cycad stem (*Antarcticycas*, Smoot et al. 1985), several types of ovules (Taylor et al. 1987; Perovich and Taylor 1989), sphenophytes (Osborn and Taylor 1989), ferns, (Millay and Taylor, in press), and a *Pteruchus*-like pollen organ (De Vore and Taylor 1988). Stratigraphically, the locality occurs in the upper Fremouw Formation, and is considered Early–Middle Triassic in age (Taylor et al. 1986; Smoot et al. 1987).

Material was studied by a combination of techniques. External morphological features were studied from dégaged weathered surfaces by a modification of the "microjackhammer" technique as described by Pigg et al. (1987). Serial sections were prepared in several planes with cellulose acetate peels, using hydrofluoric acid to etch the silicified matrix (Basinger and Rothwell 1977).

Description

Glossopteris

Two anatomically distinct forms of leaves assignable to *Glossopteris* occur at the Permian Skaar Ridge locality. Both forms are characterized by a reticulate venation pattern as seen in surface view (Figs. 12.1 and 12.4). The first form (Figs. 12.1–12.3) has a narrow-meshed venation pattern similar to that found in many species of *Glossopteris* known from the compression/impression fossil record (Fig. 12.1). Based on details of the venation pattern and the typically small size (1.5–2 cm wide and probably not exceeding 10 cm in length), these leaves most closely resemble *G. angustifolia* and related forms (e.g., Banerjee 1978; Chandra and Surange 1979). In transverse section, leaves of this type are characterized by prominent vascular bundles containing a region of mesarch xylem and a lacuna that presumably represents the former position of the phloem (Fig. 12.2). The bundle is delimited by a bundle sheath 2–3 cells thick that contains numerous dark fibers. Mesophyll is typically poorly preserved but, when present, it is relatively undifferentiated and aerenchymatous, with no evidence of palisade and spongy layers. A 2–3 cell-layered hypodermis is present on both surfaces. Commonly, epidermal cells are characterized by a single median papilla and straight-margined anticlinal walls (Fig. 12.3). Leaves are hypostomatic with stomata surrounded by a ring of 5–6 relatively undifferentiated subsidiary cells (Fig. 12.3).

In contrast, the second species of *Glossopteris* at Skaar Ridge (Figs. 12.4–12.6) is characterized by a coarser, or broader, mesh pat-

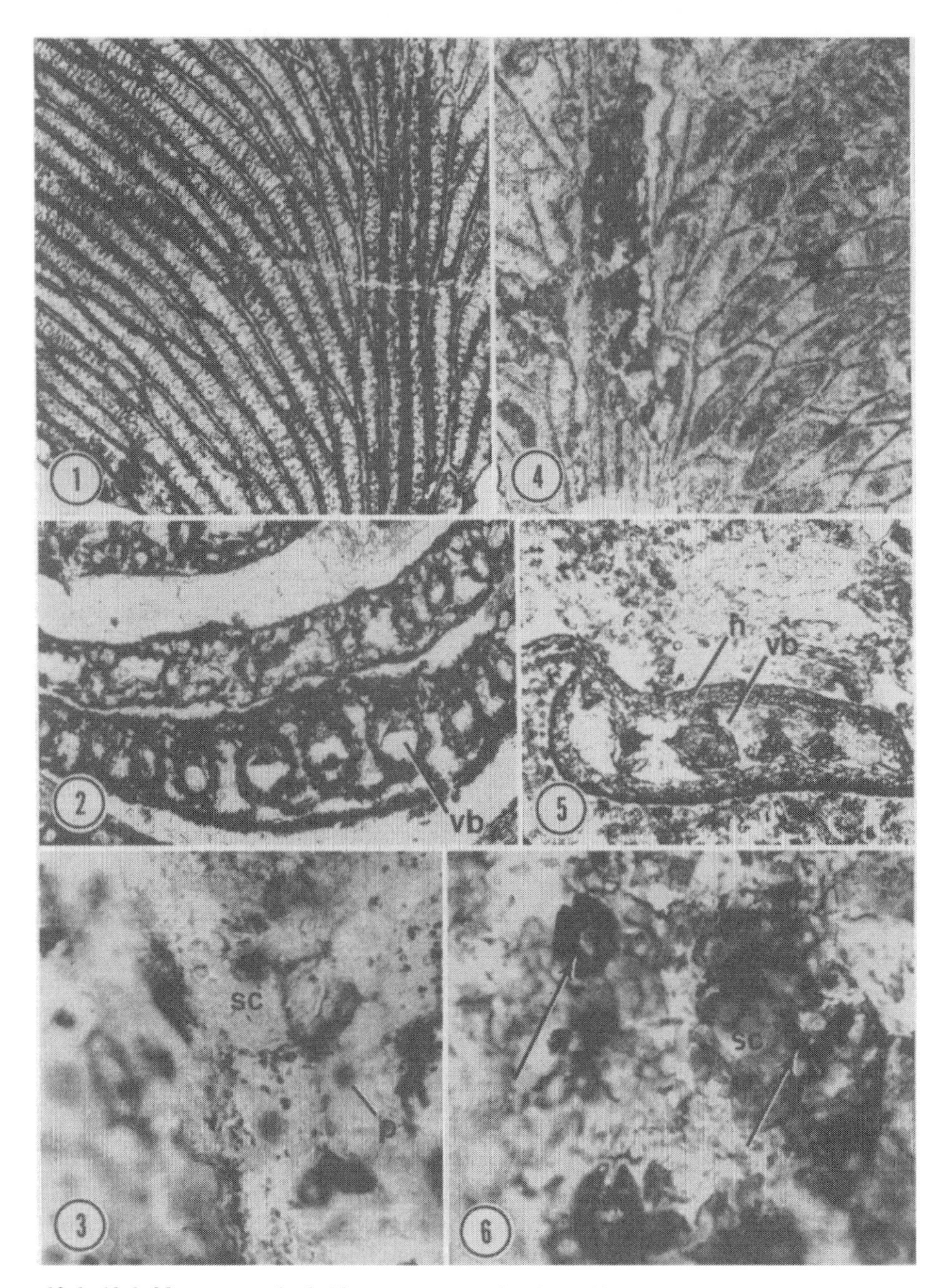

Figures 12.1–12.3. Narrow-meshed *Glossopteris* species from Skaar Ridge, Transantarctic Mountains.

Figure 12.1. Paradermal section to show venation pattern of leaf. Note narrow, elongate meshes, parallel veins of the midrib, and various types of reticulations. 435#2α×8.

Figure 12.2. Transverse section through several leaves. Note prominent vascular bundles (vb). 533 A bot # 8a × 32.

Figure 12.3. Stoma surrounded by five subsidiary cells (sc). Note centrally positioned papillae on subsidiary cells (p). 435 # 4α×400.

Figures 12.4–12.6. Broad-meshed *Glossopteris* species from Skaar Ridge, Transantarctic Mountains.

Figure 12.4. Paradermal section showing general features of venation. Note midrib and broader, polygonal meshes. 465 C4 side # 2×7.

Figure 12.5. Transverse section through leaf. Note prominent hypodermis (h) and oval vascular bundles (vb). 484 C bot #1 g ×40

Figure 12.6. Lower cuticle showing two sunken stomata (arrows). Note thickened guard cells. Stoma at left is sectioned at a more internal level such that subsidiary cells are out of the plane of section, while stoma at right is surrounded by ring of subsidiary cells (sc). 465 C4 side # 13×363

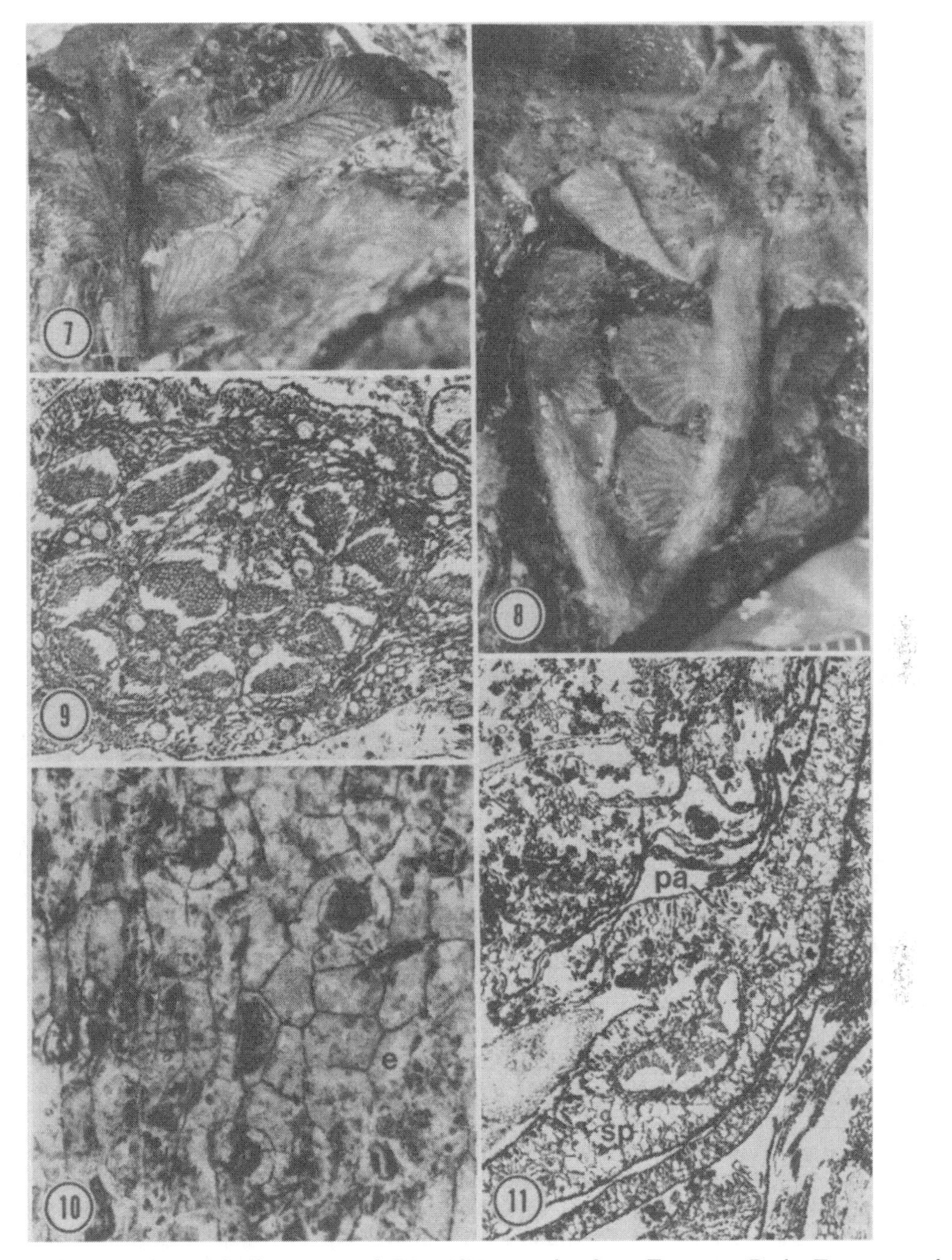

Figures 12.7–12.11. Anatomically preserved *Dicroidium* species from Fremouw Peak, Transantarctic Mountains.

Figure 12.7. Weathered surface showing fragment from distal part of frond. Note bipinnatifid pinnule with lobed margins and dichotomizing venation. D14, 10,235 B2 surface ×2.6.

Figure 12.8. Weathered surface showing frond fragment with proximal bifurcation and entire basal pinnule with dichotomizing venation. D15, 10,235 B2 surface ×2.6.

Figure 12.9. Transverse/oblique section of frond rachis showing abaxial ring of eight bundles and an adaxial line of five bundles. Note resinous cells. 10,109 B2a side b#1β ×22.

Figure 12.10. Cuticle of lower epidermis. Note polygonal and elongate epidermal cells (e) and stomata. 10,200 C3 pull α #1×214.

Figure 12.11. Transverse section through laminar pinnule to show prominent midvein and palisade (pa) and spongy (sp) mesophyll. 10,033 B bot # 13 ×22.

tern, with individual meshes appearing elongate–polygonal in outline (Fig. 12.4). Leaves of this type resemble several coarse-meshed glossopterid species of the *G. retifera* type, including *G. conspicua, G. shirleyi,* and *G. retifera* (Lacey et al. 1975; Banerjee 1978; Rigby 1978). [Rigby et al. (1980) have synonomized *G. retifera* with *G. conspicua*.] Additionally, similar venation is sometimes seen in species of *Belemnopteris* (e.g., *B. pellucida;* Pant and Choudhury 1977). In transverse section, the coarse-meshed leaves are strikingly distinct from the first form in several respects. They lack the fibrous bundle sheath characteristic of the first form but demonstrate a more prominent hypodermis of several layers in thickness (Fig. 12.5). Vascular bundles contain terete xylem strands composed of small tracheids with helical thickenings. In contrast to the first form, they lack a conspicuous phloem lacuna. Cuticular features are also markedly different, with hypostomatic leaves characterized by epidermal cells with sinuous anticlinal walls, and more elaborate stomatal structures (Fig. 12.6). Stomata are sunken, with thickened guard cells, and are surrounded by a ring of 5–6 subsidiary cells (Fig. 12.6).

The two new species of *Glossopteris* from Skaar Ridge are anatomically distinct from one another and are also unlike the putative glossopterid vegetative leaves described in association with *Glossopteris* fructifications from the Bowen Basin (Gould and Delevoryas 1977). Although the Australian leaves are characterized by a distinctive hypodermis similar to that of the broad-meshed species from Antarctica, they differ in possessing a well-differentiated mesophyll with palisade and spongy layers. External morphological features and cuticular details are presently unknown for the Australian specimens.

Dicroidium

Frond fragments of *Dicroidium* at Fremouw Peak represent a bipinnatifid frond with two intergrading types of pinnules (Figs. 12.7–12.11). Basally, near the frond dichotomy, pinnules are entire (Fig 12.8); more distally, they become lobed with up to 10 lobes/pinnule (Fig. 12.7). In transverse section, the frond rachis demonstrates a complex vascular pattern composed of 5–8 bundles arranged in an abaxial ring, and 4–6 in an adaxial line (Fig. 12.9). This pattern continues throughout the length of the frond, with the adaxial line of bundles producing strands that diverge laterally to vascularize the pinnules. Histologically, the frond rachis contains scattered resinous cells and a palisade-like adaxial layer (Fig. 12.9). Pinnules are characterized by a well-differentiated mesophyll with palisade and spongy layers (Fig. 12.11), and dichotomizing venation (Figs. 12.7 and 12.8). Leaves are amphistomatic with a relatively thin, but resistant cuticle characterized by stomata with a dicyclic arrangement of subsidiary cells (Fig. 12.10).

In comparison to the compression/impression species of *Dicroidium,* the fronds from Fremouw Peak most closely resemble *D. dubium* and *D. dubium* var. *tasmaniense* (Retallack 1977, 1980). The present material also bears resemblance to Lele's (1961) *D.* sp. cf. *feistmanteli, D. feistmanteli* (Jain and Delevoryas 1967) and other bipinnatifid fronds of the *D. dubium* type, including *D. dubium* of Anderson and Anderson (1983), and *D. dubium* var. *australe* (Retallack 1977).

Discussion

Anatomically preserved material of the present type provides a new level of resolution in the description of *Glossopteris* and *Dicroidium* leaves. This in turn allows researchers to address a number of biological questions about the leaf forms that characterized Gondwana vegetation, but which up until now have been known as artificial foliage forms, with little context to the plants that bore them.

Glossopteris

Three distinctive anatomical types can now be recognized for the genus *Glossopteris,* including the two species from Antarctica and a third form from Australia (Gould and Delevoryas 1977). Although the venation patterns of *Glossopteris,* as known from compressions and impressions, show a great deal of variety (e.g.,

Chandra and Surange 1979), the degree of anatomical diversity recognized from the petrifactions could not be predicted from morphology alone. Conversely, anatomical features are sufficiently distinct that without external morphology there is little to unite these forms.

For some time, researchers have recognized that the unusual fertile structures found in attachment and association with *Glossopteris* leaves show a great deal of diversity (e.g., Plumstead 1956; Surange and Chandra 1975; Schopf 1976; Taylor 1987). Some workers have identified within this plexus several distinct lineages, rather than one cohesive group (e.g., Surange and Chandra 1975; Rigby 1978). The anatomical diversity demonstrated among vegetative leaves supports the hypothesis that glossopterids were a diverse assemblage of plants, perhaps not all that closely related to one another, that bore leaves with a similar venation pattern. Although the selective pressures that led to the widespread distribution of leaves with glossopterid venation are unclear, the common occurrence of this pattern suggests that it had some adaptive significance. The continued study of permineralized vegetative and reproductive remains will help to establish a better frame of reference for variation among glossopterids and perhaps offer clues as to why this leaf form was so common in the Permian.

Dicroidium

In Triassic sediments, the genus *Dicroidium* represents highly variable frond-like leaves thought to characterize a group of Mesozoic seed ferns, the corystosperms (Townrow 1957; Archangelsky 1968; Petriella 1981). The affinities of the corystosperms with other Mesozoic pteridosperm groups, with presumed Paleozoic ancestral forms, with other gymnosperms, and even with the angiosperms are not well defined (e.g., Thomas 1955; Doyle and Donoghue 1986; Crane 1988). The internal anatomical structure of *Dicroidium* provides significant new information with which to address the question of corystosperm phylogeny.

Dicroidium fronds are characterized by a complex vascular architecture with a dorsiventral organization, composed of an abaxial ring of bundles and an adaxial linear arrangement. Of major gymnospermous groups with pinnate fronds, this architecture shows the greatest resemblance to forms that occur among the cycads (e.g., *Cycas,* Worsdell 1906). Petioles of the cycads are commonly multivascularized and show a range of variation in their arrangement of bundles. Typically, bundles form a U-shaped configuration flanked on either side by a vertical row of bundles that extends into an adaxial linear group (Bierhorst 1971). A slight modification of this arrangement (i.e., loss of the vertical bundles) would result in the configuration characteristic of the *Dicroidium* petiole (Fig. 12.9).

On the other hand, the configuration in *Dicroidium* is unlike that which typifies the Paleozoic seed ferns (Thomas 1955; Petriella 1981). Among Paleozoic forms, the calamopityans (Galtier 1974), the lyginopterids (Taylor and Millay 1977), and *Callistophyton* (Rothwell 1975) are all characterized by relatively simple frond vasculature composed of a single strand or several individual strands (Galtier 1988). The other major group of Paleozoic seed ferns, the medullosans, has a complex vascular architecture, but one which demonstrates a different pattern. In medullosans, petioles are characterized by an outer ring of numerous vascular bundles and an inner region of either scattered bundles (Ramanujam, et al, 1974) or bundles arranged linearly (Hamer and Rothwell 1988). Histological characters, including the abundance of transfusion tracheids, tracheids with reticulate-scalariform wall thickening patterns, and resinous cells within the ground tissue of fronds suggests additional features similar to cycads (Worsdell 1906; Bierhorst 1971).

Upon further evaluation, the relationships between *Dicroidium,* early cycadophytes, and pteridosperms may prove to be more complex than previously thought. The presence of a cycad stem (*Antarcticycas;* Smoot et al. 1985), several types of ovules and ovulate fructifications (Taylor et al. 1987; Perovich and Taylor 1989), a *Pteruchus*-like pollen organ containing bisaccate grains (De Vore and Taylor 1988), and many other gymnospermous organs in this florisitically rich peat may provide the necessary data base with which to characterize additional information about gymnosperm diversification during the Mesozoic.

Summary

Permineralized peat from the Transantarctic Mountains provides the opportunity to characterize the three-dimensional anatomical features of the leaves *Glossopteris* and *Dicroidium,* forms which dominate the Permian and Triassic megafossil records, respectively. Although Permian leaves with a prominent midrib and lateral reticulate venation pattern can be readily assigned to the form genus *Glossopteris,* the anatomical diversity now known demonstrates that the internal anatomical structure may be more variable than can be predicted from external morphological features alone. Based on this anatomical variability, the anastomosing venation that characterized the glossopterid leaf form may represent a convergence in leaf form among plants not all that closely related to one another (e.g., Delevoryas 1973).

From Triassic strata, the internal structure of *Dicroidium* leaves, characterized for the first time, reveals that corystosperms have a surprising similarity to cycads in their frond architecture. Further investigations of this type of anatomically preserved material may be invaluable in reassessing the evolutionary relationships of this presently poorly known group of Mesozoic seed ferns.

References

Anderson JM, Anderson HM (1983) Palaeoflora of Southern Africa Molteno Formation (Triassic), Volume 1. A.A. Balkema, Rotterdam

Anderson JM, Anderson HM (1985) Palaeoflora of Southern Africa. Prodromus of South African megafloras. Devonian to Lower Cretaceous. A.A. Balkema, Rotterdam

Archangelsky S (1957) Las Glossopterideas del Bajo de la Leona. Rev. Assoc. Geol. Argentina 12: 135–164

Archangelsky S (1968) Studies on Triassic fossil plants from Argentina. IV. The leaf genus *Dicroidium* and its possible relation to *Rhexoxylon* stems. Palaeontology 11: 500–512

Banerjee M (1978) Genus *Glossopteris* Brongniart and its stratigraphic significance in the Palaeozoic of India. Part 1—a revisional study of some species of the genus *Glossopteris*. Bulletin of the Botanical Society of Bengal 32:81–125

Barrett PJ, Elliot DH (1973) Reconnaissance geologic map of the Buckley Island Quadrangle, Transantarctic Mountains, Antarctica. U.S. Geol. Survey Map A-3

Basinger JF, Rothwell GW (1977) Anatomically preserved plants from the Middle Eocene (Allenby Formation) of British Columbia. Canad. J. Bot. 55:1984–1990

Bierhorst DW (1971) Morphology of Vascular Plants. Macmillan, NY

Chandra S, Surange KR (1979) Revision of the Indian Species of *Glossopteris*. Monograph No. 2. Birbal Sahni Institute of Palaeobotany. Lucknow

Crane PR (1988) Major clades and relationships in the "higher" gymnosperms. In Beck CB (ed) Origin and evolution of gymnosperms. Columbia Univ. Press, NY, pp. 218–272

Delevoryas T (1973) Postdrifting Mesozoic floral evolution. In Meggers BJ, Ayensu ES, and Duckworth WD (eds) Tropical Forest Ecosystems in Africa and South America: a comparative review. Smithsonian Inst. Press. Washington, DC, pp. 9–19

DeVore ML, Taylor TN (1988) Mesozoic seed plants: a pollen organ from the Triassic of Antarctica. Amer. J. Bot. 75 (6, part 2): 106

Doyle JA, Donoghue MJ (1986) Seed plant phylogeny and the origin of angiosperms: an experimental cladistic approach. Bot. Rev. 52: 321–431

Galtier J (1974) Sur l'organisation de la fronde des *Calamopitys,* Ptéridospermales probables du Carbonifère inférieur. C.R. Acad. Sc. Paris sér D. 279: 975–978

Galtier J (1988) Morphology and phylogenetic relationships of early pteridosperms. In Beck CB (ed) Origin and evolution of gymnosperms. Columbia Univ. Press, NY, pp. 135–176

Gould RE (1975) The succession of Australian pre-Tertiary megafossil floras. Bot. Rev. 41: 453–483

Gould RE, Delevoryas T (1977) The biology of *Glossopteris:* evidence from petrified seed-bearing and pollen-bearing organs. Alcheringa 1: 387–399

Hamer J, Rothwell GW (1988) The vegetative structure of *Medullosa endocentrica* Baxter (Pteridospermopsida). Canad. J. Bot. 66: 375–387

Jain RK, Delevoryas T (1967) A Middle Triassic flora from the Cacheuta Formation, Minas de Petroleo, Argentina. Palaeontology 10: 564–589

Lacey WS, van Dijk DE, Gordon-Gray KD (1975) Fossil plants from the Upper Permian in the Mooi River district of Natal, South Africa. Annals Natal Museum 22: 349–420

Lele KM (1961) Studies in the Indian Middle Gondwana flora: 1. On *Dicroidium* from the South Rewa Gondwana Basin. Palaeobotanist 10: 48–68 (issued 1962)

Millay MA, Taylor TN (in press). New fern stems from the Triassic of Antarctica. Rev. Palaeobot. Palynol.

Osborn JM, Taylor TN (1989) Structurally preserved sphenophytes from the Triassic of Antarctica: vegetative remains of *Spaciinodum* gen. nov. Amer. J. Bot. 77: in press

Pant DD (1958) The structure of some leaves and fructifications of the *Glossopteris* flora of Tanganyika. Bull. Brit. Museum (Nat. History) Geology 3: 127–175

Pant DD (1977) The plant of *Glossopteris*. J. Indian Bot. 56: 1–23

Pant DD, Choudhury A (1977) On the genus *Belemnopteris* Feistmantel. Palaeontographica 164B: 153–166

Pant DD, Singh KB (1971) Cuticular structure of some Indian Lower Gondwana species of *Glossopteris* Brongniart, Part 3. Palaeontographica 135B: 1–40

Perovich NE, Taylor EL (1989) Structurally preserved fossil plants from Antarctica. IV. Triassic ovules. Amer. J. Bot. 76: 992–999

Petriella B (1979) Sinopsis de las Corystospermaceae (Corystospermales, Pteridospermophyta) de Argentina. I. Hojas. Ameghiniana 16: 81–102

Petriella B (1981) Sistematica y vinculaciones de las Corystospermaceae H. Thomas. Ameghiniana 18: 221–234

Plumstead EP (1956) Bisexual fructifications borne on *Glossopteris* leaves from South Africa. Palaeontographica 100B: 1–25

Pigg KB, Taylor TN (1985) Anatomically preserved *Glossopteris* from the Beardmore Glacier area of Antarctica. Ant. J.U.S. 19: 8–10

Pigg KB, Taylor TN (1987a) Anatomically preserved *Glossopteris* from Antarctica. VII Simposio Argentino de Paleobotanica y Palinologica, Actas: 177–180

Pigg KB, Taylor TN (1987b) Anatomically preserved *Dicroidium* from the Transantarctic Mountains. Ant. J.U.S. 22 (5): 28–29

Pigg KB, Taylor TN, Stockey RA (1987) Studies of Paleozoic seed ferns: *Heterangium kentuckyensis* sp. nov., from the Upper Carboniferous of North America. Amer. J. Bot. 74: 1184–1204

Ramanujam CGK, Rothwell GW, Stewart WN (1974) Probable attachment of the *Dolerotheca* campanulum to a *Myeloxylon-Alethopteris* type frond. Amer J. Bot. 61:1057–1066

Retallack G (1977) Reconstructing Triassic vegetation of eastern Australasia: a new approach for the biostratigraphy of Gondwanaland. Alcheringa 1: 247–277, and Alcheringa-fiche 1: G1–j17, ISSN0311–5518

Retallack G (1980) Middle Triassic megafossil plants and trace fossils from Tank Gully, Canterbury, New Zealand. J. Royal Soc. New Zealand 10: 31–63

Rigby JF (1978) Permian glossopterid and other cycadopsid fructifications from Queensland. Geol. Surv. Queensland Publ. 367 (Palaeontol. Paper 41): 1–21

Rigby JF, Schopf JM (1969) Stratigraphic implications of Antarctic palaeobotanical studies. Gondwana Stratigraphy Earth Sci. 2: 91–106

Rigby JF, Maheshwari HK, Schopf JM (1980) Revision of Permian plants collected by J. D. Dana during 1839–1840 in Australia. Queensland Dept. of Mines, Geol. Survey Queensland Pub. 376, Pal. pap. 471–485

Rothwell GW (1975) The Callistophytaceae (Pteridospermopsida): 12. Vegetative structures. Palaeontographica 151B: 171–196

Schopf JM (1970) Petrified peat from a Permian coal bed in Antarctica. Science 169: 274–277

Schopf JM (1976) Morphologic interpretation of fertile structures in glossopterid gymnosperms. Rev. Palaeobot. Palynol. 21: 25–64

Schopf JM, Askin RA (1980) Permian and Triassic floral biostratigraphic zones of southern land masses. In Dilcher DL, Taylor TN (eds) Biostratigraphy of Fossil Plants. Dowden, Hutchinson, and Ross, Stroudsberg, Pennsylvania

Smoot EL, Taylor TN (1986) Structurally preserved fossil plants from Antarctica: II. A Permian moss from the Transantarctic Mountains. Amer. J. Bot. 73: 1683–1691

Smoot EL, Taylor TN, Delevoryas T (1985) Structurally preserved Permian plants from Antarctica. I. *Antarcticycas,* gen. nov., a Triassic cycad stem from the Beardmore Glacier area. Amer. J. Bot. 72: 1410–1423

Smoot EL, Taylor TN, Collinson JW, Elliot DH (1986) Structurally preserved Permian plants from Skaar Ridge, Beardmore Glacier region. Ant. J.U.S. 21(5): 27–28

Smoot EL, Taylor TN, Collinson JW (1987) Lower Triassic plants from Antarctica: Diversity and paleoecology. VII Simposio Argentino de Paleobotanica y Palinologia, Actas: 193–196.

Srivastava PN (1956) Studies in the *Glossopteris* flora of India-4. *Glossopteris, Gangamopteris,* and *Palaeovittaria* from the Raniganj coalfield. Palaeobotanist 5: 1–45

Surange KR, Chandra S (1975) Morphology of the gymnospermous fructifications of the *Glossopteris* flora and their relationship. Palaeontographica 149B: 153–180

Taylor EL (1987) *Glossopteris* reproductive organs: an analysis of structure and morphology. XIVth

International Botanical Congress, Berlin. Int. Bot. Congr. Abstr. 17: 287
Taylor TN, Millay MA (1977) Morphologic variability of Pennsylvanian lyginopterid seed ferns. Rev. Palaeobot. Palynol. 32: 27–62
Taylor TN, Smoot EL, Collinson JW (1986) Paleoenvironment of Lower Triassic plants from the Fremouw Formation. Ant. J.U.S. 21(5):26–27
Taylor TN, Taylor EL, Millay MA (1987) Mesozoic seed plants from Antarctica: Multiovulate cupules. Amer. J. Bot. 74: 691–692
Thomas HH (1955) Mesozoic pteridosperms. Phytomorphology 5: 177–185
Townrow J (1957) On *Dicroidium*, probably a pteridospermous leaf, and other leaves now removed from this genus. Trans. Geol. Soc. Afr. 60: 21–60
Worsdell WC (1906) The structure and origin of the Cycadaceae. Ann. Bot. 20: 129–159

13—Comments on the Role of Cycadophytes in Antarctic Fossil Floras

T. Delevoryas

Introduction

As in previous works, references to cycadophytes include members of both the Cycadales (or Nilssoniales) and Cycadeoidales (Bennettitales). Even though the two groups are distinct enough to be placed into separate divisions (Delevoryas 1975; Taylor 1981), it continues to be useful to refer to them together. They very often occur together in Mesozoic sediments, and both apparently reflect similar kinds of environments. Both the Cycadales (now represented by only 10 genera) and the Cycadeoidales (which are extinct) were important components of Mesozoic biotas worldwide. What appear to be the same, or closely related, genera occur in widely separated localities.

Cycadales

Extant genera occur worldwide (Fig. 13.1), but with a tendency to be more concentrated in equatorial regions. Species of the genus *Cycas* are found in northern Australia, islands of the southwest Pacific, southeast Asia, India, and Madagascar and nearby islands. *Encephalartos* is exclusively African, occurring in the southern two-thirds of the continent. *Stangeria* is confined to southern Africa. The New World *Zamia* is found in the southeastern United States, Mexico, Central America, and northwest South America. *Dioon* and *Ceratozamia* are primarily Mexican. Exclusively Australian are the genera *Macrozamia, Lepidozamia,* and *Bowenia*. *Macrozamia* is widespread throughout the country, with *Lepidozamia* concentrated in the southeast. *Bowenia* is relatively restricted to a small area in the northeast. Finally, *Microcycas* appears relictual in Cuba.

On the basis of present-day distribution, little can be said with certainty concerning the center of radiation for cycads if, indeed, there was but a single center in the past. It was hoped that a survey of fossil cycads, including Antarctic and other Gondwana members, might provide some insight into the paleobiogeography of the Cycadales.

Remains of the Cycadales are not frequently encountered in Antarctic sediments. The earliest and, incidentally, best preserved Antarctic cycad is *Antarcticycas schopfii* from the Early–Middle Triassic of the Beardmore Glacier area (Smoot et al. 1985) (Fig. 13.2). Anatomically, *Antarcticycas* appears to have more in common with *Bowenia* than any other cycad, although no relationship is implied. While such evidence of cycads in the early Mesozoic of Antarctica is interesting, that occurrence, in itself, does not serve as evidence for a center of radiation of the order. Other ancient cycads are known from widely separated parts of the world.

Well known are the reports of Mamay (1969, 1976) of *Phasmatocycas* and *Archaeocycas,* both of which are considered to have had entire megasporophylls bearing ovules. On the basis of these finds, Mamay postulates

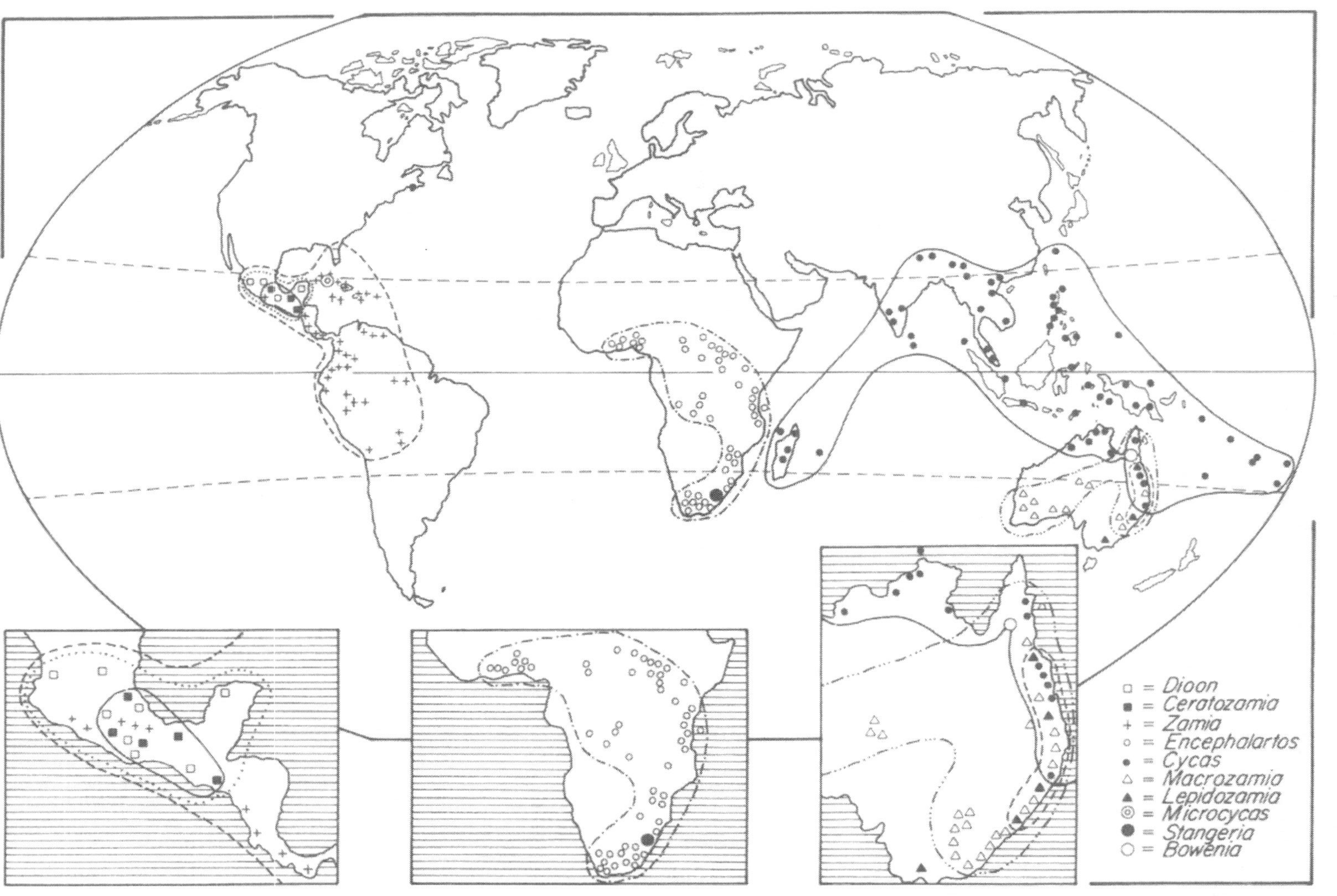

Figure 13.1. Present-day global distribution of cycad genera (modified from Greguss 1968).

Figure 13.2. Cross section of *Antarcticycas schopfii.*

that the cycad megasoporophyll was derived from an entire taeniopterid leaf. Delevoryas (1982) suggested that there may, perhaps, have been a different source of cycad megasporophylls, probably from dissected pteridosperm leaves. More cycadlike are megasporophylls from the Lower Permian of Shanxi, China, described by Zhu and Du (1981) (Fig. 13.3). These sporophylls are more like those traditionally considered primitive (e.g., megasporophylls of *Cycas*) with ovules borne along two sides of a rachis with a distal, dissected lamina. These sporophylls have been named *Primocycas* and occur in association with taeniopterid foliage. *Phasmatocycas, Archaeocycas,* and *Primocycas* all indicate that even at the time of earliest appearance of cycads, it is not possible to recognize a center of distribution.

Michelilloa is a Gondwana cycad stem from the Upper Triassic of Argentina (Archangelsky and Brett 1963). *Michelilloa,* with a stem diameter estimated to be about 10 cm, has some structural features in common with stems of members of the Cycadeoidales, but the authors felt it was more similar to cycadalean stems, mentioning *Dioon spinulosum* as an example where there are many similarities.

Late Triassic cycads were not, however, confined to Gondwana lands. The Upper Triassic *Lyssoxylon gribsbyi* was found in the southwestern United States and first named by Daugherty (1941), who considered it to be cycadeoidalean. Later reinvestigated in detail by Gould (1971), it was shown to have had a number of typical cycadalean features, including girdling leaf traces. Another early cycad, *Leptocycas gracilis,* from Upper Triassic deposits of North Carolina, is known from com-

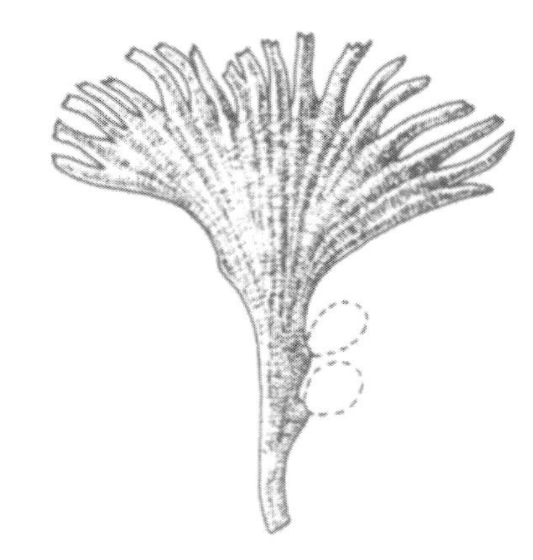

Figure 13.3. Reconstruction of *Primocycas muscariformis,* a presumed Permian cycad megasporophyll (from Zhu and Du 1981).

pressions of stems, leaves, and a microsporangiate cone (Delevoryas and Hope, 1971). It is probably the earliest example of a cycad that demonstrates the habit of the entire plant.

Florin's (1933) *Bjuvia simplex,* also from Upper Triassic deposits in Sweden, is well known, but the reconstruction is largely hypothetical. Megasporophylls, called *Palaeocycas,* that were presumed to have been borne by the plant are clearly cycadalean, however.

Thus, it is apparent that the early record of Cycadales gives no clue about the origin of the order. Furthermore, except for the unusual sporophylls of *Phasmatocycas* and *Archaeocycas,* the nature of the late Paleozoic and early Mesozoic cycads is amazingly uniform. Plumstead (1962) wrote that "Cycadophyte fronds and fruits are so common in Triassic rocks throughout the world that their existence in East Antarctica is less significant than the occurrence of less widely distributed plants."

Plumstead reported the occurrence of leaves of *Nilssonia,* generally conceded to be cycadalean, from a nunatak at the head of Upper Taylor Glacier. The age is assumed to be not later than Middle Triassic (Plumstead 1962). From nearby beds about 16 miles north of Mount Beehive, Taylor Glacier, a structure determined by Plumstead to represent a cycad megasporophyll (somewhat like that of the Jurassic *Beania*), was found. The details in Plumstead's illustration (Plate 21, Fig. 4) are so vague, however, that no definite determination is possible from the figure.

Lacey and Lucas (1981) described a collection of Triassic plants from the Williams Point area of Livingston Island, South Shetland Islands. Among the plant remains were five specimens of *Doratophyllum,* a *Taeniopteris*-type leaf customarily included in the Cycadales.

Perhaps the best known Antarctic Mesozoic flora is that from Hope Bay, described by Halle (1913) and revised by Gee (1989). Halle considered the flora to have many elements in common with the Yorkshire Jurassic flora and suggested that the Hope Bay plants were Middle Jurassic. Later workers (e.g., Florin 1940) felt that the Hope Bay flora had a "decidedly southern composition" and that the comparison with the Yorkshire flora was unwarranted. Farquharson (1984) is more inclined to suggest an Early Cretaceous age of the Hope Bay plants. The Mt. Flora Formation, the source of the plants, is conformably overlain by agglomerates and tuffs radiometrically dated as Early Cretaceous. Furthermore, on the South Orkney Islands, where there exists a sequence similar to the Mt. Flora Formation, there is a marine intercalation with ammonites dated as Neocomian (Thomson 1981).

Cycadaleans from Hope Bay are represented by leaves of two genera, *Nilssonia* and *Pseudoctenis,* both forms with a worldwide distribution at that time.

In summary, it may be stated that members of the Cycadales were not dominant components of Antarctic fossil floras. Remains are primarily in the form of leaf fragments, with only one questionable fertile structure having been reported. Antarctic cycads include genera found in all other parts of the world in rocks of the same age. (*Antarcticycas* is a conspicuous exception.) No information concerning the place of origin and subsequent radiation is evident at this time.

Cycadeoidales

Cycadeoidales, which were coexistent with Mesozoic cycadaleans, have certain features in common that make them a fairly natural group and recognizable in the fossil record. Typically, cycadeoidaleans have ovulate strobili consisting of a fleshy receptacle, on the surface of which are ovules, separated by interseminal scales. The shape of the receptacle may vary, the presence or absence of ovule stalks may differ, and some forms may have accompanying microsporophylls, while others do not. However, there is a fundamental cohesiveness in the structure of the ovule-bearing member of the cycadeoidaleans.

Microsporangiate structures also have much in common among all members. Basically, there is a whorl of microsporophylls, modified or fused in various ways, bearing microsporangia. These microsporangia are usually compound structures, each bearing a number of elongated, tubular sporangia surrounded by a thick outer wall.

Leaves of cycadeoidaleans (bennettitaleans), of course, have cuticular features that stand out from those of members of the Cycadales.

From the localities that yielded remains of *Nilssonia,* Plumstead (1962) also described fragments of leaves assignable to *Zamites,* normally considered cycadeoidalean. She also figured a specimen that she interpreted to be *Williamsonia,* an ovulate cone. Both, of course, have an extensive geographic and geologic range.

Lacey and Lucas (1981) indicate the occurrence of *Pterophyllum* from the Upper Triassic of Livingston Island of the South Shetland Islands.

In Gee's revision of the Hope Bay Flora (Gee 1989), she recognized two genera of cycadeoidalean leaves: *Zamites* and *Otozamites*. As Gee indicated, Seward (1917) recommended assigning leaves named *Zamites* by Halle (1913) to *Ptilophyllum* (another cycadeoidalean leaf genus). She, as well as Anderson and Anderson (1985), feel that *Zamites* is the appropriate identification. Gee also figured a possible *Weltrichia* (a microsporangiate cycadeoidalean structure) from Hope Bay. Two specimens were located, neither of which had been observed by Halle. Assignment to *Weltrichia* is tentative, but the specimens appear to bear whorled sporophylls. No microsporangia were observed, but Gee suggests that the surface of the cone exposed is the abaxial one, with the adaxial sporangia probably buried in the matrix.

One specimen of an ovulate cone of *Williamsonia* is also present in the Hope Bay assemblage. A fractured surface reveals fleshy receptacle from which radiate interseminal scales. Ovule detail is not evident.

Cycadolepis, a scale leaf with cycadophytic affinities, was also reported from Hope Bay.

A Word of Caution

The theme of this paper has been to emphasize the conspicuous uniformity among fossil cycadophytes from all parts of the world. Perhaps there are species differences in various localities, but a *Zamites* from Antarctica appears little different from a *Zamites* from Montana, and a *Williamsonia* ovulate cone from India is very much like one from Argentina.

However, unless information about the entire plant is available, there are built-in uncertainties. For example, *Antarcticycas* from the Antarctic Triassic is cycadalean in all respects. Comparison may even be made with certain extant genera. However, although the stem is apparently cycadalean, leaves of cycadophytes are conspicuously absent from the silicified peat. On the other hand, *Dicroidium* leaves are extremely abundant, yet no known stems of *Dicroidium* have been found. Could *Antarcticycas* be the stem of *Dicroidium*?

Another example demonstrates the need to exercise caution in assigning affinity to isolated plant parts. *Taeniopteris* is a very widespread leaf type, ranging from the Late Paleozoic to the end of the Mesozoic. When cuticle is present it tends to be cycadalean. Various kinds of possible cycadophyte fructifications have been associated with *Taeniopteris*. For example, both *Phasmatocycas* and *Archaeocycas* were believed to have been borne on plants with *Taeniopteris* leaves. The Late Permian *Primocycas* is also supposed to have been borne on plants with *Taeniopteris* leaves. On the other hand, Drinnan and Chambers (1985) described leaves referred to *Taeniopteris daintreei* from Lower Cretaceous rocks of Victoria, Australia. On the basis of close association of the *Taeniopteris* with seed-bearing organs referred to *Carnoconites* and pollen-bearing axes assignable to *Sahnia laxiphora,* the authors consider the leaves to be pentoxylalean. White (1981) actually figured leaves of *Taeniopteris spatulata* in connection with pentoxylalean cones. It is obvious, then, that *Taeniopteris* leaves are not exclusively cycadophytic, but are found on plants among various kinds of unrelated or only distantly related kinds of biological entities.

Even though much is known of the fossil record of the Cycadales and Cycadeoidales, much remains to be learned about their biology, distribution, relationships, and evolution. Antarctica is still one of the relatively unexplored areas of the earth. Perhaps significant answers may emerge with continuing vigorous research on the fossil plants from there.

Summary

The present-day distribution of the Cycadales gives no clue concerning the center of radiation. Fossil cycadophytes are not a major part of the floras in Antarctica, and those that are known are similar to forms that occur in widely separated parts of the earth. Most of the fossil remains of cycadophytes are in the form of leaves, and instances are cited where if more of the plant were known some of the plants bearing one type of leaf could conceivably be unrelated to others with the same type of leaf. Because of the still relatively unexplored nature of the Antarctic continent, it is hoped that further efforts will yield valuable information concerning the fossil history of the cycadophytes.

References

Anderson JM, Anderson HM (1985) Palaeoflora of Southern Africa. Prodromus of South African Megafloras Devonian to Lower Cretaceous. A.A. Balkema, Rotterdam

Archangelsky A, Brett DW (1963) Studies on Triassic plants from Argentina. II. *Michelilloa waltonii* nov. gen. et sp. from the Ischigualasto Formation. Annals of Botany (New Series) 27:147–154

Daugherty LH (1941) The Upper Triassic flora of Arizona, with a discussion of its geologic occurrence by H.R. Stagner. Carnegie Institution of Washington Publication 526:108 pp

Delevoryas T (1975) Mesozoic cycadophytes. In Campbell KSW (ed) Gondwana Geology. Papers from the Third International Gondwana Symposium Canberra, Australia, 1973. Australian National University Press, Canberra, A.C.T., Section 2. Gondwana flora 15:173–191

Delevoryas T (1982) Perspectives on the origin of cycads and cycadeoids. Review of Palaeobotany and Palynology 37:115–132

Delevoryas T, Hope RC (1971) A new Triassic cycad and its phylogenetic implications. Postilla 150:1–21

Drinnan AN, Chambers TC (1985) A reassessment of *Taeniopteris daintreei* from the Victorian Early Cretaceous: a member of the Pentoxylales and a significant Gondwanaland plant. Australian Journal of Botany 33:89–100

Farquharson GW (1984) Late Mesozoic, nonmarine, conglomeratic sequences of northern Antarctic Peninsula (the Botany Bay Group). British Antarctic Survey Bulletin 65:1–32

Florin R (1933) Studien über die Cycadales des Mesozoikums nebst Erörterungen über die Spaltöffnungsapparate der Bennettitales. Kungl. Svenska Vetenskapsakademiens Handligar 3rd ser. 12 (5):1–134

Florin R (1940) The Tertiary fossil conifers of south Chile and their phytogeographical significance with a review of the fossil conifers of southern lands. Kungl. Svenska Vetenkapsakademiens Handligar 3rd ser. 19:1–107

Gee CT (1989) Revision of the late Jurassic/early Cretaceous flora from Hope Bay, Antarctica. Palaeontographica (in press)

Gould RE (1971) *Lyssoxylon grigsbyi,* a cycad trunk from the Upper Triassic of Arizona and New Mexico. American Journal of Botany 58:239–248

Greguss P (1968) Xylotomy of the Living Cycads with a Description of their Leaves and Epidermis. Akadémiai Kiadó, Budapest

Halle TG (1913) The Mesozoic flora of Graham Land. Wissensch. Ergebn. Schwed. Südpolar-Exped., 1901–1903, 3(14):1–123

Lacey WS, Lucas RC (1981) The Triassic flora of Livingston Island, South Shetland Islands. British Antarctic Survey Bulletin 53:157–173

Mamay SH (1969) Cycads: fossil evidence of late Paleozoic origin. Science 164:295–296

Mamay SH (1976) Paleozoic origin of the cycads. U.S. Geological Survey Professional Paper 934:1–48

Plumstead EP (1962) Fossil floras of Antarctica. Trans-Antarctic Expedition 1955–1958. Scientific Reports No. 9, Geology 2:1–154

Seward AC (1917) Fossil Plants, Vol. III. Cambridge University Press, Cambridge

Smoot EL, Taylor TN, Delevoryas T (1985) Structurally preserved fossil plants from Antarctica. I. *Antarcticycas,* gen. nov., a Triassic cycad stem from the Beardmore Glacier area. American Journal of Botany 72:1410–1423

Taylor TN (1981) Paleobotany. An Introduction to Fossil Plant Biology. McGraw-Hill Book Company, New York

Thomson MRA (1981) Late Mesozoic stratigraphy and invertebrate palaeontology of the South Orkney Islands. British Antarctic Survey Bulletin 54:65–83

White ME (1986) The Greening of Gondwana. Reed Books Pty. Ltd., Frenchs Forest, N.S.W., 256 pp

Zhu Jia-Nan, Du Xian-Ming (1981) A new cycad—*Primocycas chinensis* gen. et sp. nov. from the Lower Permian in Shanxi, China and its significance. Acta Botanica Sinica 23:401–404

14—Antarctic and Gondwana Conifers

Ruth A. Stockey

In Antarctica, the study of fossil plants in general and the conifers in particular is in its infancy, while the potential for work in this area is enormous. Because of its unique position in Gondwana, the information that we can extract from the fossil record there will add greatly to our knowledge of conifer evolution, especially with respect to the major Southern Hemisphere families. Among the extant conifer families, two in particular, the Podocarpaceae and Araucariaceae, are dominant in Gondwana; however, other unique conifers have also been discovered in the Southern Hemisphere, some of which are now included in their own extinct families. These fossils promise to add much to our knowledge of the early evolution of the Coniferales. This chapter discusses the coniferalean remains from Antarctica and Gondwana, excluding ginkgophytes, and is based primarily on megafossil evidence.

Form Genera

Most of the genera of Coniferales described in Mesozoic floras are form genera of foliage and include *Elatocladus* Halle, *Pagiophyllum* Heer, and *Brachyphyllum* Brongniart. The remains are mostly compression/impression fossils or molds and casts that lack cuticle or internal preservation. The genus *Elatocladus* (Fig. 14.1) has been reported by Plumstead (1962, 1964) and Townrow (1967e) from the Carapace Nunatak in south Victoria Land (Jurassic) and at several localities in the South Shetland Islands dated as Upper Triassic (Banerji and Lemoigne 1987, Lemoigne 1987) and Lower Cretaceous (Jefferson 1982a, b, Laudon et al. 1985). *Pagiophyllum* (Fig. 14.2) also occurs at these localities and was reported on Livingston Island (Triassic) by Lacey and Lucas (1981). The genus *Brachyphyllum,* attributed by most workers to the Araucariaceae, has been more widely reported, sometimes as *Araucaria* or *Araucarites,* in Tertiary age sediments (Dusen 1908, Halle 1913, Barghoorn 1961, Plumstead 1962, Orlando 1963, Barton 1964, Townrow 1967e, Gee 1987). Only one specimen of *Brachyphyllum,* to date, has been found as a permineralization on Alexander Island (Thomson and Tranter 1986) with marine Jurassic fossils (Fig. 14.3). This material should be reinvestigated for internal structure.

The problems of identification of these form taxa have been discussed by Harris (1979), who points out that some species of *Brachyphyllum* may belong to the Araucariaceae while others are referrable to the Cheirolepidiaceae based on cuticular features. The three foliage genera often intergrade. In southern Gondwanaland, however, when cuticular remains are known, the Araucariaceae and Podocarpaceae appear to be the dominant forms (Townrow 1967c). As Townrow (1967c) points out, some of these leaves belong to the Taxodiaceae and others are not referrable to an extant family. One of the reasons for confusion of these foliage types can be seen in

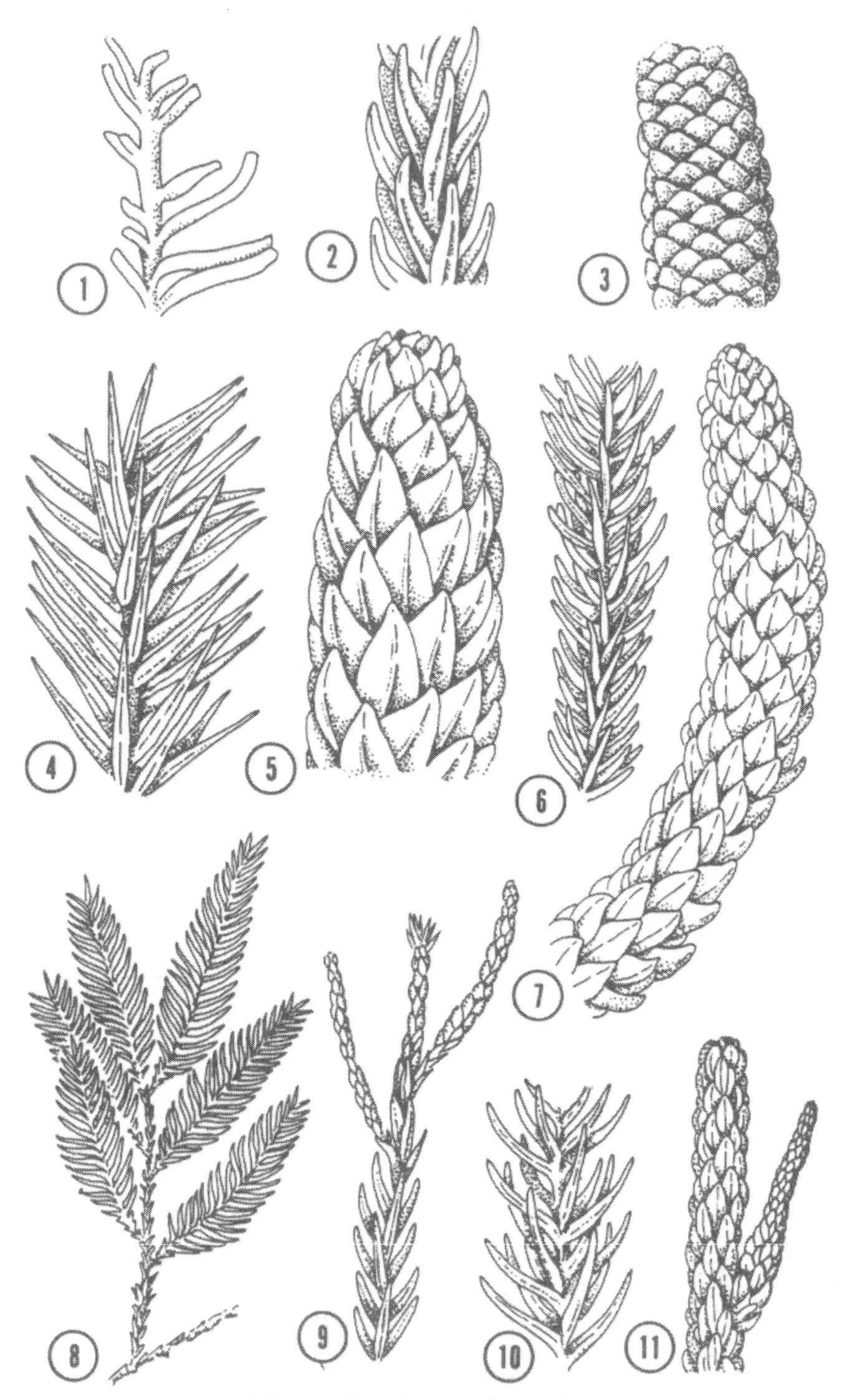

Figure 14.1–14.11. Conifer foliage.

examining growth in the extant Coniferales. For example, in extant *Araucaria,* Section *Eutacta,* there are marked differences between juvenile and adult foliage (de Laubenfels 1972). The transition from juvenile to adult foliage is also extended over several years in taxa such as *Araucaria muelleri* (Carrière) Brongn. et Gris. (Figs. 14.4 and 14.5) and *A. laubenfelsii* (Figs. 14.6 and 14.7), resulting in different leaf sizes, morphologies, and angles of attachment (Stockey 1982). These are the types of artificial criteria used to distinguish *Elatocladus, Pagiophyllum,* and *Brachyphyllum.* The differences in juvenile versus adult foliage are no less striking in the Podocarpaceae (de Laubenfels 1972), for example, in genera such as *Dacrycarpus* (Figs. 14.8 and 14.9) and *Dacrydium* (Figs. 14.10 and 14.11). Some authors (Zastawniak 1981, Zastawniak et al. 1985) have noted similar variability in Tertiary fossil leaves and have used names of modern genera of podocarps for Antarctic fossil leaf remains. However, as Florin (1940a, 1940b) and Townrow (1967c) have noted, cuticular remains for coniferalean taxa are usually quite diagnostic, and cuticular remains have not been reported for most Antarctic fossil conifers.

The genus *Araucarioxylon* is common throughout Gondwana from the Lower Carboniferous to the Tertiary and has often been thought to represent wood of araucarian conifers (e.g., Cortemiglia et al. 1981, Stockey 1982, Nishida and Nishida 1985). This wood type, however, is similar to the genus *Dadoxylon,* and the two are often confused in the literature (Stockey 1982). Such wood may have belonged to cordaites, araucarian conifers, and even glossopterids and is of little use by itself in conifer studies. Several specimens of *Araucarioxylon* wood, however, have provided the basis for studies on fungal decay (white pocket rot) of Antarctic wood (Stubblefield and Taylor 1986).

Permian Conifers

Walkomiella australis Florin (=*Brachyphyllum australe* O. Feistm.) was described from the Upper Permian of Bowenfels near Lithgow and the Newcastle Series (Upper Coal Measures) of New South Wales, Australia (Florin 1940b, 1944). Florin believed that these vegetative remains of branches with triangular, imbricate, spirally arranged leaves terminated in tufts of branchlets (Fig. 14.12). Subsequent workers (White 1981a, 1986) have claimed to have terminal cones at the ends of branches. So far, pollen however, has not been described, and the numbers and positions of "attached seeds" have not been convincingly demonstrated. Based on cuticle, Florin (1940b) distinguished these leaves from the contemporary genera *Paranocladus* Florin and *Lebachia* Florin by their toothed margins, wavy epidermal cell outlines, hairs 1–3 cells long on adaxial leaf bases, 5–9 papillate subsidiary cells, and adaxial stomata. The genus *Walkomiella* has also been reported from the

Figure 14.1. Elatocladus sp.,×2 (Redrawn from Banerji and Lemoigne 1987.).

Figure 14.2. Pagiophyllum sp., ×5.4 (Redrawn from Lacey and Lucas 1981.

Figure 14.3. Brachyphyllum sp., ×1.5. (Redrawn from Thomson and Tranter 1986.)

Figure 14.4. Araucaria muelleri (Carr.) Brongn. et Gris., juvenile to transition foliage, ×0.6. (Redrawn from de Laubenfels 1972).

Figure 14.5. A. muelleri. adult foliage, ×0.6. (Redrawn from de Laubenfels 1972).

Figure 14.6. A. laubenfelsii Corbasson, juvenile foliage, ×0.6. (Redrawn from de Laubenfels 1972).

Figure 14.7. A. laubenfelsii, adult foliage, ×0.6. (Redrawn from de Laubenfels 1972).

Figure 14.8. Dacrycarpus viellardii (Parl.) de Laub., juvenile foliage, ×0.6. (Redrawn from de Laubenfels 1972).

Figure 14.9. D. vieillardii, adult foliage with pollen cones, ×2. (Redrawn from de Laubenfels 1972).

Figure 14.10. Dacrydium araucarioides Brongn. et Gris., transition foliage, ×2. (Redrawn from de Laubenfels 1972).

Figure 14.11. D. araucarioides, adult foliage with pollen cone, ×2 (Redrawn from de Laubenfels 1972).

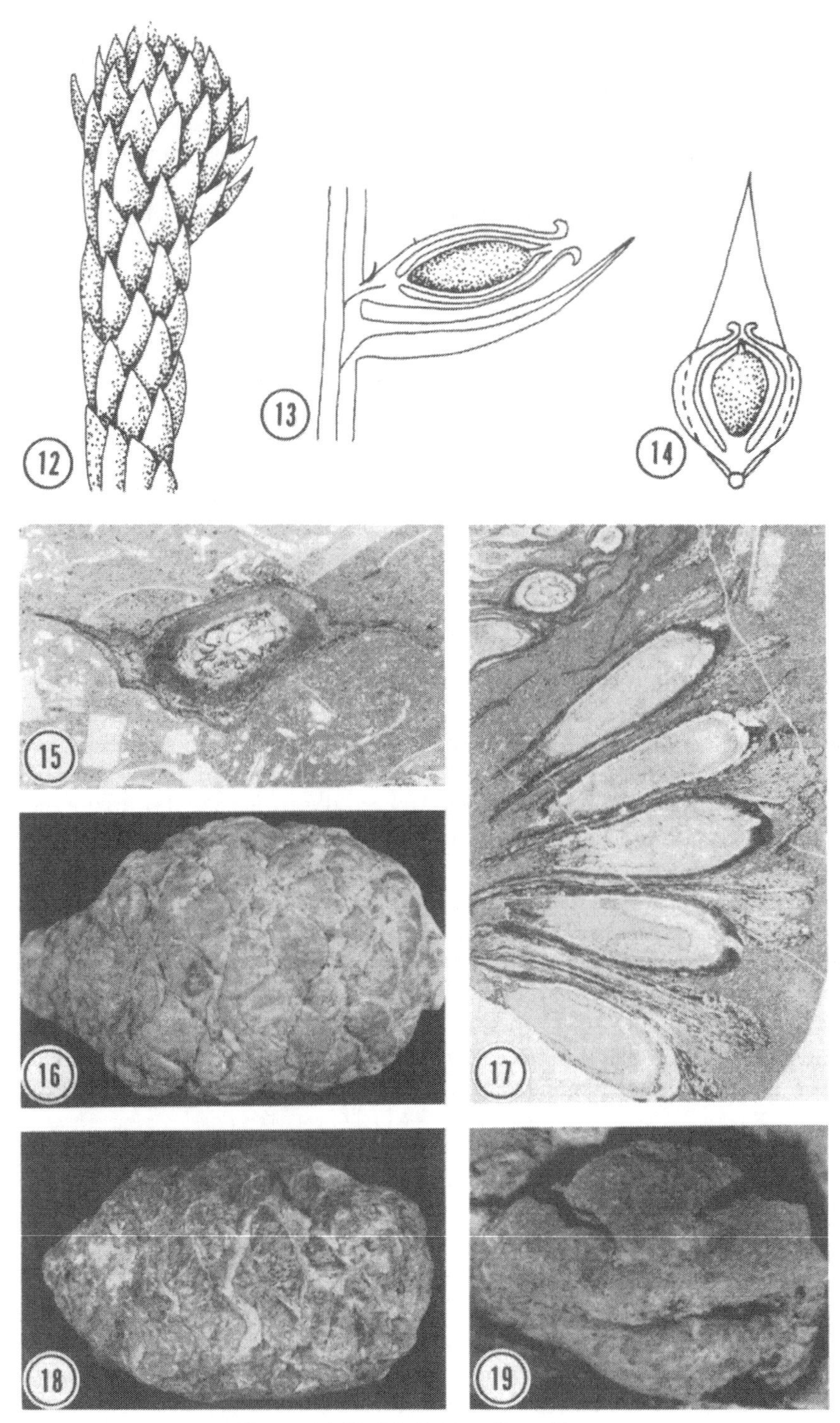

Figure 14.12–14.19. Fossil conifers

Permian of India (Surange and Singh 1951, 1953) and South Africa (LeRoux 1964); however, Pant (1982) suggests that the affinities of some of these remains with *Walkomiella* may be in doubt.

Voltziopsis Potonié from the Upper Permian and Lower Triassic of Africa and Australia (Townrow 1967d) is reported to show leaf dimorphism. Attached ovulate cones are known with five-lobed ovuliferous scales, each with its own anatropous ovule. This material is treated by Miller (1977) in the Voltziaceae, a predominantly Northern Hemisphere family.

Recently, Archangelsky and Cuneo (1987) have described a new conifer family with two genera from the Permian of Argentina. The Ferugliocladaceae, containing *Ferugliocladus* (Figs. 14.13 and 14.14) and *Ugartecladus*, shows foliage of the well-known and variable *Paranocladus* type (Florin 1940b, 1963; Fittipaldi and Rösler 1978) and seeds originally described as *Eucerospermum*. *Ferugliocladus* is distinguished from *Ugartecladus* by its larger winged ovules with bifid apices and five orders of branching versus four. Both genera show ovulate cones with simple bracts, orthotropous ovules, and no documented evidence of ovuliferous scale tissue (Archangelsky and Cuneo 1987). These cones have been compared with those of *Buriadia* and *Genoites* by Archangelsky (1985) and Archangelsky and Cuneo (1987), who believe that they may have had a common ancestor. Pollen cones have been reported; however, the number of pollen sacs is in question. *Cannanoropollis* Potonié and Sah is the most common monosaccate grain associated with the cone material. Further specimens should clarify many aspects of the family delimitation.

Araucariaceae

Pollen referred to the Araucariaceae has been reported throughout Gondwana and also extended into the Northern Hemisphere (Cookson 1947, Archangelsky and Romero 1974, Stockey 1982). *Araucariacites* Cookson (1947) grains have been reported from the Tertiary lignites of the Kerguelen Archipelago and as early as the Lower–Upper Triassic from Timber Peak and Section Peak (Beacon and Ferrar Groups) in northern Victoria Land (Cookson 1947, Gair et al. 1965).

Of the several genera that we now know to represent "the *Brachyphyllum crassum* complex," Townrow (1967c) described *Araucaria crassa* from Ipswich, Queensland (?Tertiary). Even though he recognized the various specimens lumped into this group in different genera and families based on cuticular evidence, Townrow reported small, hollow papillae on the outer cuticle of epidermal cells of *A. crassa*, a character not seen in living Araucariaceae (Stockey and Ko 1986, Stockey and Atkinson 1988), suggesting perhaps affinities elsewhere.

Recently, several araucarian leaves with cuticular remains have been described from the Eocene of Tasmania (Bigwood and Hill

◁

Figure 14.12. *Walkomiella australis* (O. Feistm.) Florin, leafy twig with swollen apex, ×3.25 (Redrawn from Florin 1940b).

Figure 14.13. *Ferugliocladus patagonicus* (Feruglio) Arch., longitudinal section of bract and ovule, ×4 (Redrawn from Archangelsky and Cuneo 1987).

Figure 14.14. *F. patagonicus*, adaxial view of bract and ovule, ×3. (Redrawn from Archangelsky and Cuneo 1987).

Figure 14.15. Araucarian cone scale, transverse section near micropylar end of ovule, 840208 C top #9, Chiba University, Japan, ×5.

Figure 14.16. Longitudinal section of crushed araucarian cone, 860961 A #1, Chiba University, Japan, ×2.5.

Figure 14.17. Conifer cone from Cerro Cuadrado (Alma Gaucha-Santa Cruz), La Plata Museum #122-1, ×1.3.

Figure 14.18. Conifer cone from Cerro Cuadrado (Alma Gaucha-Santa Cruz), La Plata Museum # 125-1. ×1.3.

Figure 14.19. Cone from Cerro Cuadrado (Alma Gaucha-Santa Cruz), La Plata Museum #122-2, ×3.5.

1985). One of the leaf forms referred to as *Araucarioides* apparently has intermediate cuticular features between *Agathis* and *Araucaria*. This evidence coupled with the large numbers of "araucarian" cone-scales seen in Northern Hemisphere localities perhaps suggests that the Araucariaceae may have included an additional genus in the past. Recent finds of araucarian cones and isolated scales from the Upper Cretaceous of Hokkaido (Figs. 14.15 and 14.17) also suggest that this might be the case (Stockey et al. 1986). Several cones have been found with araucarian structure, the single seeds are embedded in scale tissue, but no free ovuliferous scale tip is present (Fig. 14.15). Integumentary anatomy is identical to that described for araucarian cones (Stockey 1975, 1978, 1982). The genus *Agathis,* while it does show a fused bract and scale, has winged seeds, and these Japanese araucarians shed their scales rather than their seeds at maturity.

Cones described as "araucarian" or *Araucarites* have also been found in Cretaceous age deposits of New Zealand (Mildenhall and Johnston 1971), South Africa (Seward 1903, Brown 1977), the Jurassic to Cretaceous of India (Singh 1957, Bose and Jain 1964, Pant and Srivastava 1968, Bose and Maheshwari 1973, Suk-Dev and Bose 1974, Sharma and Bohra 1977, Suk-Dev and Zeba-Bano 1978), Antarctica (Halle 1913), and Argentina (Archangelsky 1966). These remains need to be reexamined in the future for structure and affinities.

The best known araucarian remains are those of the Jurassic Cerro Cuadrado petrified forest of Patagonia (Stockey 1975, 1978, 1982; Stockey and Taylor 1978). Cones of the most common taxon, *Araucaria mirabilis,* have been studied developmentally from the free nuclear phase of gametophyte development to the mature dicotyledonous embryo phase. The structure of these cones most closely resembles that of extant *A. bidwillii* from Queensland, and they have been placed into the section *Bunya* of the genus *Araucaria*. Similar permineralized cones have been reported from the Jurassic of India (Vishnu-Mittre 1954). Fossil seedlings that also probably belong to *A. mirabilis* have been described as having swollen hypocotyls and hypogeal germination (Stockey and Taylor 1978). The remains of foliage and wood, some of which have been found attached to cones, have not yet been thoroughly studied, but future work promises to add much to our knowledge of these araucarians as whole plants.

The second species described based on cones from the Cerro Cuadrado, *Pararaucaria patagonica,* combines features of several conifer families, especially the Taxodiaceae and Pinaceae, and shows little affinity to the Araucariaceae (Stockey 1977). These cones also exhibit well-preserved embryo tissues and probably had epigeal germination with 6–8 photosynthetic cotyledons.

Also from the Jurassic Cerro Cuadrado, several larger cylindrical cones resembling *Pararaucaria* have been found (Figs. 14.16 and 14.18). Most of these cones have badly scorched surfaces and were thought to be large *Pararaucaria* cones (Stockey 1977). However, several well-preserved cones up to 6 cm long and 3.5 cm wide are now known with large bracts and three-lobed ovuliferous scales (Fig. 14.19). Future sectioning of this material should provide anatomical details. The three-lobed nature of the ovuliferous scale is reminiscent of the voltzialean conifers *Pseudovoltzia* and *Tricranolepis* and several of the strobiloid podocarps both recent and extinct (Miller 1977; Townrow 1967a, 1967b).

Taxodiaceae

The one genus of the Taxodiaceae widespread in Gondwana is *Athrotaxis* Don, today confined to Tasmania. *Athrotaxis tasmanica* (Townrow 1965b, 1967c) from the Lower Tertiary of Australia was once also part of "the *Brachyphyllum crassum* complex." *Athrotaxis ungeri* (Halle) Florin is known from the Lower Cretaceous Ticó Flora of Patagonia (Florin 1940c, Archangelsky 1966). Leaves of this genus are spirally arranged, imbricate and scale-like, lack papillae, and have a scarious or winged edge, collars around stomatal pits, and irregular subsidiary cell and stomatal arrangements (Fig. 14.20 and 14.21). Thus, the genus was more widespread in the past, and as more cuticular remains are known, identification of other fossils as *Athrotaxis* and their separation from other *Brachyphyllum*-like taxa will be possible.

Podocarpaceae

By far, the largest number of coniferous remains in Gondwana are referrable to the Podocarpaceae. The best known of these are strobiloid forms that belong to extinct genera. *Nothodacrium warrenii* (Townrow 1967b, 1967e) from the Carapace Nunatak, Jurassic Beacon Group in East Antarctica, is represented by branches bearing helically arranged rhomboidal leaves with decurrent bases and terminal spike-like ovulate cones with free simple bracts, trilobed ovuliferous scales, and one anatropous seed/scale (Figs. 14.22–14.25). Both juvenile and adult foliage types were documented based on cuticular remains (Townrow 1967c). Associated pollen cones are referred to *Masculostrobus warrenii* (Townrow 1967c) and bear two pollen sacs/microsporophyll with trisaccate pollen probably referrable to *Tsugaepollenites trilobatus* (Balme) Dettmann. Pollen cones lack cuticle and have not yet been found attached to branches. Townrow (1967c) suggests similarities of *Nothodacrium* to certain *Dacrydium* species. Preliminary studies of cuticles of *Dacrydium,* Gruppe B of Florin (1931) by Stockey and Ko (1989), however, do not show epidermal papillae as in *Nothodacrium*.

The genus *Rissikia* Townrow is one of the most widespread of the early podocarps occurring in the Triassic of Madagascar, South Africa, Australia, and Antarctica from the Allan Hills in Victoria Land (Townrow 1967a, 1967e, 1969). *Rissikia* species are characterized by foliar spurs that were perhaps deciduous (Fig. 14.29), bearing thick flattened leaves with decurrent bases, acute apices, and a rhomboidal shape in transverse section (Townrow 1967a). Attached pollen cones are round with 2 pollen sacs per microsporophyll and bisaccate, striate pollen. Ovulate cones are terminal and spike-like with trifurcate bracts and trilobed scales with 2–6 anatropous seeds/scale, only 1 or 2 of which probably matured (Figs. 14.26 and 14.27). Townrow (1967a) suggested that *Elatocladus planus* (Feistmantel) Seward from the Talbragar (?Jurassic) Fish Beds of New South Wales, Halle's (1913) specimens from Hope Bay, Antarctica, and *E. australis* Frenguelli (1940) from Argentina may also represent species of *Rissikia*. White (1981b, 1986) figured an elongate, spike-like cone attached to *E. planus*-like foliage in the Talbragar Fish Beds that she called *Rissikia talbragarensis*. It is not clear at the present time which of the numerous "*Elatocladus planus*" leaves at this locality belong to this cone. Cuticular remains in this instance would be invaluable. Cones of *R. talbragarensis* are more complete than any of those described by Townrow (1967a, 1969) and should be reinvestigated.

Mataia podocarpoides (Ett.) Townrow (1967a), also formerly in part known as a species of *Elatocladus,* from the Jurassic of New Zealand and Australia, has bifacial, spirally arranged leaves that are twisted into a two-ranked arrangement, with abaxial midribs and acute apices. Spike-like ovulate cones 3 cm long have free bracts and ovuliferous scales that curved back adaxially to partially cover the two anatropous ovules (Fig. 14.28). Townrow (1967a) suggests the evolution of the podocarp epimatium from such infolded ovuliferous scale tissue.

In addition to these well-preserved cone and leaf remains, several other reproductive structures have been reported. An unnamed specimen that may represent a podocarp cone has been briefly described from the Early Cretaceous of New Zealand (Mildenhall 1976). *Mehtaia* (Vishnu-Mittre 1958), twigs with scale-like leaves and attached cones from the Jurassic of India, have erect ovules, lack an aril, epimatium, or other sterile tissue, and have been compared to *Pherosphaera*. These cones should probably be compared to the Ferugliocladaceae. *Nipaniostrobus* Rao (Vishnu-Mittre 1958) includes ovulate cones (Jurassic, India) on branches with small imbricate leaves. Seeds are anatropous and partially

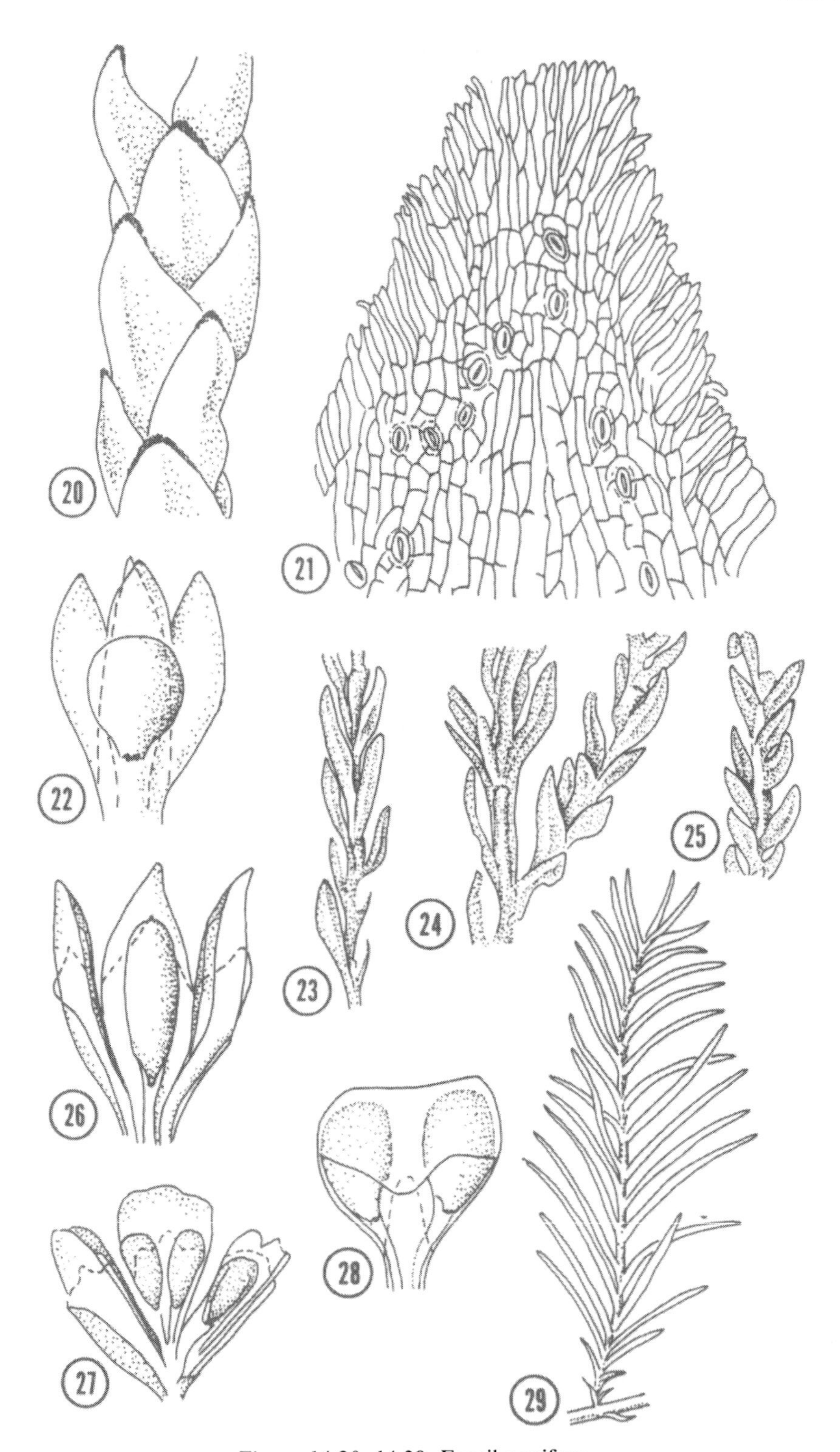

Figure 14.20–14.29. Fossil conifers.

enclosed in recurved ovuliferous scale tissue as in *Mataia* and *Nipanioruha* Rao (Vishnu-Mittre 1958). Pollen cones in *Nipaniostrobus* have trisaccate pollen as in several living podocarps (Miller 1977). *Podostrobus* Rao and Bose (1970) has also been described from the Jurassic of India and Antarctica from permineralized material with bisaccate and trisaccate pollen grains. *Sitholyea* also from the Jurassic of India is reported to have an ovuliferous scale with one ovule terminal on a leafy shoot.

Other genera in the family include several South American taxa. *Trisaccocladus, Apterocladus, Podocarpus dubius,* and *Brachyphyllum tigrense* from the Cretaceous of Argentina have now all been included in the Podocarpaceae (Gamerro 1965, Archangelsky 1966, Archangelsky and del Fueyo 1987). All except *P. dubius* are known from leaf cuticles as well as pollen cones with bisaccate and trisaccate pollen.

Tertiary Gondwana podocarp taxa have usually been put into extant genera and new species have usually been erected. *Acmopyle,* confined to New Caledonia and Fiji today, has been described from the Late Tertiary of Seymour and King George Islands and northern Patagonia (Berry 1938, Florin 1940a, Zastawniak 1981) from megafossils and pollen (Cranwell 1959, Couper 1960).

Other genera that appear to have been more diverse and widespread also include *Microstrobus, Microcachrys,* and *Phyllocladus* based on pollen evidence (Cookson 1947, Cranwell 1959, Couper 1960, Mildenhall 1978). Wood of the *Phyllocladus* type has been described from the Cretaceous of Alexander Island, Antarctica, from stumps referrable to *Circoporoxylon* Kräusel (Jefferson 1982a). Megafossils (cuticular pieces) and pollen have also been found from the Oligocene of Australia and New Zealand (Couper 1960) showing a wider past distribution than at present.

Several *Podocarpus* species, *P. stzeleckianus, P. tasmanicus, P. setiger, P. goedei, P. acicularis, P. inopinatus, P. araucoensis,* and *P. dubius* (Florin 1940a, Townrow 1965a, Archangelsky 1966), have all been compared to sections of the living genus. However, even after major revisions to the large (100+ species) extant genus (de Laubenfels 1969, 1972), comparisons are difficult. Future work on elucidating the cuticular structure of the Podocarpaceae will greatly aid in comparisons of fossil and extant podocarps (Stockey and Ko 1988a, 1989).

An extinct genus, *Coronelia* (Florin 1940a, Townrow 1965a), is known from the Eocene of Concepción, Chile, and from Buckland, Tasmania. The distinct features of this genus include very constricted leaf bases, epidermal papillae, uni- or multicellular hairs in the two abaxial stomatal zones fusing to form Florin rings, and marginal tooth-like projections on

Figure 14.20. Athrotaxis tasmanicus Townrow, twig with leaves, ×10 (Redrawn from Townrow 1965b).

Figure 14.21. A. tasmanicus leaf showing wing and stomatal orientations, ×135. (Redrawn from Townrow 1965b).

Figure 14.22. Nothodacrium warrenii Townrow, cone-scale complex (bract dashed), ×3.8 (Redrawn from Townrow 1967b).

Figure 14.23. N. warrenii leaves, ×7. (Redrawn from Townrow 1967b).

Figure 14.24. N. warrenii, variable leaf morphology, ×7. (Redrawn from Townrow 1967b).

Figure 14.25. N. warrenii leaves, ×7 (Redrawn from Townrow 1967b).

Figure 14.26. Rissikia apiculata Townrow, cone-scale complex (bract dashed), ×3.8 (Redrawn from Townrow 1969).

Figure 14.27. R. media (Tenison-Woods) Townrow, cone-scale complex (bract dashed), ×3.8 (Redrawn from Townrow 1969).

Figure 14.28. Mataia podocarpoides (Ett.) Townrow, cone-scale complex showing ovuliferous scale tissue partially covering ovules (bract dashed), ×3.8. (Redrawn from Townrow 1969).

Figure 14.29. R. media, portion of a leafy shoot, ×1. (Redrawn from Townrow 1969).

the leaves. The Chilean leaves show swollen tips on the multicellular hairs, while the Tasmanian leaves do not.

Incertae Sedis

Leaves that have been formerly referred to as *Podozamites lanceolatus* from the Talbragar Fish Beds of Australia, renamed by White (1981b) *Agathis jurassica,* may also prove to be podocarp taxa with foliar spurs similar to those described for *Rissikia* (Townrow 1967a). The assignment of associated cone scales described by White (1981b) to *Agathis* has been questioned (Stockey 1982). Assignment to *Araucaria* Section *Eutacta* cannot be ruled out at the present time.

The genus *Allocladus* Townrow (1967c), although known from cuticular data and found in the Jurassic of India and Australia, does not fit into any extant family. These leaves have been referred to as *Echinostrobus, Brachyphyllum, Pagiophyllum, Walchia,* and *Elatocladus* and appear similar to *Araucaria crassa* and *Athrotaxis tasmanica.* This form, along with *Pagiophyllum feistmantelii* Halle from the Middle Jurassic of Australia, is also part of the "*Brachyphyllum crassum* complex" (Townrow 1967c).

Tomaxellia Archangelsky (1963, 1966) from the Upper Cretaceous of Argentina is known from long, narrow, heteromorphic leaves that have been compared with *Elatides, Elatocladus,* and extant members of the Podocarpaceae, Araucariaceae, and Taxodiaceae. Compressed cones of this genus shed their scales, retained their seeds, and had two seeds/scale.

Summary

Most Antarctic conifer material described to date has been based on poorly preserved foliage form genera. Based on our knowledge of Northern Hemisphere and other Gondwana taxa with similar morphologies, several conifer families probably occur. Gondwana floras of similar ages have revealed well-preserved cuticular remains of the Araucariaceae, Podocarpaceae, and Taxodiaceae and several distinct Permian taxa, including the Ferugliocladaceae and Voltziaceae. Based on the Gondwana distributions of these families, their presence in Antarctica should be confirmed by future collecting.

One new genus and species, *Nothodacrium warrenii* Townrow, was based primarily on Antarctic fossil remains. The presence of cuticle on both adult and juvenile leaves, attached ovulate cones, and associated pollen-bearing cones from the Carapace Nunatak (Jurassic-Beacon Group) indicates that this locality is one that will enable whole plant reconstructions and should be collected extensively in the future.

The future identification of fossil conifer material from Antarctica and elsewhere, whether compressed or permineralized, requires a broad data base. Further studies of extant conifer growth, anatomy, and cuticular structure, in particular, will greatly help the paleobotanist in making taxonomic decisions and in understanding these remains as once parts of whole plants. Only then will our ideas on evolution of taxa be meaningful.

Because of Antarctica's unique position in the former continent of Gondwana, it probably played a vital role in the dispersal and present-day distribution of the extant Coniferales. The fossil record, while very incomplete, shows much promise for future investigation.

Acknowledgments. I thank Dr. Gar W. Rothwell for photographic assistance and Rebecca Sampson, Ohio University, for providing the line drawings. This chapter was supported in part by Natural Sciences and Engineering Research Council of Canada Grant A-6908.

References

Archangelsky S (1963) A new Mesozoic flora from Ticó, Santa Cruz Province, Argentina. Bull. British Museum (Natural History) Geol. 8:45–92

Archangelsky S (1966) New gymnosperms from the Ticó Flora, Santa Cruz Province, Argentina. Bull. British Museum (Natural History) Geol. 13:259–295

Archangelsky S (1985) Aspectos evolutivos de las coníferas gondwánicas del Paleozóico. Bull. Sect. Sci. Com. Trav. Hist. Sci. 8:115–124
Archangelsky S, Cúneo RC (1987) Ferugliocladaceae, a new conifer family from the Permian of Gondwana. Rev. Palaeobot. Palynol. 51:3–30
Archangelsky S, del Fueyo G (1987) Sobre una Podocarpacea fertíl del Cretácico Inférior de la Provincia Santa Cruz, República Argentina. VII Simposio Argentino de Paleobotánica y Palinologia Actas. Buenos Aires, pp. 85–87
Archangelsky S, Romero EJ (1974) Polen de gimnospermas (coníferas) del Cretácico supérior y Paleoceno de Patagonia. Ameghiniana 11:217–236
Banerji J, Lemoigne Y (1987) Significant additions to the Upper Triassic flora of Williams Point, Livingston Island, South Shetlands (Antarctica). Géobios 20:4:469–487
Barghoorn, ES (1961) A brief review of fossil plants of Antarctica and their geologic implications. Natl. Acad. Sci. Publ. 839:5–9
Barton CM (1964) Significance of the Tertiary fossil floras of King George Island, South Shetland Islands. In Adie RJ (ed) Antarctic Geology (Proc. 1st Int. SCAR Symp. on Antarctic Geol., Cape Town 1963), North-Holland Publ. Co., Amsterdam, pp. 603–608
Berry EW (1938) Tertiary flora from the Río Pichileufu, Argentine. Spec. Pap. Geol. Soc. Amer. 12:40
Bigwood AJ, Hill RS (1985) Tertiary araucarian macrofossils from Tasmania. Aust. J. Bot. 33: 645–656
Bose MN, Jain KP (1964) A megastrobilus belonging to the Araucariaceae from the Rajmahal Hills, Bihar, India. Palaeobotanist 12:229–231
Bose MN, Maheshwari HK (1973) Some detached seed-scales belonging to Araucariaceae from the Mesozoic rocks of India. Geophytology 3:205–214
Brown JT (1977) On *Araucarites rogersii* Seward from the Lower Cretaceous Kirkwood Formation, Algoa Basin, Cape Province, South Africa. Palaeont. Afr. 20:47–51
Cookson IC (1947) Plant microfossils from the lignites of Kerguelen Archipelago. B.A.N.Z. Antarct. Res. Expd. Rep. Ser. A 2:129
Cortemiglia GC, Gastaldo P, Terranova R (1981) Studio di piante fossili trovate nella King George Island delle Isole Shetland del sud (Antartide). Atti Soc. ital. Sci. Nat. Museo civ. Stor. nat. Milano 122:37–61
Couper RA (1960) Southern Hemisphere Mesozoic and Tertiary Podocarpaceae and Fagaceae and their palaeogeographic significance. Royal Society of London, Proceedings, Ser. B, 152:491–500
Cranwell LM (1959) Fossil pollen from Seymour Island, Antarctica. Nature 184:1782–1785
de Laubenfels DJ (1969) A revision of the Malesian and Pacific rainforest conifers, I. Podocarpaceae, in part. J. Arnold Arbor. 50:274–369
de Laubenfels DJ (1972) Flore de la Nouvelle-Caledonie et dépendances. 4. Gymnospermes. Mus. Natl. Hist. Nat., 168 pp
Dusén P (1908) Über die tertiare Flora der Seymour-Insel. Wiss. Ergebn. schwed. Südpolarexped. 3:27
Fittipaldi FC, Rösler O (1978) *Paranocladus ?fallax* (Conifera). Estudios cuticulares. Bolm. IG Sao Paulo 9:109–113
Florin R (1931) Untersuchungen zur Stammesgeschichte der Coniferales und Cordaitales. Kungl. Svenska Vetenskapsakademiens Handlingar, Ser. 5, Bd. 10:1–588
Florin R (1940a) The Tertiary conifers of south Chile and their phytogeographical significance with a review of the fossil conifers of southern lands. Kungl. Svenska Vetenskapsakademiens Handlingar, Ser. 3, Bd. 19:1–103
Florin R (1940b) On *Walkomia* n. gen. a genus of Upper Paleozoic conifers from Gondwanaland. Kungl. Svenska Vetenskapsakademiens Handlingar, Ser. 3, Bd. 18:1–23
Florin R (1940c) Die heutige und fruhere Verbreitung der Koniferengattung *Acmopyle* Pilger. Svensk Botanisk Tidskrift 34:119–140
Florin R (1944) Die Koniferen des Oberkarbons und des unteren Perms. Palaeontographica B85:365–456
Florin R (1963) The distribution of conifer and taxad genera in time and space. Acta Horti Bergiani Bd. 20:121–312
Frenguelli J (1940) Contributions al concimiento de la flora del Gondwana supérior en la Argentina XXIX. *Elatocladus australis*. Nat. Mus. La Plata 9:543–548
Gair HS, Norris G, Ricker J (1965) Early Mesozoic microfloras from Antarctica. N.Z.J. Geol. Geophys. 8:231–235
Gamerro JC (1965) Morfologia del polen de *Apterocladus lanceolatus* Archang. (Coniferae) de la Formación Baqueró, Provincia de Santa Cruz. Ameghiniana 4:133–136
Gee CT (1987) Revision of the Early Cretaceous flora from Hope Bay, Antarctica. XIV International Botanical Congress, Berlin (West) Germany. Abstracts, p. 333
Halle TG (1913) Some Mesozoic plant bearing deposits in Patagonia and Tierra del Fuego, and their floras. Kungl. Svenska Vetenskapsakademiens Handlingar. 51:1–58
Harris TM (1979) The Yorkshire Jurassic Flora. V. Coniferales. Brit. Mus. (Nat. Hist.) 803:1–166
Jefferson TH (1982a) Fossil forests from the Lower Cretaceous of Alexander Island, Antarctica. Palaeontology 25:681–708
Jefferson TH (1982b) The preservation of fossil leaves in Cretaceous volcaniclastic rocks

from Alexander Island, Antarctica. Geol. Mag. 119:291–300

Lacey WS, Lucas RC (1981) The Triassic flora of Livingston Island, South Shetland Islands. Br. Antarct. Surv. Bull. 53:157–173

Laudon WS, Lidke DJ, Delevoryas T, Gee CT (1985) Sedimentary rocks of the English Coast, eastern Ellsworth Land, Antarctica. Antarctic Jour. of the U.S. 20(5):38–40

Lemoigne Y (1987) Confirmation de l'existence d'une flore triasique dans l'île Livingston des Shetland du sud (Ouest Antarctique). C.R. Acad. Sc. Paris t. 304. sér. II. 10:543–546

LeRoux SF (1964) *Walkomiella transvaalensis*, a new conifer from the Middle Ecca beds of Vereeniging Transvaal. Trans. Geol. Soc. S. Afr. 66:1–8

Mildenhall DC (1976) Early Cretaceous podocarp megastrobilus. N.Z.J. Geol. Geophys. 19:389–391

Mildenhall DC (1978) *Cranwellia costata* n. sp. and *Podosporites erugatus* n. sp. from Middle Pliocene (?Early Pleistocene) sediments, South Island, New Zealand, J. Roy. Soc. N. Z. 8:253–274

Mildenhall DC, Johnston MR (1971) A megastrobilus belonging to the genus *Araucarites* from the Upper Motuan (Upper Albian), Wairarapa, North Island, New Zealand. N.Z.J. Bot. 9:67–79

Miller, Jr. CN (1977) Mesozoic conifers. Bot. Rev. 43:217–280

Nishida M, Nishida H (1985) Petrified woods from the Upper Cretaceous of the Quirquina Island. In Nishida M (ed) A Report of the Botanical Survey to Bolivia and Southern Chile by a Grant-in-Aid for Overseas Scientific Survey, 1983. Faculty of Science, Chiba University. Hasedate Printing Co., Chiba, Japan., pp. 27–36

Orlando HA (1963) La flora fósil en las inmediaciones de la Peninsula Ardley, Isla 25 de Mayo, Islas Shetland del Sur. Contr. del Instituto Antártico Argentino 79:1–17

Pant DD (1982) The lower Gondwana gymnosperms and their relationships. Rev. Palaeobot. Palynol. 37:55–70

Pant DD, Srivastava GK (1968) On the cuticular structure of *Araucaria (Araucarites) cutchensis* (Feistmantel) comb. nov. from the Jabalpur series, India. J. Linn. Soc. (Bot.) 61:201–206

Plumstead EP (1962) Fossil floras of Antarctica (with an appendix on Antarctic fossil wood by Richard Kräusel). Trans-Antarctic Exped. 1955–1958. Sci. Rep. Geol. 9:1–154

Plumstead EP (1964) Palaeobotany of Antarctica. In Adie RJ (ed) Antarctic Geology, North-Holland Publ. Co., Amsterdam, pp. 637–654

Rao AR, Bose MN (1970) *Podostrobus* gen. nov., a petrified podocarpaceous male cone from the Rajmahal Hills, India. Palaeobotanist 19:83–85

Seward AC (1903) Flora of the Uitenhage Series. Fossil floras of the Cape Colony. Ann. S. Afr. Mus. 4:1–22

Sharma BD, Bohra, DR (1977) Petrified araucarian megastrobili from the Jurassic of the Rajmahal Hills, India. Acta Palaeobot. 18:31–36

Singh G (1957) *Araucarites nipaniensis* sp. nov.—a female araucarian cone-scale from the Rajmahal Series. Palaeobotanist 5:64–65

Stockey RA (1975) Seeds and embryos of *Araucaria mirabilis*. Amer. J. Bot. 62:856–868

Stockey RA (1977) Reproductive biology of Cerro Cuadrado (Jurassic) conifers: *Pararaucaria patagonica*. Amer. J. Bot. 64:733–744

Stockey RA (1978) Reproductive biology of Cerro Cuadrado fossil conifers: ontogeny and reproductive biology of *Araucaria mirabilis* (Spegazzini) Windhausen. Palaeontographica B166:1–15

Stockey RA (1982) The Araucariaceae: an evolutionary perspective. Rev. Palaeobot. Palynol. 37:133–154

Stockey RA, Atkinson IJ (1988) Cuticle micromorphology of Araucariaceae: *Agathis* Salisbury. Amer. J. Bot. 75:117–118

Stockey RA, Ko H (1986) Cuticle micromorphology of *Araucaria* de Jussieu. Bot. Gaz. 147:508–548

Stockey RA, Ko H (1988) Cuticle micromorphology of some New Caledonian podocarps. Bot. Gaz. 149:240–252

Stockey RA, Ko H (1989) Cuticle micromorphology of *Dacrydium* (Podocarpaceae) from New Caledonia. Bot. Gaz. (in press)

Stockey RA, Taylor TN (1978) On the structure and evolutionary relationships of the Cerro Cuadrado fossil conifer seedlings. J. Linn. Soc. London. Bot. 76:161–176

Stockey RA, Nishida M, Nishida H (1986) Permineralized araucarian remains from the Upper Cretaceous of Japan. Amer. J. Bot. 73:78

Stubblefield SP, Taylor TN (1986) Wood decay in silicified gymnosperms from Antarctica. Bot. Gaz. 147:116–125

Suk-Dev, Bose MN (1974) On some conifer remains from Bansa, South Rewa Gondwana Basin. Palaeobotanist 21:59–69

Suk-Dev, Zeba-Bano (1978) *Araucaria indica* and two other conifers from the Jurassic–Cretaceous rocks of Madhya Pradesh, India. Palaeobotanist 25:496–508

Surange K, Singh P (1951) *Walkomiella indica*, a new conifer from the lower Gondwanas of India. J. Indian Bot. Soc. 30:143–147

Surange K, Singh P (1953) The female dwarf shoot of *Walkomiella indica*—a conifer from the lower Gondwanas of India. Palaeobotanist 3:5–8

Thomson MRA, Tranter TH (1986) Early Jurassic fossils from central Alexander Island and their geological setting. Br. Antarct. Surv. Bull. 70:23–39

Townrow JA (1965a) Notes on some Tasmanian pines. I. Some lower Tertiary podocarps. Roy. Soc. Tasmania Papers Proc. 99:87–108

Townrow JA (1965b) Notes on Tasmanian pines. II. *Athrotaxis* from the lower Tertiary. Roy. Soc. Tasmania Papers Proc. 99:109–113

Townrow JA (1967a) On *Rissikia* and *Mataia* podocarpaceous conifers from the lower Mesozoic of southern lands. Roy. Soc. Tasmania Papers Proc. 101:103–136

Townrow JA (1967b) On a conifer from the Jurassic of East Antarctica. Roy. Soc. Tasmania Papers Proc. 101:137–148

Townrow JA (1967c) The *Brachyphyllum crassum* complex of fossil conifers. Roy. Soc. Tasmania Papers Proc. 101:149–172

Townrow JA (1967d) On *Voltziopsis*, a southern conifer of Lower Triassic age. Roy. Soc. Tasmania Papers Proc. 101:173–188

Townrow JA (1967e) Fossil plants from Allan and Carapace Nunataks, and from the Upper Mill and Shackleton Glaciers, Antarctica. N.Z.J. Geol. Geophys. 10:456–473

Townrow JA (1969) Some lower Mesozoic Podocarpaceae and Araucariaceae. In Gondwana Stratigraphy: IUGS Symposium, Mar del Plata, 1–15 Oct. 1967, UNESCO, Sec. I. (Biochronologia) pp. 159–184

Vishnu-Mittre (1954) *Araucarites bindrabunensis* sp. nov., a petrified megastrobilus from the Jurassic Rajmahal Hills, Bihar. Palaeobotanist 3:103–108

Vishnu-Mittre (1958) Studies on the fossil flora of Nipania (Rajmahal Series) Bihar-Coniferales. Palaeobotanist 6:82–122

White ME (1981a) The cones of *Walkomiella australis* (Feist.) Florin. Palaeobotanist 28–29:75–80

White ME (1981b) Revision of the Talbragar Fish Bed flora (Jurassic) of New South Wales. Records of the Australian Museum 33:695–721

White ME (1986) The greening of Gondwana. Reed Books Pty. Ltd., Frenchs Forest, N.S.W. 256 p.

Zastawniak E (1981) Tertiary leaf flora from the Point Hennequin Group of King George Island (South Shetland Islands, Antarctica). Preliminary Report. Studia Geologica. Polonica. Warszawa 72:97–108

Zastawniak E, Wrona R, Gazdzicki A, Birkenmajer K (1985) Plant remains from the top part of the Point Hennequin Group (Upper Oligocene) King George Island (South Shetland Islands, Antarctica). Studia Geologica Polonica. Warszawa. 81:143–164

15—Cretaceous Paleobotany and Its Bearing on the Biogeography of Austral Angiosperms

Andrew N. Drinnan and Peter R. Crane

Introduction

Plant fossils from Antarctica have played a major role in developing concepts of the geological history of the Southern Hemisphere. Disjunct austral distributions of extant taxa (e.g., Hooker 1853) and the strong floristic similarities between the Permo-Triassic *Glossopteris* floras on the austral land masses (e.g., Halle 1937) were important in establishing that now separate continental areas were once contiguous. Since widespread acceptance of continental drift, paleobotanical contributions to ideas of continental realignment have emphasized the delimitation of floristic provinces ("phytochoria" sensu Meyen 1987) and assessments of overall similarities among fossil floras from different areas (e.g., Chaloner and Meyen 1973, Chaloner and Lacey 1973, Vakhrameev et al. 1978, Chaloner and Creber 1988). An alternative approach, made possible by recent developments in systematics, uses patterns of phylogenetic relationship in different taxa to develop and test hypotheses about relationships between different areas and patterns of geographical fragmentation (Brundin 1966, Rosen 1978, Nelson and Platnick 1981, Nelson and Rosen 1981, Platnick and Nelson 1978, Humphries and Parenti 1986). Angiosperms are particularly significant for the application of this technique in the Southern Hemisphere because they were undergoing a major evolutionary diversification as component areas of Gondwana were beginning to separate. If the appearance of major geographical discontinuities influenced patterns of speciation, then repeated patterns of systematic relationship may reflect spatial relationships established between different austral land masses as they separated from each other (Fig. 15.1). Obvious limitations on the present vegetation of Antarctica dictate that paleobotanical studies are the only means of incorporating this critical land mass into such interpretations of austral biogeography.

In this chapter, we focus on the significance of paleobotanical data for understanding Cretaceous plant evolution in the Southern Hemisphere and the biogeography of austral angiosperm groups. Tertiary biogeography and plant ecology, with particular reference to Antarctica, are covered elsewhere in this volume (Truswell 1989). First, we review current knowledge of Cretaceous floras from the southern continents and briefly discuss the Cretaceous history of austral angiosperms in the context of global patterns of angiosperm evolution. We then consider both neobotanical and paleobotanical data on angiosperm biogeography in terms of current ideas on the breakup of Gondwana.

Austral Cretaceous Paleobotany

Southern Africa

Preangiosperm Early Cretaceous floras from southern Africa are reviewed by Anderson and Anderson (1985), who describe 40 paleo-

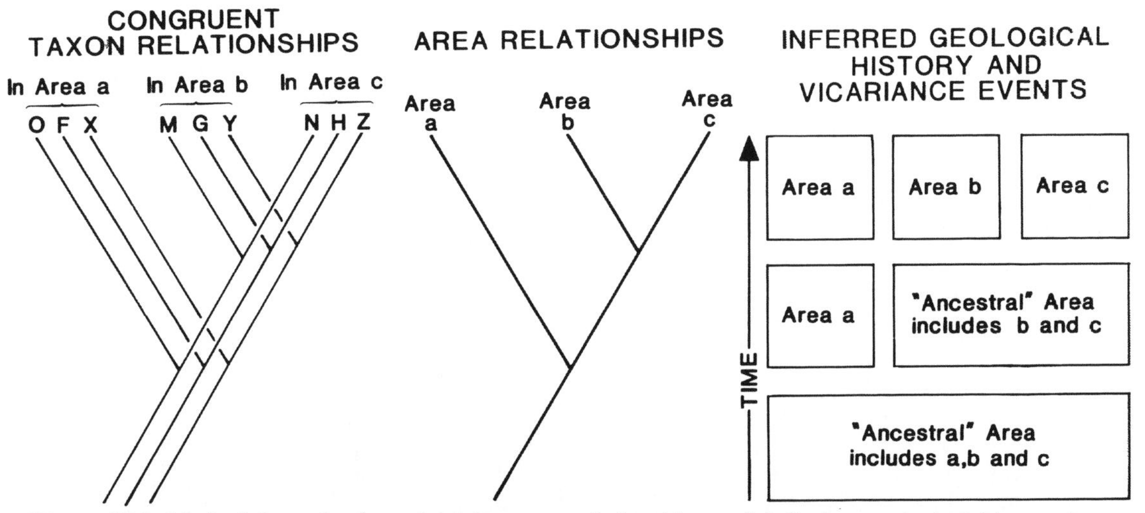

Figure 15.1. Methodology for hypothesizing area relationships and inferring geological history from congruent taxon relationships. See text for further discussion of this technique.

botanical species representing a total of about 30 "whole plants" from 3 geological formations. The Makatini Formation outcrops in Zululand (Fig. 15.2) along the eastern coast of southern Africa and is thought to be of Barremian to earliest Aptian age; the Mngazana Formation occurs in a limited outcrop in the Transkei between Durban and Port Elizabeth and is thought to be of late Valanginian age (McLachlan et al. 1976); and the Kirkwood Formation (Fig. 15.2) outcrops in the Algoa, Gamtoos, and Pletmos basins around and particularly to the east of Port Elizabeth (McLachlan and McMillan 1976). The Kirkwood Formation is regarded as of Berriasian to early Valanginian age. Knowledge of these floras is based almost entirely on leaf remains and fossil assemblages are dominated by leaves of ferns and Bennettitales (Anderson and Anderson 1985; Seward 1903, 1907). The cycadophyte leaves in these floras seem larger than those in floras of similar age from southern South America (Archangelsky 1967) and southeastern Australia (Douglas 1969). Palynofloras from the Kirkwood Formation in the Algoa Basin are dominated by pollen of cheirolepidiaceous conifers (*Classopollis*), with lesser amounts of inaperturate and bisaccate conifer pollen and pteridophyte spores (Scott 1971, 1972, 1976; see also Martin 1960 for palynology in the Gamtoos Basin).

There are few other palynological studies of Cretaceous sediments from southern Africa. Stapleton and Beer (1977) described the palynoflora from the marine Brenton Beds, which they interpret as probably correlative with the Sundays River Formation overlying the Kirkwood Formation. They suggest a probable Valanginian to Hauterivian age for the Brenton Beds. The palynoflora does not include angiosperm pollen and is dominated by trilete spores with smaller amounts of *Classopollis* and bisaccate pollen. Early Cretaceous (undifferentiated Aptian–Albian) palynological samples from DSDP site 361 off the southwestern tip of Africa (Fig. 15.2) contain a few monosulcate angiosperm grains, but the palynoflora is of low diversity and is dominated by *Classopollis*, with smaller amounts of saccate pollen and pteridophyte spores (McLachlan and Pieterse 1978).

There is little information on the Late Cretaceous paleobotany of southern Africa, and the Late Cretaceous was largely a period of erosion in this area, although a few megafloras are preserved in the vents of kimberlite pipes

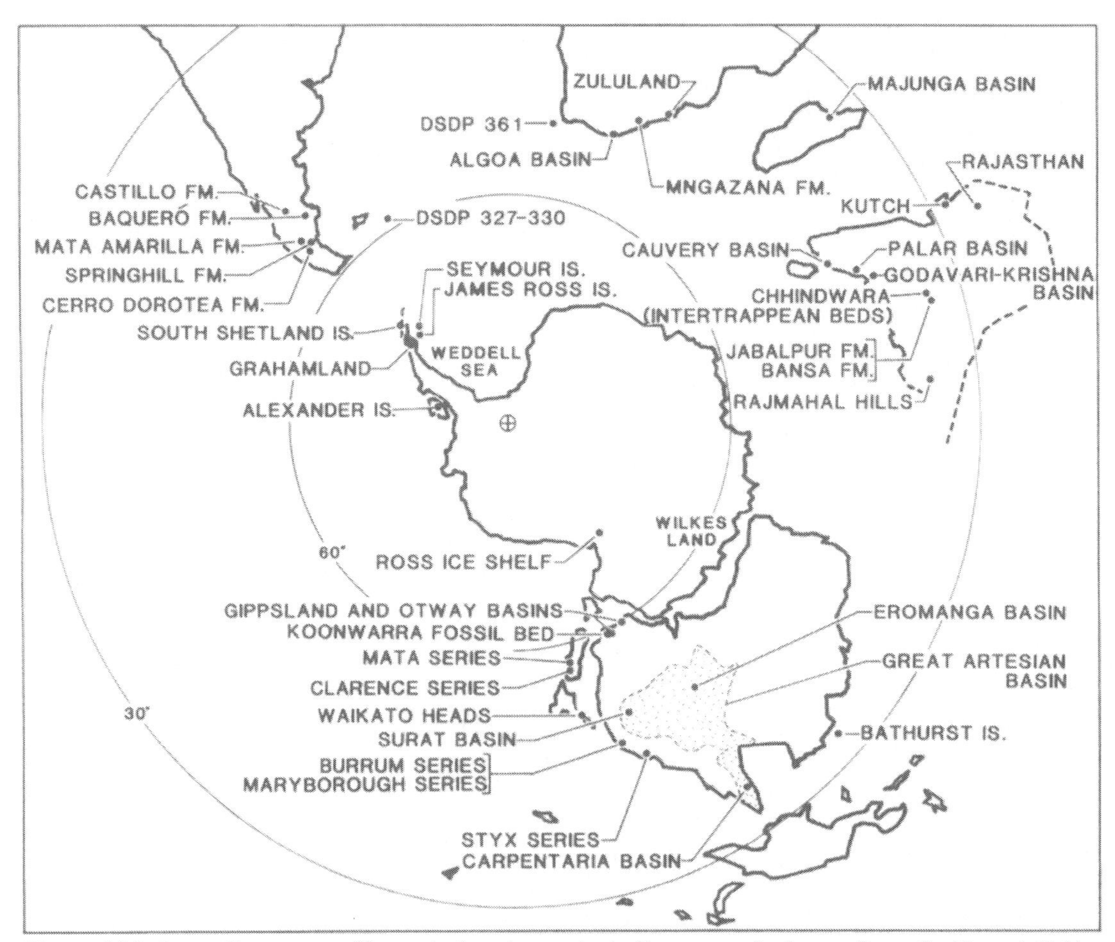

Figure 15.2. Late Cretaceous (Santonian) paleocontinental map of middle and high latitudes in the Southern Hemisphere showing the major paleobotanical localities/geological units mentioned in the text. Base map (redrawn from Smith and Briden 1977) shows only present coastlines. See Barron (1987) for alternative estimates of plate positions at this time.

(Anderson and Anderson 1985). The pollen flora from the Arnot Pipe, Banke, Namaqualand, originally thought to be of Late Cretaceous age (Kirchheimer 1934 and references therein), is regarded as of probable Miocene age by Axelrod and Raven (1978) based on radiometric dates from similar structures in the region (Kröner 1973). Cenomanian palynofloras from DSDP site 361 contain a diverse angiosperm component along with fern spores and high levels (typically 25%) of *Classopollis*. Post-Cenomanian Cretaceous palynofloras (precise age uncertain) contain very little *Classopollis* and much more abundant fern spores. Axelrod and Raven (1978) infer the presence of a more or less continuous subtropical rain forest in the Late Cretaceous of southern Africa based on inferences from contemporary floras in India. They also suggest that *Nothofagus*–podocarp forest was present in extreme southern Africa during the Late Cretaceous, but recent assessments of the fossil history of *Nothofagus* (Romero 1986b, Tanai 1986) provide no indication of this genus in southern Africa, Madagascar, or India. Indeed, one of the difficulties encountered by McLachlan and Pieterse (1978) in correlating palynofloras from DSDP 361 was that most of the common pollen taxa that occur in Australia were not present.

Madagascar

In the Majunga Basin of northeastern Madagascar, pollen floras are known from mid-Cretaceous sediments ranging in age from early Aptian to early Cenomanian (Doyle et al. 1982, p.66). Detailed palynological analyses have not been published but the pollen floras contain diverse angiosperm pollen, including *Afropollis*, tricolpates, and tricolporates. Other elements are typical of southern Gondwana palynofloras (Brenner 1976) and include common *Classopollis*, few ephedroid grains, abundant putative gleicheniaceous trilete spores, and bi- and trisaccate conifer pollen (Doyle et al. 1982). Other mid-Cretaceous palynofloras from northern Madagascar have a similar composition (Herngreen and Chlonova 1981).

India

The extensive "Upper Gondwana" sequences of India have provided a wealth of information on the mid-Mesozoic biota of the subcontinent (see Bose et al. 1989). The plant megafossil remains were first described in the latter half of last century, and numerous studies of both the plants and associated fauna have been completed since. The utility of these data in a global context has been lessened by the long, and still lively, dispute as to the age of the "Upper Gondwana" strata in India, which has centered on the different age indications from plant megafossil and faunal evidence. Most Indian paleobotanists followed Feistmantel's early lead in equating the Indian floras with the well-known Yorkshire Jurassic flora of England, whereas evidence from ammonites clearly indicated a Lower Cretaceous age, as did later investigations of foraminifera. Since 1970, extensive palynological investigations have added further information on subsurface and outcrop strata and in virtually all cases have been in general agreement with the faunal evidence. Together with radiometric dates obtained from lava flows in the Rajmahal Hills (Fig. 15.2), the biotic evidence overwhelmingly points to an Early Cretaceous age for the "Upper Gondwana"strata of India.

The floras of the Rajmahal Hills in western Bengal, northeastern India (Fig. 15.2), are perhaps the most well known of Indian Mesozoic plant assemblages, and certainly the ones from which most comparisons are drawn. The Rajmahal sequence consists of a number of plant beds associated with lava flows, which may be divided into two main "floras" (Zeba-Bano et al. 1979). The older flora (Zone I) commonly contains *Brachyphyllum* and *Elatocladus* conifer foliage as well as diverse and abundant *Ptilophyllum* (particularly *P. acutifolia* and *P. cutchense*), *Pterophyllum*, *Dictyozamites* and other bennettitalean fronds. Taeniopterid leaves (*Taeniopteris spatulata*) that are clearly pentoxylalean (Sahni 1948, Vishnu-Mittre 1957, Bose et al. 1985) are common, and leaves of *Ginkgoites rajmahalensis* are known from several localities. Fern-like foliage of *Thinnfeldia indica* is locally abundant, and sphenopterid and cladophleboid fern foliage is widespread. This flora is typical of leaf assemblages found throughout the Indian "Upper Gondwanas". In the slightly younger flora (Zone II), conifers and ferns are dominant, while Bennettitales are rare. The Rajmahal flora was first described by Oldham and Morris (1863) and was long considered to be of Middle Jurassic age, a view still cited in some recent publications, especially those concerned with plant megafossils. The spore/pollen evidence that Sah and Jain (1965) used to support a Jurassic determination has been reinterpreted by Playford and Cornelius (1967), Filatoff (1975), and Herngreen and Chlonova (1981) as more indicative of an Early Cretaceous age. This is supported by Maheshwari and Jana (1984). Potassium–argon dates of the lava traps of 100–105 myr (McDougall and McElhinny 1970) and 109.1 ± 2.6 myr (Agrawal and Rama 1976) also clearly point to an Early Cretaceous age, although the correlation of the K–Ar samples with the paleofloras is not clear. It is likely that the Rajmahal flora is at least as young as the Neocomian, is quite possibly Aptian, and may even extend into the Albian.

Similar considerations apply to the megaflora of Kutch in Gujarat State, western India (Fig. 15.2). This *Ptilophyllum*-dominated flora from the Bhuj Beds (="Umia plant beds", Bhuj Series) was first described by Feistmantel (1876). Bose and Banerji (1984), in the most recent and comprehensive treatment of the flora, noted that the most distinctive taxa are a few species of pteridophytes, *Sagenopteris*, species of *Taeniopteris*, *Nilssoniopteris*, *Ptilophyllum*, *Pterophyllum*, *Otozamites*, *Elatocladus*, *Pagiophyllum*, and, to a lesser extent, *Brachyphyllum*. Most of these taxa occur in other Indian Early Cretaceous floras, but the presence of *Linguifolium*, *Sagenopteris*, and *Nilssoniopteris* is distinctive of the Kutch assemblage (Bose and Banerji 1984). Bose and Banerji (1984) maintain a Jurassic age based on a megaflora comparison with the Yorkshire and Greenland fossil assemblages, but other evidence suggests an Early Cretaceous age. The stratigraphically older Umia Beds (also Bhuj Series) contain Neocomian species of ammonites and the bivalve *Trigonia* (Spath 1933, Arkell 1956, Bhalla 1972). The Umia Beds are considered post-Jurassic by Venkatachala and Kar (1970) and appear to correlate with the uppermost strata of the East Coast "Upper Gondwanas" on faunal evidence (Bhalla 1972). One of the *Trigonia* species (*T. ventricosa*) has been recorded from the Early Cretaceous of South Africa and Neocomian of Tanganyika (Bhalla 1972). The Bhuj Beds themselves are included in the youngest of the three palynological zones described by Venkatachala and Kar (1970), which makes them Aptian, or at least post-Neocomian, in age. This is consistent with the megaspore flora, which Banerji et al. (1984) consider to be pre-Aptian, but certainly Early Cretaceous. It appears that an approximate Barremian age is most probable for the Bhuj plants.

Studies of the East Coast Gondwanas also strongly suggest that the Indian "Upper Gondwana" sequences are Early Cretaceous. Most of the critical evidence has been gained from investigations of several basins in Tamil Nadu and Andhra Pradesh, southeastern India. Exposures of the Terani and Sivaganga Beds in the Cauvery Basin (Fig. 15.2) have yielded an assemblage of plants typical of the Rajmahal flora (Feistmantel 1879, Gopal et al. 1957, Chowdhury 1958, Mamgain et al. 1973). Maheshwari (1986) lists the familiar species *Ptilophyllum acutifolia, P. cutchense, Dictyozamites indicus, Otozamites* sp., *Thinnfeldia indica, Taeniopteris spatulata, Elatocladus heterophylla, E. conferta, Pagiophyllum marwarensis,* and *Araucarites cutchensis*. The associated palynoflora, which is indicative of a Neocomian age, is dominated by the coniferous taxa *Araucariacites* and *Inaperturopollenites* (47% together), *Callialasporites dampieri* (24.5%), and *Podocarpites* (7%). *Alisporites* accounted for 9.5%, and *Vitreisporites pallidus* almost 2%. The Sivagana Beds are overlain by the Dalmiapuran Formation, which contains a rich ostracod, foraminifera, and palynofossil assemblage of Aptian to early Albian age, constraining the upper age limit for these beds (Venkatachala 1977).

The Sriperumbudur Beds of the Palar Basin (Fig. 15.2) similarly contain a megaflora typified by *Ptilophyllum cutchense, P. acutifolia, Dictyozamites, Pterophyllum,* and *Taeniopteris* (Feistmantel 1879, Suryanarayana 1954, 1955, Ramanujam and Varma 1977). Venkatachala (1977), commenting on their age, notes evidence of an underlying Upper Jurassic sequence. This is consistent with records of Early Cretaceous ammonites and foraminifera in the Sriperumbudur Beds (Murthy and Sastri 1962). Ramanujam and Varma (1977) investigated the palynology of the Sriperumbudur Beds and equated them with the Jabalpur and Bhuj Beds and Cauvery and Godavari Basins (Fig. 15.2) on the consistent occurrence of a suite of stratigraphically significant spore/pollen taxa that are strongly indicative of a Neocomian–Aptian age (also supported by Venkatachala 1977). Ramanujam and Varma (1977) consider all these strata to be closely comparable with the Neocomian–Aptian "Stylosus" assemblage of southeastern Australia (Dettman 1963), the *Microcachryidites antarcticus* and "microflora IIB" subzones of western Australia (Balme 1957, 1964), and the Baqueró Formation of Argentina (Archangelsky and Gamerro 1965, 1966a, 1966b).

The Vemavaram and Raghavapuram Shales in the Godavari–Krishna Basin (Fig. 15.2) have also yielded a typical Rajmahal flora (Feistmantel 1879; Suryanarayana 1954, 1955; Bose 1966; Baski 1966, 1967) and have also been determined as Early Cretaceous on the basis of palynology (Neocomian, Venkatachala et al. 1974) and foraminifera (Bhalla 1965, 1968; Baksi 1966). Bhalla equates the foraminifera with those of the Early Cretaceous of the Great Artesian Basin of Australia. Sastri and Mamgain (1971) correlate these two shales with the Palar Basin and assigned a Neocomian–Aptian age to the East Coast Gondwanas. According to the stratigraphy of Mitra et al. (1971), the entire East Coast Gondwanas in the Cauvery and Palar basins are younger than in the Rajmahal Hills.

The palynological investigations, reviewed in detail by Venkatachala (1977) and Herngreen and Chlonova (1981), reveal a dominance of coniferous pollen throughout the Indian "Upper Gondwanas". The Cauvery Basin has been subdivided into three palynostratigraphic zones (Venkatachala et al. 1972, Venkatachala and Sharma 1974a, Ramanujam and Varma 1977). The oldest zone, of uppermost Jurrasic age, contains up to 40% *Callialasporites segmentatus,* but *Microcachryidites* is not present. Some geographic variation in the abundance of the major elements is evident throughout this zone, with *Araucariacites australis* accounting for up to 70% of the pollen and *Callialasporites* as little as 14% (Ramanujam and Srisailam 1974, Maheshwari 1975). The middle *Microcachryidites antarcticus* zone is of Neocomian age. *Callialasporites* and *Alisporites* are recorded as most abundant, together with the other coniferous types, *Araucariacites, Inaperturopollenites, Classopollis, Podosporites,* and *Podocarpites*. These conifers remain in similar abundances into the youngest Aptian–Albian zone.

Three similar zones are distinguishable at Kutch (Venkatachala and Kar 1970), although there are some differences in the abundance of the major elements. *Microcachryidites* is recorded in the older, supposedly Late Jurassic, portion (=upper Katrol Series), and *Platysaccus* and *Vitreisporites* are present in significant quantities, although *Araucariacites* is less prevalent. In the middle Neocomian (Katrol–Bhuj Transition) zone, pteridophyte spores and coniferous pollen of *Impardecispora, Callialasporites,* and *Araucariacites* form the major elements, to the virtual exclusion of saccate pollen (*Podocarpites, Podosporites, Microcachryidites*). Although *Araucariacites* (20–70%) remains most common in the Aptian zone (=Bhuj Series), *Podosporites, Podocarpites, Callialasporites,* and *Microcachryidites* are also significant in these floras.

Pollen assemblages from the Bansa and Jabalpur Formations (central India) are similarly dominated by *Araucariacites,* but *Classopollis* is rare. In the Bansa Formation, *Callialasporites* is more abundant than *Cyadopites,* while the latter does not occur at Jabalpur. Differences are also obvious in the megaflora; *Ptilophyllum* and *Pachypteris* are common in Jabalpur beds, but *Weichselia* characterizes the Bansa flora (Maheshwari 1974). The Bansa Formation is considered of Early Cretaceous age, while the Jabalpur Formation is possibly slightly older: both are probably younger than the Rajmahal Hills plant beds. Maheshwari (1974) considers all three to comprise one biostratigraphic zone with Rajmahal the oldest and Bansa the youngest. One distinct trend in all of these Indian Early Cretaceous microfloras is a decrease in bi- and trisaccate conifer pollen (mostly podocarpaceous conifers, and particularly *Microcachryidites*) from south to north India. *Microcachryidites* accounts for up to 15% in southern Indian palynofloras, but only 3% at Kutch, and less that 1% in the Bansa Formation.

Little is known about the mid- or Late Cretaceous floras from India, or the diversification of angiosperms on the subcontinent. Venkatachala and Sharma (1974b) record a late Albian to early Cenomanian palynoflora from southeastern India containing two species of *Liliacidites* and four species of *Tricolpites*; and Lukose (1974) notes *Clavatipollen-*

ites and *Tricolpites* from the Goru Formation (?Albian–Cenomanian), northwestern India. In the latest Cretaceous, a rich flora of tropical aspect from the Deccan (Chhindwara) Intertrappean Cherts (Fig. 15.2) (Maastrichtian–Paleocene) contains remains of a variety of families, including Anacardiaceae, Arecaceae, Burseraceae, Combretaceae, Datiscaceae, Elaeocarpaceae, Euphorbiaceae, Flacourtiaceae, Lecythidaceae, Rutaceae, Sapindaceae, Simaroubaceae, Tiliaceae, Vitaceae, and others (Axelrod and Raven 1978). Interestingly, there are also records of Araucariaceae and Podocarpaceae, which today occur only in southern and eastern India.

Southern South America

Several preangiosperm floras have been described from the Cretaceous of southern South America of which the most intensively studied are those from the Baqueró Formation of Santa Cruz Province (Fig. 15.2), southern Argentina (late Barremian–early Aptian; Archangelsky 1963, 1967; Archangelsky et al. 1986; Archangelsky and Taylor 1986; Archangelsky and del Fueyo 1987; Taylor and Archangelsky 1985). Frequently, the plants from this formation are very well preserved in volcanoclastic sediments, and the flora is dominated by a variety of cycadophytes (including true cycads, e.g., Archangelsky and Petriella 1971) and conifers. It also includes diverse ferns and a significant component of pteridosperms with complex pinnately divided foliage (e.g., *Ruflorinia,* Taylor and Archangelsky 1985), which may be related to corystosperms from the earlier Mesozoic (Archangelsky and Gamerro 1967; see also Archangelsky et al. 1981, 1983, 1984). Palynofloras from the Baqueró Formation are dominated by conifer pollen, among which *Classopollis* and putative podocarpaceous saccate grains are the most conspicuous components.

The Springhill Formation in the Austral (Magellanes) Basin of southern Chile and Argentina is thought to be of Berriasian–Valanginian age (Baldoni and Archangelsky 1983, Riccardi 1987) and is very similar in composition to that of the Baqueró Formation (Archangelsky 1976, Baldoni 1979, Baldoni and Taylor 1983). Palynological analyses (Baldoni and Archangelsky 1983) have demonstrated that the palynofloras are dominated by saccate conifer pollen and pteridophyte spores with *Classopollis* as a smaller, but still significant (<20%), component.

To the north of the Baqueró Formation, exposures in Santa Cruz Province sediments preserved in the Neuquén Embayment (Basin) of Neuquén Province have also provided extensive information on pollen floras ranging from the Jurassic–Cretaceous boundary (lower Tithonian–lower Valanginian, e.g., Vaca Muerta Formation) to Hauterivian–Barremian (Agrio Formation) and Albian (e.g., Huitrin Formation) (see Archangelsky 1980 for review). Palynological assemblages from the Neuquén Embayment (e.g., Volkheimer and Salas 1975, 1976; Volkheimer and Quatrocchio 1975a,b; Volkheimer et al. 1975) resemble those from the Baqueró Formation. They are frequently dominated by *Classopollis* and saccate conifer pollen.

Pollen floras from the tip of the Falklands Plateau (DSDP 327-330), which are dated as of Neocomian through Albian age, are similar to those from southern South America, but no angiosperm pollen is recorded (Harris 1977, Hedlund and Beju 1977).

In terms of the changing proportions of major palynomorph types, the palynological succession in southern South America is basically similar to that in northwest Europe. Early Cretaceous floras have abundant *Classopollis,* bisaccate conifer pollen, and fern spores. During the mid-Cretaceous, spores of Gleicheniaceae and other ferns diversify, and angiosperm pollen appears. In the Late Cretaceous, angiosperm pollen diversifies, typical mid-Cretaceous fern spores decline, while saccate conifer (podocarpaceous) pollen remains important.

Fossil angiosperm remains from southern South America first appear in the Early Cretaceous. The pollen taxa *Clavatipollenites* and *Liliacidites* are recorded from the Barremian of Argentina (Archangelsky 1980), and the former type is the only angiosperm pollen associated with unidentified, but clearly angiosperm, leaves in Aptian deposits (part of the Baqueró Formation, Romero and Archangelsky 1986). *Stephanocolpites* and *Tricolpites* occur in the mid-Albian upper part of the Huitrin Formation of Argentina and constitute the earliest unequivocal records of "higher" (nonmagnoliid) dicotyledons in southern South America (Volkheimer and Salas 1975).

Romero and Arguijo (1982) provide a detailed review of Late Cretaceous floras from southern South America (see also Romero 1978). In the Castillo Formation (Fig. 15.2, late Albian–Senonian, Riccardi 1987), the Cerro San Bernardo megaflora is dated as late Albian–Senonian, while the Cerro Cachetamán flora is dated as Turonian–Senonian (Romero and Arguijo 1982). Both floras contain predominantly angiosperm leaves, although the Cerro Cana flora from the same formation contains only *Onychiopsis* (Romero personal communication). Angiosperms are the dominant floristic component in all megafloras from the Mata Amarilla Formation (Turonian–Santonian, Riccardi 1987) of the Austral Basin (Fig. 15.2) and are dated as of Coniacian age (Romero and Arguijo 1982). Maastrichtian megafossils and palynofloras have been descibed from the Cerro Dorotea Formation of the Austral Basin (Fig. 15.2), and other Maastrichtian pollen assemblages have been described from the Rio Blanco and Jaguel formations (Archangelsky and Romero 1974a,b; Romero 1973).

In southern South America, as elsewhere, the transition to the latest Cretaceous and early Tertiary involved increased representation of more modern angiosperm taxa. All three types of *Nothofagus* pollen are known from the Maastrichtian (Romero 1986b) and early Tertiary (Baldoni 1987), and pollen representative of the families Myrtaceae, Proteaceae, Myricaceae, Loranthaceae, and Gunneraceae is evident during the Eocene (Romero and Castro 1986, see also Romero 1986a). Romero and Hickey (1976) assign a fossil leaf from the Paleocene of Argentina to Akaniaceae, a family that is extant only in Australia.

Australia

Cretaceous sediments in Australia are preserved in several major sedimentary basins. Although these basins outcrop in all mainland states, by far the most important information on the floras of this age comes from the closely comparable Otway and Gippsland Basins in southeastern Australia, and the Great Artesian Basin (encompassing the Carpentaria, Eromanga, and Surat Basins) of Queensland in the northeast (Fig. 15.2). Additional palynofloras have been described for western Australia (Balme 1957, Ingram 1968, Kemp 1976).

The Australian Cretaceous has been comprehensively reviewed with respect to palynology, stratigraphy (Dettmann and Playford 1969) and general floristics (Dettmann 1981). Biostratigraphic schemes based on detailed studies of the microfloras (Balme 1957, 1964; Dettmann 1963; Evans 1966, Dettmann and Playford 1969; Burger 1973, 1980, 1988; Stover and Evans 1973; Stover and Partridge 1973; Norvick and Burger 1975; Dettmann and Douglas 1976; Morgan 1980; Helby et al. 1987) and megafloras (Douglas 1969) are well correlated and dated by both marine faunal evidence (see references above) and radiometric methods (Gleadow and Duddy 1980). These factors make the Australian Cretaceous floras important benchmarks for the interpretation of other Gondwana Cretaceous floras.

Douglas (1969, 1973) divides the Victorian Early Cretaceous into four successive megaflora zones (A–D). The oldest (Zone A) does not outcrop and is poorly understood, but the remaining three zones provide a well-supported indication of floristic changes. The southeast Australian flora in the Neocomian is dominated by small-leaved *Brachyphyllum*- and *Elatocladus*-type conifers of presumed araucariaceous and podocarpaceous affinity, Bennettitales (*Ptilophyllum* and *Otozamites*), and taeniopterid-leaved plants of the Pentoxylales. Sphenopterid and cladophleboid ferns are well represented, and there are several peculiar leaf taxa of possible pteridosperm affinity (*Rienitsia variabilis* Douglas and *Pachypteris austropapillosa* Douglas) that are distinctive of Zone B. In general, however, it is the small-leaved, thick-cuticled Bennettitales that are distinctive of this earlier part of the Cretaceous (Zones A and B). Palynofloras of this age are dominated by saccate, presumably podocarpaceous pollen with subsidiary representation of araucariaceous conifers, and cycad and pteridosperm-like pollen *Cycadopites* and *Alisporites* (Dettmann 1963). Ferns include probable Osmundaceae, Schizaeaceae, Gleicheniaceae, and Cyatheaceae.

The succeeding late Aptian (Zone C) flora almost totally lacks the formerly prominent Bennettitales. Occasional *Ptilophyllum* fronds are found at a few of the older Zone C localities, but they are absent from most sites, including the well-sampled Koonwarra Fossil Bed (Douglas 1969, Drinnan and Chambers 1986). The most notable addition to the Zone C flora is *Ginkgo australis,* which is sufficiently common and consistently represented to be considered a guide fossil for Zone C. The Pentoxylales are perhaps more abundant than in Zone B, possibly reflecting the absence of the Bennettitales. The ferns from Zones B and C show little difference in terms of abundance, diversity, or species composition. Podocarpaceous and araucariaceous conifers remain abundant.

The disappearance of *Gingko australis* and the decline of the Pentoxylales in the Albian (Zone D) are accompanied by the diversification of conifers and widespread abundance of the osmundaceous *Phyllopteroides dentata* (Cantrill and Webb 1987). The sphenopterid and cladophleboid ferns of Zones B and C persisted, but were less common. Angiosperm megafossil evidence is rare and fragmentary, but the pollen record reveals their first appearance and initial slow diversification during the late Aptian and Albian.

Walkom's (1918, 1919) treatments of the late Aptian and Albian Maryborough, Burrum, and Styx megafloras allow comparison with Victorian floras and those from other areas of Gondwana. All of these plant assemblages are similarly rich in araucariaceous and podocarpaceous conifers, spatulate taeniopterid leaves, bennettitalean fronds, ginkgoalean leaves, and sphenopterid and cladophleboid fern foliage. The presence of pinnate cycadophyte fronds (*Zamites* and *Nilssonia*) is exceptional. The only other striking difference between the Queensland and Victorian floras is the association of ginkgoalean leaves with bennettitalean fronds–two components of the Victorian Lower Cretaceous flora that are mutually exclusive. Ginkgoalean leaves and bennettitalean fronds occur together regularly in other Cretaceous floras (e.g., Rajmahal Hills, India).

Palynological data for the late and post-Aptian are extensive from eastern Australia, and provide a detailed insight into the introduction and diversification of angiosperms in the Australian flora (Burger 1970, 1973, 1974, 1975, 1976, 1980, 1988; Dettmann 1973, 1986a; Dettmann and Playford 1968, 1969; Stover and Partridge 1973; Playford et al. 1975). The earliest records of dispersed angiosperm pollen known from Australia are rare occurrences of *Clavatipollenites* from the Barremian of the Eromanga Basin, southern Queensland (Burger 1988) and the late Aptian Koonwarra Fossil Bed in the Gippsland Basin (Dettmann 1986a). It is not until the mid-Albian that *Clavatipollenites* and *Asteropollis* become significant components of palynoflora counts, together accounting for a little over 3% of the spore/pollen flora (Burger 1988). In the Surat Basin of southeastern Queensland (Burger 1980), *Clavatipollenites hughesii* and *Asteropollis asteroides* make their first appearance in the early Albian, together with *Ephedrites multicostatus*. By the mid-Albian, *Clavatipollenites* and *Asteropollis* are consistently present, and are joined by *Retimonocolpites peroreticulatus* and the tricolpates *Rousea georgensis* and *Tricolpites variabilis*. Playford et al. (1975) describe nine angiosperm pollen taxa from a latest Albian or early Cenomanian drill sample in the Carpentaria Basin (Allaru Mudstone), including two species of *Liliacidites, Striatopollis,* and the tricolporoidate *Cupuliferoidaepollenites*.

Norvick and Burger (1975) investigate two drill samples from Bathurst Island, northern Australia (Carpentaria Basin) (Fig. 15.2) that span the late Albian to early Turonian. The palynofloras are dominated by fern spores of the *Cyathidites-Gleicheniidites* type (30–45% of the total spores/pollen). Schizaeaceous spores are only a small component (1–7%), while other trilete spores are abundant. Conifer pollen is less abundant than spores, with saccate grains (mostly the podocarpaceous *Microcachryidites antarcticus*) typically accounting for 6–20% of the palynofloras. *Classopollis* is usually sparse (1-7%), although rises locally to 39% apparently at the expense of various spore types. Angiosperm pollen abundance (mostly tricolpates) gradually increases from 1–4% in the late Albian to 9–10% in the Turonian. Tricolporate grains are first encountered in the early Cenomanian.

Although angiosperm pollen comprises only a minor component of Albian to Turonian palynofloras of eastern Australia (Dettmann 1973), it nonetheless shows a pattern of first appearance and diversity that is consistent with the rise of angiosperms elsewhere, starting with monosulcates in the Barremian–Aptian, tricolpates in the Albian, tricolporates in the early Cenomanian and tri- and polyporates in the Turonian. The systematic affinities of these early angiosperm grains are uncertain, but through the Albian similar forms are recorded throughout eastern Australia with the maximum diversity in the north (Dettman 1981). By the early Turonian, there is already considerable differentiation between the angiosperm floras of northern and southern Australia. In southeastern Australia, podocarpaceous conifers increase in abundance, while araucariaceous and cheirolepidiaceous conifers become less important. Cyatheaceous and hydropteridaceous fern spores remain a significant component of the flora, but schizaeaceous and gleicheniaceous forms become less abundant and diverse. In the north and west of Australia, nonsaccate araucariaceous pollen and *Classopollis* dominate podocarps in the pollen record, and ephedroid types become a minor component. Gleicheniaceous and schizaeaceous ferns remain well represented (Dettmann 1981).

Pollen with recognizable affinity to the modern flora of Australia begins to appear in the Senonian, and includes *Nothofagus* and proteaceous types (Dettmann and Playford 1969, Stover and Partridge 1973). The abundance and diversity of angiosperm pollen also increases markedly during the Senonian and Maastrichtian. Podocarpaceous conifers remain significant in the palynofloras, consistently occurring with *Nothofagus*. This provides early evidence of a plant association that was dominant in the southern Australian flora in the Early Tertiary, and still persists to a much lesser extent in parts of Australasia and South America.

New Zealand

Early Cretaceous sediments in New Zealand are almost entirely absent, leaving a virtual gap in the fossil record between the latest Jurassic and Albian. Current information on Cretaceous floras comes from the Albian–mid Cenomanian Clarence Series, late Cenomanian–mid Senonian Raukumara Series, and the Campanian–Maastrichtian Mata Series (McQueen 1956; Couper 1953, 1960; Raine 1984).

The fossil evidence from the latest Jurassic suggests a comparable flora to eastern Australia around the Jurassic–Cretaceous boundary. Arber (1917) and Bartrum (1921) note a typical mid-Mesozoic flora from Waikato Heads comprising *Elatocladus* and *Araucarites* conifers, *Cladophlebis* and *Coniopteris* ferns, and *Taeniopteris* leaves. Harris (1961) describes pentoxylalean fruiting structures (*Carnoconites cranwelli*) from this flora. Blashke and Grant-Mackie (1976) and Drinnan and Chambers (1985) consider the *Taeniopteris* and *Carnoconites* taxa from Waikato conspecific with those that persist through to the Albian in southeastern Australia. Although not included in Arber's list, bennettitalean *Ptilophyllum* and *Pterophyllum* fronds are known from other Jurassic localities and are reported as a component of the flora as late as the early Senonian (McQueen 1956). McQueen describes several fern and cycadophyte leaf taxa from a series of Upper Cretaceous localities, and summarizes floristic changes in the New Zealand Cretaceous based on megafossil evidence. Further insights into mid–Late Cretaceous and younger floras are gained from the detailed palynological investigations of Couper (1953, 1960), Mildenhall (1980), and Raine (1984).

Couper's spore/pollen tables indicate a dominance of podocarpaceous conifers (particularly *Microcachryidites* pollen) from the Albian through the Senonian. *Podocarpidites* and *Dacrydiumites*-type pollen then become abundant. *Nothofagus* pollen first appears in the early Senonian, and proteaceous pollen in the mid-Senonian, both slightly later than their first appearance in southeastern Australia (Fleming 1975). The Maastrichtian sees a sharp rise in angiosperm pollen diversity, with about 30 taxa recorded for the Haumurian stage of the Mata Series. While several of these taxa can be related to elements of the modern flora (e.g., *Nothofagus*, Proteaceae), the affinities of the majority of these palynomorphs are unknown.

Floristic changes in the Cretaceous flora of New Zealand are very similar to those that occur in Australia, and the same palynozones and datum planes are applicable (Mildenhall 1980). Many pollen, spore, and megafossil taxa are shared, particularly prior to the tectonic separation of the two areas in the mid–Senonian. The sequence of floristic changes through the Late Cretaceous also seems to be essentially similar. During the mid-Cretaceous the araucariaceous and podocarpaceous conifers become more important in megafloras and dominate floristic assemblages until the Campanian. Monosulcate and tricolpate angiosperm pollen first appears in the late Albian of the Clarence Series in the northeast of the South Island (Couper 1953, 1960) and steadily becomes more diverse throughout the Late Cretaceous. Palynofloras with a floristic composition similar to Recent New Zealand vegetation appear during the Campanian–Maastrichtian. Raine (1984) presents an outline palynological zonation of the west coast South Island from the Barremian to Maastrichtian, recognizing four biostratigraphic zones that are consistent with, but provide less resolution than, those for equivalent strata in southeastern Australia. Raine (1984) does not record pre-Turonian angiosperm pollen, although this has been recognized by the end of the Albian elsewhere in New Zealand (Couper 1960, Fleming 1975).

Mildenhall (1980, Table 1) presents a list of the first appearances of the families and genera of New Zealand vascular plants in the fossil pollen record, along with a list (1980, Table 2) of a number of New Zealand Tertiary pollen and spores with extant relatives in neighboring regions. Among these are several prominent extant Australian taxa (e.g., *Casuarina, Eucalyptus, Acacia,* and *Banksia.*) In addition, a number of taxa occurring in New Zealand assemblages occur first in Austrailia, suggesting the possibility of trans-Tasman dispersal. For example, the disjunct *Nothofagus* pollen taxa *Nothofagidites flemingii* and *N. falcatus*, which range in New Zealand from the lower Eocene and Oligocene respectively, are known from the late Paleocene and later Eocene of Australia. Similarly pollen of *Acacia,* which has a continuous record from the early Miocene in Australia, is known from New Zealand only during a relatively short period from the early Pliocene to the middle Pleistocene.

Antarctica

The plant fossil record from Anarctica is poor, but does allow subdivision of the continent into East and West Antarctica along broadly tectonic lines. The larger East Antarctica has yielded a considerable range of land plants from the Middle Devonian until the Middle Jurassic (Plumstead 1963), the most prominent being Permian *Glossopteris* floras. Most of the collecting localities are concentrated along the Transantarctic Mountains, which reflects both the bias in exploration in this area and the favorable rock exposures. No outcrops containing fossil floras younger than the Jurassic are known from East Antarctica.

In contrast, no pre-Jurassic plant fossils have been reported from West Antarctica. The first described, and one of the best known, West Antarctic fossil floras consists of latest Jurassic or earliest Cretaceous conifers, cycads, bennettitaleans, pteridosperms, ferns, and equisetaleans described by Halle (1913) from Hope Bay, Graham Land (see also

Gee 1987). Plant-bearing localities of similar age occur at Ablation Point and Fossil Bluff on Alexander Island (Jefferson 1982a,b). Megafossils from the Early Cretaceous of the Byers Peninsula, Livingston Island, South Shetland Islands, have been described by Hernandez and Azcarate (1971). This post-Barremian but apparently preangiosperm flora consists of conifers identified as podocarpaceous and araucariaceous, cycads, bennettitaleans, *Taeniopteris*, some possible pteridosperms, and ferns. It was compared by Hernandez and Azcarate (1971) to the flora of the Baqueró Formation of southern South America. A significant gap in our knowledge of Antarctic floras is the absence of Late Cretaceous megafloras from this continent.

The Early Cretaceous megafloras, together with in situ palynomorph studies by Askin (1981) of pre-Aptian Lower Cretaceous sediments also from the Byers Peninsula and by Domack et al. (1980) from the continental shelf, are consistent with other Gondwana floras, indicating a continuation of predominantly conifer- and pteridophyte-dominated vegetation from the Jurassic into the Early Cretaceous. Although the evidence is sparse, it is clear that by the late Senonian a typical austral flora had developed. Podocarpaceous conifers and *Nothofagus* (mostly *brassii* type) pollen are the most common components of the Campanian–Maastrichtian of the Antarctic Peninsula, and are accompanied by pteridophytes, rare proteaceous pollen, and various simple tricolpate, tricolporate, and triporate angiosperm grains (Askin 1983, 1988). Askin (1985) considers this spore/pollen assemblage to correlate closely with the subzone PM2 of the *Phyllocladidites mawsonii* assemblage of the west coast of New Zealand's South Island (Raine 1984) and the *Tricolporites lilliei* and *Tricolpites longus* zones from the Gippsland basin of southeastern Australia (Stover and Partridge 1973). These correlative intervals also range from the Campanian through the Cretaceous–Tertiary boundary. Dinoflagellates from these sequences provide age estimates consistent with the evidence from spores and pollen (Askin 1983).

Truswell (1983) compiles previous palynological information from the study of Antarctic offshore bottom sediments, mostly from the Ross Ice Shelf area, the Weddell Sea, and the coast of Wilkes Land (Kemp 1972a, 1972b; Truswell 1982; Truswell and Drewry 1984). Truswell (1983) maps palynomorph distribution and densities off the Ross Ice Shelf against patterns of ice drainage (Robin 1975). Two results are quite clear. First, samples with relatively low densities of Paleozoic and Mesozoic palynomorphs are probably derived from the Transantarctic Mountains, with rare Early Tertiary taxa probably derived from previously reported McMurdo Sound erratics. Second, samples with high concentrations of Late Cretaceous and Early Tertiary pollen originate from sediments of Marie Byrd Land, West Antarctica. The exact location of these latter sediments is unclear, but they are presumably eroded from sub-ocean or sub-ice rocks.

The tables of reworked spores and pollen from the Antarctic coastline presented by Truswell (1983, Tables 4 and 5) show a similar trend to that suggested by the results of Askin (1981, 1983, 1985, 1988). Of the 64 Early Cretaceous spore/pollen taxa listed by Truswell (1983, Table 4), the only one of angiosperm affinity is *Asteropollis asteroides*, which also appears in the Australian sequence in the Aptian–Albian. The remainder of the palynoflora consists of Araucariaceae (5 spp.), Podocarpaceae (4 spp.) Cheirolepidiaceae (2 spp.), pteridophytes (24 ferns plus 8 lycopods), hepatics (8 spp.), 1 *Sphagnum* spore type, and 11 taxa of unknown affinity. In contrast, of the 70 spore pollen taxa listed for the Late Cretaceous and Early Tertiary, 51 are angiosperms. Although the affinity of 17 of these could not be determined, the others demonstrate the presence of typical austral plant groups. These include *Nothofagus* (9 spp.); Proteaceae (10 spp.), Myrtaceae (4 spp.), Casuarinaceae (1 sp.), and Restionaceae (1 sp.) The remainder of the palynoflora shows a predominance of podocarpaceous conifers (9 spp.) over Araucariaceae (1 sp.), the absence of *Classopollis*, and a sharp reduction in the diversity of ferns (5 spp.) and lycopods (2 spp.).

More details are provided by a study of in situ palynomorphs from the James Ross Island area of the Antarctic Peninsula (Fig. 15.5, Dettmann and Thomson 1987). This study documents a clear change from a vegetation dominated by ferns and mixed gymnosperms (Podocarpaceae 29–35%, Araucariaceae, *Classopollis, Alisporites*) in the early Albian to one dominated by angiosperms and podocarpaceous conifers in the mid-Maastrichtian. The earliest angiosperm pollen is the monosulcate *Clavatipollenites*, accounting for up to 1% of the early Albian palynoflora that has a close affinity to the spore/pollen assemblages of the same age from Australia. The mid–late Albian palynofloras have components that indicate similarities to either Australasia or the Falklands Plateau. Although still essentially a mixed gymnosperm/fern dominated flora, the diversity of the infrequent angiosperms increases with the appearance of *Asteropollis asteroides, Rousea georgensis, Tricolpites* spp., and *Phimopollenites pannosus* in the mid-Albian followed by *Liliacidites, Tricolporites lilliei*, and *Nyssapollenites* spp. toward the end of the stage.

Similar floristic associations persist until the early Campanian. Angiosperm diversity continues to increase with the addition of *Nothofagidites senectus, Tricolpites gillii, Periporopollenites*, and *Triporopollenites*, but without a corresponding increase in their abundance. This palynoflora in general, and the coniferous elements in particular, closely resembles coeval floras from Australia and New Zealand. *Classopollis* pollen disappears from the sequence at this stage. The mid-Campanian samples show a general similarity to those of Australia and New Zealand, but several Antarctic taxa are unknown from the Campanian of Australia. Two species of *Proteacidites* and *Myrtaceidites eugenioides* make their first appearance, and *Nothofagidites* is represented by five taxa.

The mid-Maastrichtian sample documents a sharp rise in the abundance of angiosperm pollen. Twenty-two angiosperm pollen taxa account for 44% of all pollen and spores. *Nothofagus* pollen is the largest component (22%), with *Tricolpites* spp. (mostly *T. gillii*) contributing 15%. Podocarpaceous pollen (40%) is far more significant than Araucariaceae (1%), while spores of ferns and lycopods account for less than 15% of the palynoflora. This association can be correlated with palynofloras of southeastern Australia. Dettmann and Jarzen (1988) describe 10 species of tricolpate and tricolporate angiosperm pollen from Vega Island, Antarctic Peninsula, and the Otway Basin, southeastern Australia, three of which range from the Antarctic Peninsula to Australia and New Zealand.

The paleobotanical data currently available from Antarctica emphasize the strong floristic similarities with southern South America and Australasia, but also provide evidence of local floristic differentiation and Antarctic endemics (Dettmann and Thomson 1987). In addition, comparison of palynological results from southern South America and the Antarctic Peninsula document the existence of a strong latitudinal floristic gradient over what appears to be relatively short distances (Dettmann and Thomson 1987). The Late Cretaceous development in Antarctica of a distinctly southern flora (i.e., *Nothofagus*, Podocarpaceae) with links to Australia and to the Falklands Plateau is paralleled by the development of a recognizably austral faunal realm (Dettmann and Thomson 1987). These faunal biogeographic links were similarly between the Antarctic Peninsula and southern South America on one hand and between the Antarctic Peninsula and Australasia on the other. It is also of biostratigraphic interest that the ages for these Late Cretaceous sediments indicated by both dinoflagellates and megafaunas are consistent, while those suggested by the spore/pollen record are partly out of synchrony. Dettmann and Thomson (1987) interpreted this as reflecting a delayed stepwise migration of the parent plants, and that consequently spore/pollen taxa have diachronous stratigraphic ranges across Gondwana (see also Dettmann 1981, 1986a,b).

Austral Angiosperm Biogeography

Early Angiosperm Evolution

The earliest unequivocal remains of angiosperms occur in Hauterivian sediments from southern Israel (Brenner 1984) and southern England (Hughes and McDougall 1987), and a major diversification of angiosperms through the mid-Cretaceous is well established based on analyses of both megafloras and palynofloras (Brenner 1963, Doyle 1969, Muller 1970, Beck 1976, Hickey and Doyle 1977, Friis et al. 1987, Lidgard and Crane 1988). Recent phylogenetic analyses suggest that angiosperms are most closely related to the Gnetales and Bennettitales, and because both of these groups were present during the Triassic, it is possible that the lineage leading to angiosperms may have diverged from related gymnosperm groups early in the Mesozoic. However, even though the precise time of angiosperm origin is uncertain, it is clear from current palynological and megafossil evidence that the group was neither diverse nor extensively differentiated into extant taxa prior to the mid-Cretaceous.

In the Northern Hemisphere, the angiosperm radiation is well documented, and angiosperm pollen first appears and becomes both diverse and abundant at low paleolatitudes. In terms of diversity and abundance, angiosperms only subsequently become important in middle and high paleolatitudes (Brenner 1976). This pattern is seen most clearly for triaperturate pollen, which is diagnostic of the "higher" nonmagnoliid dicotyledons (Hickey and Doyle 1977, fig. 64; Crane and Lidgard manuscript, 1). Although there is much less data for the Southern Hemisphere, it is clear that the same latitudinally diachronous pattern occurs (Fig. 15.3). As in high latitude areas of the Northern Hemisphere, the first appearance of angiosperms in Antarctica (early Albian, Dettmann and Thomson 1987) is after their appearance at lower latitudes. There is no support for the view that Antarctica or any other "isolated continent in the S. Hemisphere" (Darwin, p. 22, in Darwin and Seward 1903) was a location of cryptic early angiosperm evolution. The significance of Antarctica in terms of angiosperm evolution lies in its central position in the supercontinent of Gondwana and its importance for understanding austral biogeography.

Systematic Diversity of Cretaceous Angiosperms

One of the most striking results to emerge from recent paleobotanical studies is the rapid increase in the diversity of mid-Cretaceous angiosperms. By the Cenomanian, few megafossil floras contain less than 30% of angiosperm species and more typical values are about 50% (Fig. 15.4, see also Lidgard and Crane 1988; Crane and Lidgard, manuscript 2). Similar results are obtained from analyses of pollen floras (Lidgard and Crane manuscript), although in these data the numerical dominance of angiosperm species is attained slightly later (Turonian). The data of Dettmann and Thomson (1987) also clearly show the increase in diversity of angiosperms (and the decline in gymnosperms and pteridophytes) through the Late Cretaceous, although this is less marked than in other areas (Fig. 15.5). Many of the structural characteristics typical of modern angiosperms appear in the fossil record for the first time prior to the Turonian, including vessels, reticulate and parallel-veined leaves, pinnatifid, pinnate, and palmate leaves, hermaphrodite and unisexual flowers, superior and inferior and apocarpous and syncarpous gynoecia, and flowers with a great variety of different arrangements and numbers of floral parts (Friis and Crepet 1987, Crane 1989).

By the mid-Cretaceous, a variety of extant angiosperm taxa had already differentiated, and it is becoming clear that previous ideas on the rate of "modernization" of the angiosperm flora have substantially underestimated the number of extant taxa that may be recognized early in angiosperm evolution. By the Early and Late Cretaceous boundary, there is good evidence for considerable diversity in

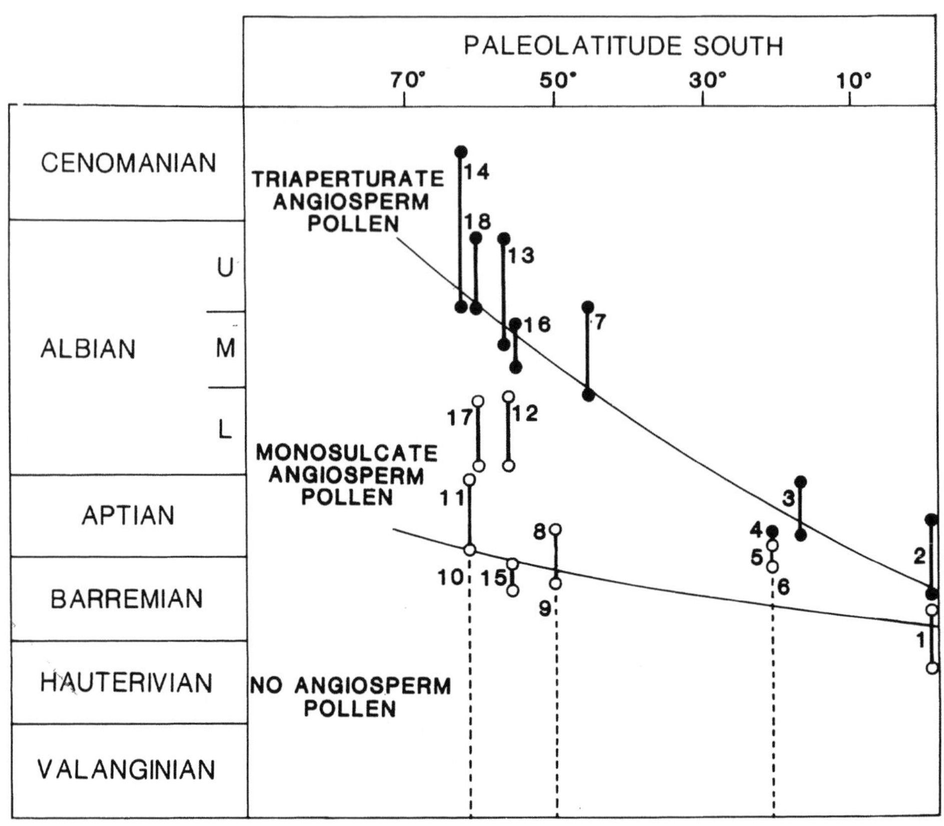

Figure 15.3. Latitudinally diachronous first appearance of angiosperms in the Southern Hemisphere based on palynofloras. Dotted lines indicate absence of angiosperm pollen. Open circles indicate presence of monosulcate or monosulcate-derived angiosperm pollen. Black dots indicate the presence of monosulcate/monosulcate-derived and triaperturate pollen diagnostic of "higher" (non-magnoliid) dicotyledons (Crane 1989). Lines joining pairs of dots indicate uncertainties of dating. 1, Israel, Helez Fm. (Brenner 1984). 2, Israel, northern Negev, Zeweira Fm. (Brenner 1976). 3, Brazil, Alagoas and Sergipe basins (Müller 1966, Brenner 1976, see also Hickey and Doyle 1977). 4, Gabon, Cocobeach Group, Zone C-VI (Doyle et al. 1982). 5, Gabon, Cocobeach Group, Zone C-VII (Doyle et al. 1982). 6, Gabon, Cocobeach Group, Zones I–IV (Doyle et al. 1982). 7, Argentina, upper Huitrin Fm. (Volkheimer and Salas 1975). 8, Argentina, Baquerό Fm. (Archangelsky et al. 1984, see also Romero and Archangelsky 1986). 9, Argentina, Vaca Muerta Fm. (Volkheimer and Quatrocchio 1975a,b), Springhill Fm. (Baldoni and Archangelsky 1983), and other Neocomian strata (Archangelsky et al. 1984). 10, Australia, Valanginian-Aptian of southern Victoria (Dettmann 1963). 11, Australia, southern Victoria, subsurface sediments correlative with Koonwarra Fossil Bed (Dettmann 1986a). 12, Australia, Surat Basin, Queensland (Burger 1980). 13, Australia, Queensland, Allaru Mudstone (Burger 1970, 1973; Dettmann 1973). 14, New Zealand, northeast South Island, Clarence Series (Couper 1953, 1960; see also Hickey and Doyle 1977). 15, Australia, Eromanga Basin (Burger 1988). 16, Australia, Eromanga Basin (Burger 1988). 17, Antarctica, James Ross Island (Dettmann and Thomson 1987). 18, Antarctica, James Ross Island (Dettmann and Thomson 1987).

the Magnoliidae (e.g., Crane and Dilcher 1984) and the presence of several extant families, including the Magnoliaceae (Dilcher and Crane 1984), Chloranthaceae (Walker and Walker 1984, Friis et al. 1986, Crane et al. 1989), Winteraceae (Walker et al. 1983), and possibly the Lauraceae (Drinnan and Crane, in preparation) and Lactoridaceae (Zavada

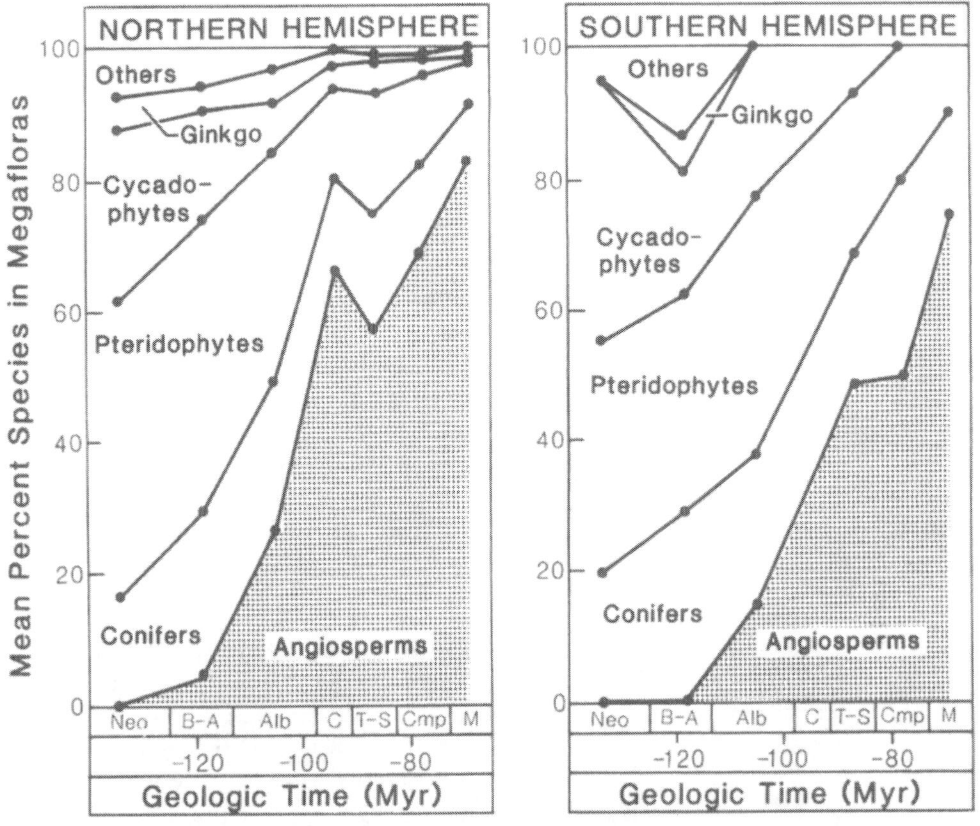

Figure 15.4. Changing floristic composition of Cretaceous fossil floras in the Northern and Southern hemispheres. Compilations include vascular plant species only (bryophytes excluded). Northern Hemisphere data based on more than 150 megafossil floras (Lidgard and Crane 1988, Crane and Lidgard, manuscript 2). Southern Hemisphere data based on a more limited sample. Neocomian: Makatini Fm., Zululand, southern Africa (Anderson and Anderson 1985); Kirkwood Fm., Algoa Basin, southern Africa (Anderson and Anderson 1985); Springhill Fm., Austral Basin, Argentina (Baldoni 1979, Baldoni and Taylor 1983). Barremian–Aptian: Koonwarra Fossil Bed, Korumburra Group, southeastern Victoria, Australia (Drinnan and Chambers 1986); Burrum Fm., Queensland, Australia (Walkom 1919); Maryborough Fm., Queensland, Australia (Walkom 1918); Anfiteatro de Tico, Baqueró Fm., Argentina (Archangelsky 1967); Baqueró, Baqueró Fm., Argentina (Archangelsky 1967); Bajo Grande, Baqueró Fm., Argentina (Archangelsky 1967). Albian: Styx Fm. Queensland (Walkom 1919). Cenomanian: no data. Turonian–Santonian: Seymour River Coal Measures, South Island, New Zealand (McQueen 1956); Mata Amarilla Fm., Argentina (Romero and Arguijo 1982). Campanian: Paparoa Coal Measures, South Island, New Zealand (McQueen 1956). Maastrichtian: Cerro Guido Fm., Argentina (Romero and Arguijo 1982).

and Benson 1987). Around the same time, there is good evidence for the Hamamelidae; for example, the Platanaceae (Crane et al. 1986, Friis et al. 1988) and probably the Trochodendrales and other groups (Crane 1989). In the Rosidae, there is evidence of several different kinds of floral structures (Basinger and Dilcher 1984). Work currently in progress on the Early Cretaceous angiosperm flora of the Atlantic Coastal Plain of North America has also revealed a considerable variety of angiosperm reproductive structures. By the Santonian–Campanian, there is evidence of further angiosperm diversity, including representatives of the Zingiberales (Friis 1988), Saxifragales (Friis and Skarby 1981, 1982),

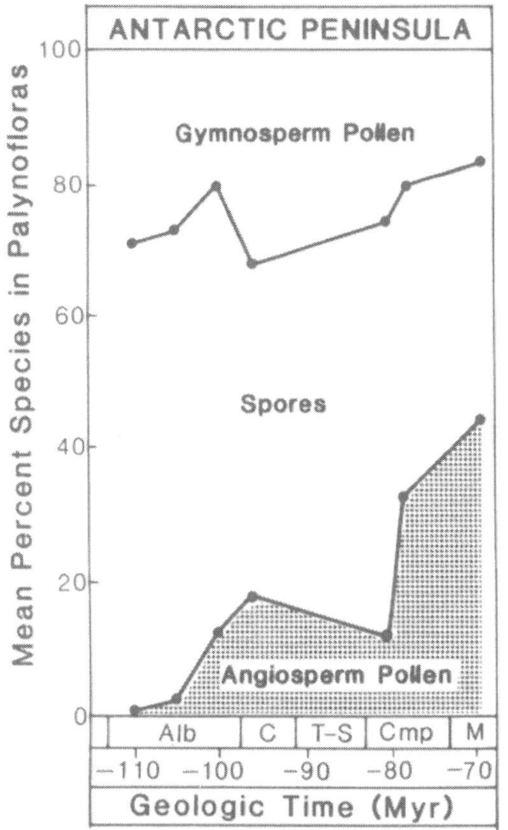

Figure 15.5. Comparison of changing floristic composition in Albian to Maastrichtian palynofloras from the James Ross Island area, Antarctica (compiled from data in Dettmann and Thomson 1987). See text for discussion.

Juglandales/Myricales (Friis 1983, Friis and Crane 1989), Fagales (Tiffney and Friis, personal communication). Palynomorphs indicate the presence of Urticales, Hippuridales, Celastrales, and Betulales (Muller 1981, 1984). *Actinocalyx bohrii* from the Santonian–Campanian of southern Sweden is closely related to extant Ericales (Friis 1985) and currently provides the earliest evidence of the Dilleniidae. By the Maastrichtian, a variety of other taxa of extant angiosperms are known on the basis of megafossils, including the Annonaceae (Chesters 1955), Arecaceae (Muller 1981, Weber 1978, Friis and Crepet 1987), and Myrtales (Sahni 1943, Chitaley 1977), as well as possibly the Icacinaceae and Passifloraceae (Chesters 1955). A variety of other orders are recorded on the basis of fossil pollen (Muller 1981, 1984).

Although these data are based largely on the Northern Hemisphere, they clearly indicate that differentiation into extant angiosperm taxa occurred rapidly through the Cretaceous and overlapped substantially with some of the major events in the breakup of Gondwana. These considerations suggest that patterns of differentiation in a large number of angiosperm taxa were influenced by tectonic events in the Southern Hemisphere and thus detailed understanding of phylogenetic relationships in these taxa is likely to provide useful information on the breakup of the supercontinent.

Major Patterns of Austral Biogeography and Comparison with Tectonic History

The dominant pattern of biogeographic relationship exhibited by extant austral angiosperms is between southern South America and Australasia. J.D. Hooker was among the first to recognize this pattern and suggest that it arose by the fragmentation "of a once more extensive southern biota." A critical evaluation of the major disjunctions in the geographic ranges of seed plants in the Southern Hemisphere by Thorne (1972, 1973) indicated that South American–Australasian distributions are the dominant pattern of austral phytogeography. Seven angiosperm families and 48 genera occur largely in South America, Australasia (Australia, New Zealand, New Guinea, New Caledonia), and some sub-Antarctic islands, with occasional range extension into Mexico or into Asia or Polynesia (Thorne 1972). The major angiosperm families with this distribution are the Epacridaceae, Stylidiaceae, and Goodeniaceae, but the same pattern is shown by other groups at lower taxonomic levels, such as the generic groups within the Monimiaceae (Martínez-Laborde 1988) and the genus *Nothofagus* (Fagaceae) (Humphries 1981, 1983; Humphries et al. 1986, Romero 1986b).

The same pattern is also seen in other groups of plants. Schuster (1976) illustrates a southeastern Australia–New Caledonia–New

Zealand–southern South America distribution for isophyllous, triradial Jungermanniales (6 orders comprising 19 genera). Similarly, Galloway (1988) emphasizes the predominantly generic level similarities between the macrolichen floras of Australasia and southern South America.

Various animal taxa also show a strong relationship across the southern Pacific. Edmunds (1981) shows a set of congruent area relationships between the genera of four subclades of temperate Southern Hemisphere mayflies (Ephemoptera), while Brundin (1966) establishes similar patterns of relationship based on the distribution and systematics of austral midges. Cracraft (1980) enumerates a variety of austral disjunctions among vertebrates of which the Australia–South America "track" is one of the most strongly supported.

In contrast, the distributions of very few taxa link Africa with either Australasia (mostly Australia) or South America (Thorne 1972). Of the few taxa that are disjunct across the Indian Ocean (18 genera common to Australasia and Africa, plus a further 11 represented in Australasia and either Madagascar, the Mascarenes or the Seychelles), many (13 genera) have other insular populations scattered throughout the Indian Ocean, and Thorne suggests dispersal as the most obvious explanation for their disjunction. Thorne also predicts that further systematic analysis will prove some of these disjuncts to be less closely related than is currently suspected.

Similarly, Thorne (1972, 1973) considers the phytogeographic patterns shared by Africa (and Madagascar) and South America to be weaker than those linking either of these areas with Australasia. Although numerous taxa are listed [12 families, 7 subfamilies, 111 genera, 108 indigenous spp.—mostly weedy, shoreline, aquatics (some naturalized) comprising only 0.6% of the total flora], on the basis of autecology and of comparative South America–Africa and South America–Africa–Madagascar distributions, Thorne again considers dispersal between these continents as the most plausible explanation. Alternative hypotheses for some taxa include penetration from the Northern Hemisphere (therefore South American and African taxa are not closely related vicariants), or that the Atlantic margins of northern South America and Africa retained a closer proximity than was previously believed. Indeed, Raven and Axelrod (1974) counter that South America and Africa parted as recently as 100 mya [although their southern extremes had separated much earlier (~125 mya)] and during the early Paleocene may have been only 800 km apart, leaving ample opportunity for transatlantic migration for a considerable period following the initial physical disjunction. Raven and Axelrod (1974) consider impoverishment of the African flora due to post-Eocene climatic change as a possible explanation. They suggest that basically similar angiosperm floras have become significantly different because of major species depletion in Africa, as opposed to Thorne's hypothesis that once different floras have become marginally similar by exchange of a small number of taxa. However, the shared taxa suggest links between the northern (equatorial) parts of Africa and South America, rather than their southern regions. Recent paleogeographical reconstructions (Ziegler et al. 1982, Barron 1987) show the coastlines of northern Brazil and west tropical Africa to be relatively close during the Maastrichtian, but the southern tip of Africa was already well separated from both the southern South American and Australian regions of the remainder of Gondwana.

Although Raven and Axelrod's considerations may well be applicable to the northern parts of Africa and South America, Thorne's explanations are supported by extant austral taxa and, where well documented, their fossil record. For example, distribution data presented by Johnson and Briggs (1975, Tables 3 and 4) for the subfamilies, tribes, and subtribes of the Proteaceae show distinct relationships between Australia and South America and between Australia and Africa, yet no close association between South America and Africa. The genus *Nothofagus* (Fagaceae), perhaps the most widely quoted example of austral Pacific area relationships, has an extant distribution in South America and Australasia, and a fossil pollen record in these areas and in Antarctica dating back to at least the early Senonian (~90 mya). In contrast, *Nothofagus* pollen is not known from the fossil record of Africa, and there is no evidence that it ever occurred there. Both *Nothofagus* and the Proteaceae are ancient in terms of modern angiosperm taxa, having fossil records extending well back into the Upper Cretaceous, although both postdate the accepted age of the final South Atlantic opening. The fact that neither of these taxa exhibit any indication of transatlantic relationship, either past or present, suggests that direct floristic connections between southern Africa and southern South America have not existed since the initiation of major angiosperm diversification.

In broad terms, these biogeographic patterns correspond with, and hence may be explained by, the current hypothesis of the breakup of Gondwana. India and Madagascar are thought to have been the first continental plates to separate from the main Gondwana land mass. Subsequently, Africa separated and Madagascar separated from India and remained in close proximity with the African plate. After the separation of Africa and India, southern South America, Australia, and New Zealand remained in connection through Antarctica. Consequently, the biotas of these regions show a strong historical relationship. Subsequently, New Zealand is thought to have separated first, leaving Australia and southern South America in direct connection via Antarctica.

Biogeographic Significance of Antarctica

Although the major biogeographical patterns exhibited by the austral biota are clear and broadly correspond to current ideas on the tectonic history of Gondwana, there are numerous complications, particularly involving Antarctica. These mainly arise because many of the critical areas are themselves tectonic composites. For example, there has been considerable discussion concerning the nature and origin of the terranes of the southern and central South American Pacific margin (Ziel 1979; James 1973; Nur and Ben Avraham 1981; Jones et al. 1982). Similarly, New Zealand consists of several tectonic fragments (Dalziel and Elliot 1982, Howell 1980), and this may account in part for the complexity of its biogeographic patterns (Craw 1982).

The tectonic situation in western Antarctica is also complex. Reconstructions of the southern Pacific land masses prior to the Gondwana breakup (e.g., Dalziel and Elliot 1982; Dalziel and Grunow 1985; Zinsmeister 1987) show southern South America, southeastern Australia, and the components of New Zealand and East Antarctica closely associated through a number of separate terranes that now comprise West Antarctica (i.e., Ellsworth and Whitmore Mountains, Antarctic Peninsula, Thurston Island–Eights Coast, and Marie Byrd Land). The behavior of these tectonic components of West Antarctica since the Jurassic is critical to clarifying the evolution of the southern Pacific land masses.

The development of the Antarctic ice sheet is thought to have commenced with the final separation of Australia from eastern Antarctica at the end of the Eocene (Zinsmeister 1987) and the opening of the Drake Passage between West Antarctica and South America. Up to this point, Antarctica is thought to have had substantial vegetational cover, and it is clear that vegetation persisted in some areas well into the late Tertiary (Truswell 1989). There may also have been some floristic continuity between southern South America and southern Australia until as late as the Miocene. Since the fossil record indicates that many modern angiosperm taxa (especially at lower taxonomic levels, i.e., family, subfamily, tribe, and even some genera) had already differentiated in the Cretaceous, the discovery of systematically informative angiosperm fossils in Antarctica could have a significant bearing on ideas of the composition and breakup of the Australasia–Antarctica–South America land mass.

Extension of the geographic ranges of certain angiosperm taxa into Antarctica would permit an evaluation of the phylogenetic relationships of the Antarctic biota with respect to related taxa in Australia, South America, and elsewhere. Congruence between the area relationships suggested by a majority of such patterns would provide important data bearing on the history of austral floras, because it is more plausible to hypothesize one set of vicariant events to explain the coincident distribution of all taxa rather than several separate, identical, dispersal events. Such argumentation requires detailed work on systematically informative fossil material, but could provide a relative chronology of vicariant events and thus a relative chronology of Gondwana fragmentation. Paleobotanical data also provide the means of recognizing extinctions and distinguishing noncongruent distribution patterns resulting from superposition of younger (on older) vicariant events (Grande 1985).

Although angiosperms are not the only biotic group bearing on the breakup of Gondwana, they have certain advantages. A detailed understanding of systematics in the groups concerned is a prerequisite for biogeographical analysis, and advances in the systematics of extant groups of austral angiosperms are providing the comparative systematic and biogeographic data with which the fossil history of these groups can be integrated. Angiosperms were also undergoing an active diversification during the critical interval in which the constituent areas of western Gondwana were beginning to separate. Although combined information from both megafossils (e.g., floral structures, leaves) and pollen would be of the greatest systematic value, pollen data alone could also provide much useful information. Pollen grains frequently provide systematically diagnostic characters (synapomorphies) that can be used to establish the presence of monophyletic groups, and studies of pollen in situ within floral structures (e.g., Crane et al. 1986; Friis et al. 1986, 1988) can be used to integrate palynological and megafossil records.

Conclusions

Antarctica is the only major isolated land mass for which there is no reasonably diverse extant terrestrial vegetation, and increased paleobotanical data is therefore critical both for evaluating the biotic relationships of this land mass and for a full understanding of austral biogeography. Increased knowledge of Cretaceous floras from Antarctica will provide some resolution of the geographical affinities of the extinct Antarctic biota, but studies of well-preserved angiosperm remains from this area are likely to provide more precise data for interpreting the details of austral biogeographic patterns. Antarctica is a critical element in understanding the biological and geological evolution of the Southern Hemisphere, and in particular the several tectonic components of West Antarctica are important for clarifying relationships between the southern Pacific land masses. Integration of paleobotanical data from Antarctica and related austral areas could provide a relative time control on the pattern of Gondwana fragmentation and the origin of austral biogeographic patterns. Because of their extraordinary radiation to floristic and vegetational dominance in the Late Cretaceous, angiosperms are likely to be particularly informative for these purposes.

Acknowledgments. We thank Thomas and Edith Taylor for their kind invitation to participate in this symposium. We are also grateful to S. Archangelsky, M.N. Bose, S.H. Lidgard, G. Playford, E. Romero, and E. Truswell for helpful discussions of some of the material covered in this chapter. Clara Richardson drew the illustrations. This work was supported by N.S.F. grant BSR-8708460 to P.R.C.

References

Agrawal JK, Rama (1976) Chronology of Mesozoic volcanics of India. Proceedings of the Indian Academy of Science 84A:157–179

Anderson JM, Anderson HM (1985) Palaeoflora of Southern Africa: Prodromus of South African Megafloras—Devonian to Lower Cretaceous. A.A. Balkema, Rotterdam, 423 p

Arber EAN (1917) The earlier Mesozoic flora of New Zealand. New Zealand Geological Survey Palaeontological Bulletin 6:1–80

Archangelsky S (1963) A new Mesozoic flora from Ticó, Santa Cruz Province, Argentina. Bulletin

of the British Museum (Natural History), Geology 8:47–92
Archangelsky S (1967) Estudio de la Formación Baqueró Cretácico Inferior de Santa Cruz, Argentina. Revista del Museo de la Plata (Nueva Serie), Paleontología 5:63–171
Archangelsky S (1976) Vegetales fósiles de la Formación Springhill, Cretácico en el subsuelo de la cuenca Magállenica, Chile. Ameghiniana 13:141–158
Archangelsky S (1980) Palynology of the Lower Cretaceous in Argentina. Proceedings of the Fourth International Palynological Conference, Lucknow (1976–77) 2:425–428
Archangelsky S, del Fueyo G (1987) Sobre una podocarpacea fertil del Cretácico Inferior de la Provincia Santa Cruz, Republica Argentina. VII Symposio Argentino de Paleobotanica y Palinología Actas, Buenos Aires. 1987. pp 85–97
Archangelsky S, Gamerro JC (1965) Estudio palinológico de la Formación Baqueró (Cretácico), Provincia de Santa Cruz, I. Ameghiniana 4:159–167
Archangelsky S, Gamerro JC (1966a) Estudio palinológico de la Formación Baqueró (Cretácico), Provincia de Santa Cruz, II. Ameghiniana 4:201–209
Archangelsky S, Gamerro JC (1966b) Estudio palinológico de la Formación Baqueró (Cretácico), Provincia de Santa Cruz, III. Ameghiniana 4:229–234
Archangelsky S, Gamerro JC (1967) Spore and pollen types of the Lower Cretaceous in Patagonia (Argentina). Review of Palaeobotany and Palynology 1:211–217
Archangelsky S, Petriella B (1971) Notas sobre la flora fósil de la Zona de Ticó, Provincia de Santa Cruz, IX, Nuevos datos acerca de la morfologia foliar de *Mesodescolea plicata* Arch. (Cycadales, Stangeriaceae). Boletín de la Sociedad Argentina de Botánica 14:88–94
Archangelsky S, Romero EJ (1974a) Los registros más antiguos de *Nothofagus* (Fagaceae) de Patagonia (Argentina y Chile). Boletín de la Sociedad México de Botánica 33:13–30
Archangelsky S, Romero EJ (1974b) Polen de Gimnospermas (coniferas) del Cretácico Superior y Paleoceno de Patagonia. Ameghiniana 11:217–236
Archangelsky S, Taylor TN (1986) Ultrastructural studies of fossil plant cuticles, II, *Tarphyderma* gen. n., a Cretaceous conifer from Argentina. American Journal of Botany 73:1577–1587
Archangelsky S, Baldoni A, Gamerro JC, Palamarczuk S, Seiler J (1981) Palinología estratigráfica del Cretácico de Argentina Austral. Diagramas de Grupos Polínicos del Suroeste de Chubut y Noroeste de Santa Cruz. VIII Congreso Geológico Argentino, San Luis (20–26 septiembre, 1981), Actas IV:719–742
Archangelsky S, Baldoni A, Gamerro JC, Seiler J (1983) Palinología estratigráfica del Cretácico de Argentina Austral, II, Descripciones sistemáticas. Ameghiniana 20:199–226
Archangelsky S, Baldoni A, Gamerro JC, Seiler J (1984) Palinología estratigráfica del Cretácico de Argentina Austral, III, Distribución de las especies y conclusiones. Ameghiniana 21:15–33
Archangelsky S, Taylor TN, Kurmann MH (1986) Ultrastructural studies of fossil plant cuticles: *Ticoa harrisii* from the Early Cretaceous of Argentina. Botanical Journal of the Linnean Society 92:101–116
Arkell WJ (1956) Jurassic Geology of the World. Oliver and Boyd, London, 806 p
Askin RA (1981) Jurassic–Cretaceous palynology of Byers Peninsula, Livingston Island, Antarctica. Antarctic Journal of the United States 16(5):11–13
Askin RA (1983) Campanian palynomorphs from James Ross and Vega Islands, Antarctic Peninsula. Antarctic Journal of the United States 18(5):63–64
Askin RA (1985) Palynological studies in the James Ross Island Basin, Antarctic Peninsula: a progress report. Antarctic Journal of the United States 20(5):44–45
Askin RA (1988) Campanian–Paleocene vegetation of northern Antarctic Peninsula: palynological evidence from Seymour Island. Geological Society of America, Abstracts, A227
Axelrod DI, Raven PH (1978) Late Cretaceous and Tertiary vegetation history of Africa. In Werger MJA, van Bruggen AC (eds) Biogeography and Ecology of Southern Africa, Dr. W Junk, The Hague, pp 77–130
Baksi SK (1966) On the foraminifera from Raghavapuram mudstone, West Godavari district, Andhra Pradesh, India. Bulletin of the Mineralogical and Metallurgical Society of India 37:1–19
Baksi SK (1967) Fossil plants from Raghavapuram mudstone, West Godavari district, Andhra Pradesh, India. The Palaeobotanist 16:206–216
Baldoni AM (1979) Nuevos elementos paleoflorísticos de la tafoflora de la Formación Spring Hill, limite Jurásico–Cretácico subsuelo de Argentina y Chile austral. Ameghiniana 16:103–119
Baldoni AM (1987) Nuevas descripciones palinológicas en el area de Collon Cura (Terciario Inferior) Provincia del Neuquén, Argentina. 4th Congreso Latinoamericano de Paleontología, Bolivia 1:399–414
Baldoni A, Archangelsky S (1983) Palinología de la Formación Springhill (Cretácico Inferior), Subsuelo de Argentina y Chile Austral. Revista Española de Micropaleontología 15:47–101
Baldoni AM, Taylor TN (1983) Plant remains from a new Cretaceous site in Santa Cruz, Argentina. Review of Palaeobotany and Palynology 39: 301–311

Balme BE (1957) Spores and pollen grains from the Mesozoic of Western Australia. Coal Research, CSIRO, TC 25:1–48
Balme BE (1964) The palynological record of Australian pre-Tertiary floras. In Ancient Pacific Floras, University of Hawaii Press, Honolulu, pp 49–80
Banerji J, Jana BN, Maheshwari HK (1984) The fossil floras of Kachchh, II.—Mesozoic megaspores. The Palaeobotanist 33:190–227
Barron EJ (1987) Global Cretaceous paleogeography—International Geologic Correlation Program Project 191. Palaeogeography, Palaeoclimatology, Palaeoecology 59:207–214
Bartrum JA (1921) Note on the Port Waikato Mesozoic flora. New Zealand Journal of Science and Technology 4:258
Basinger JF, Dilcher DL (1984) Ancient bisexual flowers. Science 224:511–513
Beck CB (ed) (1976) Origin and Early Evolution of Angiosperms, Columbia University Press, New York, 341 p
Bhalla SN (1965) New species of foraminifera from the Raghavapuram shales (Lower Cretaceous), Andhra Pradesh, India. Bulletin of the Geological Society of India 2:39–43
Bhalla SN (1968) Palaeoecology of the Raghavapuram shales (Early Cretaceous), East Coast Gondwanas, India. Palaeogeography, Palaeoclimatology, Palaeoecology 5:345–357
Bhalla SN (1972) Upper age limit of the East Coast Gondwanas, India. Lethaia 5:271–280
Blaschke PM, Grant-Mackie JA (1976) Mesozoic leaf genus *Taeniopteris* at Port Waikato and Clent Hills, New Zealand. New Zealand Journal of Geology and Geophysics 19:933–941
Bose MN (1966) Fossil plant remains from the Rajmahal and Jabalpur Series in the Upper Gondwana of India. Symposium on the Floristics and Stratigraphy of Gondwanaland, Birbal Sahni Institute of Palaeobotany, Lucknow, 1964:143–154
Bose MN, Banerji J (1984) The fossil floras of Kachchh, I.—Mesozoic megafossils. The Palaeobotanist 33:1–189
Bose MN, Pal PK, Harris TM (1985) The *Pentoxylon* plant. Philosophical Transactions of the Royal Society of London B310:77–108
Bose MN, Taylor TN, Taylor EL (1989) Gondwana Floras of India and Antarctica—A Survey and Appraisal. This volume, pp. 118–148
Brenner GJ (1963) The spores and pollen of the Potomac Group of Maryland. State of Maryland Department of Geology, Mines and Water Resources, Bulletin 27:1–215
Brenner GJ (1976) Middle Cretaceous floral provinces and early migrations of angiosperms. In Beck CB (ed) Origin and Early Evolution of Angiosperms, Columbia University Press, New York, pp 23–47
Brenner GJ (1984) Late Hauterivian angiosperm pollen from the Helez Formation, Israel. 6th International Palynological Conference, Calgary 1984, Abstracts, p 15
Brundin L (1966) Transantarctic relationships and their significance as evidenced by chironomid midges. Kungl. Svenska Vetenskapsakademiens Handlingar, Series 4, 11:1–472
Burger D (1970) Early Cretaceous angiospermous pollen grains from Queensland. Bulletin of the Bureau of Mineral Resources, Geology and Geophysics, Australia 116:1–110
Burger D (1973) Spore zonation and sedimentary history of the Neocomian, Great Artesian Basin, Queensland. Special Publication of the Geological Society of Australia 4:87–118
Burger D (1974) Palynology of subsurface Lower Cretaceous strata in the Surat Basin, Queensland. Bulletin of the Bureau of Mineral Resources, Geology and Geophysics, Australia 150:27–42
Burger D (1975) Cenomanian spores and pollen grains from Bathurst Island, Northern Territory, Australia. Bulletin of the Bureau of Mineral Resources, Geology and Geophysics, Australia. 151:114–167
Burger D (1976) Some Early Cretaceous plant microfossils from Queensland. Bulletin of the Bureau of Mineral Resources, Geology and Geophysics, Australia 160:1–22
Burger D (1980) Palynology of the Lower Cretaceous in the Surat Basin. Bulletin of the Bureau of Mineral Resources, Geology and Geophysics, Australia 189:1–106
Burger D (1988) Early Cretaceous environments in the Eromanga Basin: palynological evidence from GSQ Wyandra-1 corehole. Memoir of the Association of Australasian Palaeontologists 5:173–186
Cantrill DJ, Webb JA (1987) A reappraisal of *Phyllopteroides* Medwell (Osmundaceae) and its stratigraphic significance in the Lower Cretaceous of eastern Australia. Alcheringa 11:59–85
Chaloner WG, Creber GT (1988) Fossil plants as indicators of late Paleozoic plate positions. In Audley Charles MG, Hallam A (eds) Gondwana and Tethys. Geological Society Special Publication 37:201–210
Chaloner WG, Lacey WS (1973) The distribution of Late Paleozoic floras. In Hughes NF (ed) Organisms and Continents through Time, Special Papers in Palaeontology 12:271–289
Chaloner WG, Meyen SV (1973) Carboniferous and Permian floras of the northern continents. In Hallam A (ed) Atlas of Palaeobiogeography, Elsevier, Amsterdam, pp 169–186
Chesters KIM (1955) Some plant remains from the Upper Cretaceous and Tertiary of West Africa. Annals and Magazine of Natural History 12: 498–504

Chitaley SD (1977) *Enigmocarpon parijai* and its allies. Frontiers of Plant Sciences—Professor P. Parija Felicitation Volume, pp 421–429

Chowdhury A (1958) Plant fossils from Uttatur plant beds of Terany clay pit, Trichinopoly district. Quarterly Journal of the Geological, Mineralogical and Metallurgical Society of India 30:141–143

Couper RA (1953) Upper Mesozoic and Cainozoic spores and pollen grains from New Zealand. New Zealand Geological Survey Palaeontological Bulletin 22:1–77

Couper RA (1960) New Zealand Mesozoic and Cainozoic plant microfossils. New Zealand Geological Survey Palaeontological Bulletin 32:1–87

Cracraft J (1980) Biogeographic patterns of terrestrial vertebrates in the southwest Pacific. Palaeogeography, Palaeoclimatology, Palaeoecology 31:353–369

Crane PR (1989) Paleobotanical evidence on the early radiation of nonmagnoliid dicotyledons. Plant Systematics and Evolution 162:165–191

Crane PR, Dilcher DL (1984) *Lesqueria:* an early angiosperm fruiting axis from the mid-Cretaceous. Annals of the Missouri Botanical Garden 71:384–402

Crane PR, Lidgard SH (manuscript 1) Angiosperm diversification and paleolatitudinal gradients in Cretaceous floristic diversity. Science (submitted)

Crane PR, Lidgard SH (manuscript 2) Angiosperm diversity and large scale floristic change through the Cretaceous: Analysis of macrofossil data from the Northern Hemisphere. Palaeontology (submitted)

Crane PR, Friis EM, Pedersen KR (1986) Angiosperm flowers from the Lower Cretaceous: fossil evidence on the early radiation of the dicotyledons. Science 232:852–854

Crane PR, Friis EM, Pedersen KR (1989) Reproductive structure and function in Cretaceous Chloranthaceae. Plant Systematics and Evolution (in press)

Craw R (1982) Phylogenetics, areas, geology and the biogeography of Croizat: a radical review. Systematic Zoology 31:304–316

Dalziel WD, Elliot DH (1982) West Antarctica: problem child of Gondwanaland. Tectonics 1:3–19

Dalziel WD, Grunow AM (1985) The Pacific margin of Antarctica: terranes within terranes within terranes. In Howell DG (ed) Tectonostratigraphic Terranes of the Circum-Pacific Region, Circum-Pacific Council for Energy and Mineral Resources, Houston, pp 555–564

Darwin F, Seward AC (1903) More letters of Charles Darwin, Vol II, John Murray, London

Dettmann ME (1963) Upper Mesozoic microfloras from south-eastern Australia. Proceedings of the Royal Society of Victoria 77:1–148

Dettmann ME (1973) Angiospermous pollen from Albian to Turonian sediments of eastern Australia. Special Publication of the Geological Society of Australia 4:3–34

Dettmann ME (1981) The Cretaceous flora. In Keast A (ed) Ecological Biogeography of Australia, Vol 1, Dr W Junk, The Hague, pp 357–375

Dettmann ME (1986a) Early Cretaceous palynoflora of subsurface strata correlative with the Koonwarra Fossil Bed, Victoria. Memoir of the Association of Australasian Palaeontologists 3:79–110

Dettman ME (1986b) Significance of the spore genus *Cyatheacidites* in tracing the origin and migration of *Lophosoria* (Filicopsida). Special Papers in Palaeontology 35:63–94

Dettmann ME, Douglas JG (1976) Mesozoic palaeontology. Special Publication of the Geological Society of Australia 5:164–169

Dettmann ME, Jarzen DM (1988) Angiosperm pollen from uppermost Cretaceous strata of southeastern Australia and the Antarctic Peninsula. Memoir of the Association of Australasian Palaeontologists 5:217–237

Dettmann ME, Playford G (1968) Taxonomy of some Cretaceous spores and pollen grains from eastern Australia. Proceedings of the Royal Society of Victoria 81:69–93

Dettmann ME, Playford G (1969) Palynology of the Australian Cretaceous: a review. In Campbell KSW (ed) Stratigraphy and Palaeontology, Essays in Honour of Dorothy Hill, Australian National University Press, Canberra, pp 174–210

Dettman ME, Thomson MRA (1987) Cretaceous palynomorphs from the James Ross Island area—a pilot study. British Antarctic Survey Bulletin 77:13–59

Dilcher DL, Crane PR (1984) *Archaeanthus:* an early angiosperm from the Cenomanian of the western interior of North America. Annals of the Missouri Botanical Garden 71:351–383

Domack EW, Fairchild WW, Anderson JB (1980) Lower Cretaceous sediment from the East Antarctic continental shelf. Nature 287:625–626

Douglas JG (1969) The Mesozoic floras of Victoria, Parts 1 and 2. Memoirs of the Geological Survey of Victoria 28:1–310

Douglas JG (1973) The Mesozoic floras of Victoria, Part 3. Memoirs of the Geological Survey of Victoria 29:1–185

Doyle JA (1969) Cretaceous angiosperm pollen of the Atlantic Coastal Plain and its evolutionary significance. Journal of the Arnold Arboretum 30:1–35

Doyle JA, Jardiné S, Doerenkamp A (1982) *Afropollis,* a new genus of early angiosperm pollen, with notes on the Cretaceous palynostratigraphy and paleoenvironments of Gondwana. Bulletin Centres Recherches Explor.-Prod. Elf-Aquitaine 6:39–117

Drinnan AN, Chambers TC (1985) A reassessment of *Taeniopteris daintreei* from the Victorian Early Cretaceous: a member of the Pentoxylales and a significant Gondwanaland plant. Australian Journal of Botany 33:89–100
Drinnan AN, Chambers TC (1986) Flora of the Lower Cretaceous Koonwarra Fossil Bed (Korumburra Group), South Gippsland, Victoria. Memoir of the Association of Australasian Palaeontologists 3:1–77
Edmunds GF Jr (1981) Discussion. In Nelson G, Rosen DE (eds) Vicariance Biogeography: A Critique, Columbia University Press, New York, pp 287–297
Evans PR (1966) Mesozoic stratigraphic palynology in Australia. Australasian Oil and Gas Journal 12:58–63
Feistmantel O (1876) Jurassic (Oolitic) flora of Kach. Memoirs of the Geological Survey of India, Palaeontologia Indica, Series 11, 2:1–80
Feistmantel O (1879) Upper Gondwana flora of the outliers on the Madras coast. Memoirs of the Geological Survey of India, Palaeontologia Indica, Series 11, 4:192–224
Filatoff J (1975) Jurassic palynology of the Perth Basin, Western Australia. Palaeontographica 154B:1–115
Fleming CA (1975) The geological history of New Zealand and its biota. In Kuschel (ed) Biogeography and Ecology in New Zealand, Dr W Junk, The Hague, pp 1–96
Friis EM (1983) Upper Cretaceous (Senonian) floral structures of juglandalean affinity containing *Normapolles* pollen. Review of Palaeobotany and Palynology 39:161–188
Friis EM (1985) *Actinocalyx* gen. nov., sympetalous angiosperm flowers from the Upper Cretaceous of southern Sweden. Review of Palaeobotany and Palynology 45:171–183
Friis EM (1988) *Spirematospermum chandlerae* sp. nov., an extinct species of Zingiberaceae from the North American Cretaceous. Tertiary Research 9:7–12
Friis EM, Crane PR (1989) Reproductive structures of Cretaceous Hamamelidae. In Crane PR, Blackmore S (eds) Evolution, Systematics and Fossil History of the Hamamelidae, Clarendon Press, Oxford (in press)
Friis EM, Crepet WL (1987) Time of appearance of floral features. In Friis EM, Chaloner WG, Crane PR (eds) The Origins of Angiosperms and Their Biological Consequences, Cambridge University Press, Cambridge, pp 145–179
Friis EM, Skarby A (1981) Structurally preserved angiosperm flowers from the Upper Cretaceous of southern Sweden. Nature 291:485–486
Friis EM, Skarby A (1982) *Scandianthus* gen. nov., angiosperm flowers of saxifragalean affinity from the Upper Cretaceous of southern Sweden. Annals of Botany 50:569–583
Friis EM, Crane PR, Pedersen KR (1986) Floral evidence for Cretaceous chloranthoid angiosperms. Nature 320:163–164
Friis EM, Chaloner WG, Crane PR (eds) (1987) The Origins of Angiosperms and Their Biological Consequences, Cambridge University Press, Cambridge, 358 pp
Friis EM, Crane PR, Pedersen KR (1988) Reproductive structures of Cretaceous Platanaceae. Biologiske Skrifter Danske Videnskabernes Selskab 31:1–55
Galloway DJ (1988) Plate tectonics and the distribution of cool temperate Southern Hemisphere macrolichens. Botanical Journal of the Linnean Society 96:45–55
Gee CT (1987) Revision of the Early Cretaceous flora from Hope Bay, Antarctica. Dissertation Abstracts International, Section B 48:1284
Gleadow AJW, Duddy IR (1980) Early Cretaceous volcanism and the early breakup history of southeastern Australia: evidence from fission track dating of volcaniclastic sediments. In Cresswell MM, Vella P (eds) Proceedings of the Fifth International Gondwana Symposium, Wellington, New Zealand, AA Balkema, Rotterdam, pp 295–300
Gopal V, Jacob C, Jacob K (1957) Stratigraphy and palaeontology of the Upper Gondwana of the Ramnad district on the East Coast. Records of the Geological Society of India 84:477–496
Grande L (1985) The use of paleontology in systematics and biogeography, and a time control refinement for historical biogeography. Paleobiology 11:234–243
Halle TG (1913) The Mesozoic flora of Graham Land. Wissenschaftliche Ergebnisse der Schwedischen Südpolar-Expedition (1901–1903) 3:1–124
Halle TG (1937) The relation between the Late Palaeozoic floras of eastern and northern Asia. C.R. 2e Congrès l'avancement étud. stratigr. Carbonifère, Heerlen 1935, Maastricht 1:237–245
Harris TM (1961) The occurrence of the fructification *Carnoconites* in New Zealand. Transactions of the Royal Society of New Zealand (Geology) 1:17–27
Harris WK (1977) Palynology of cores from deep sea drilling sites 327, 328 and 330, south Atlantic Ocean. Initial Reports Deep Sea Drilling Project 36:761–817
Hedlund RW, Beju D (1977) Stratigraphic palynology of selected Mesozoic samples, DSDP Hole 327A and Site 330. Initial Reports Deep Sea Drilling Project 36:817–827
Helby R, Morgan R, Partridge AD (1987) A palynological zonation of the Australian Mesozoic. Memoir of the Association of Australasian Palaeontologists 5:1–94
Hernandez PJ, Azcarate V (1971) Estudio paleobotánico preliminar sobre restos de una tafoflora de

la Peninsula Byers (Cerro Negro), Isla Livingston, Islas Shetland del Sur. Antártida. Instítudo Antartido Chileno, Ser. Cientifica 2:15–50
Herngreen GFW, Chlonova AF (1981) Cretaceous microfloral provinces. Pollen et Spores 23:441–555
Hickey LJ, Doyle JA (1977) Early Cretaceous fossil evidence for angiosperm evolution. Botanical Review 43:3–104
Hooker JD (1853) Introductory essay. In The Botany of the Antarctic Voyage of H. M. Discovery, Erebus and Terror in the Years 1853–55, II, Flora Nova Zelandiae, published by the author, London
Howell DG (1980) Mesozoic accretion of exotic terranes along the New Zealand segment of Gondwanaland. Geology 8:487–491
Hughes NF, McDougall AB (1987) Records of angiospermid pollen entry into the English Early Cretaceous succession. Review of Palaeobotany and Palynology 50:255–272
Humphries CJ (1981) Biogeographical methods and the southern beeches (Fagaceae: *Nothofagus*). In Funk VA, Brooks DR (eds), Advances in Cladistics, Proceedings of the First Meeting of the Willi Hennig Society, New York Botanical Garden, Bronx, NY, pp 177–207
Humphries CJ (1983) Biogeographical explanations and the southern beeches. In Sims RW, Price JH, Whalley PES (eds) Evolution, Time and Space: The Emergence of the Biosphere, Academic Press, London, pp 335–365
Humphries CJ, Parenti LR (1986) Cladistic Biogeography. Clarendon Press, Oxford, 98p
Humphries CJ, Cox JM, Nielsen ES (1986) *Nothofagus* and its parasites: a cladistic approach to coevolution. In Stone AR, Hawksworth DL (eds), Coevolution and Systematics, Clarendon Press, Oxford, pp 56–76
Ingram BS (1968) Stratigraphical palynology of Cretaceous rocks from bores in the Eucla Basin, Western Australia. Western Australia Department of Mines, Annual Report 1967:102–105
James DE (1973) The evolution of the Andes. Scientific American 229:60–69
Jefferson TH (1982a) Fossil forests from the Lower Cretaceous of Alexander Island, Antarctica. Palaeontology 25:681–708
Jefferson TH (1982b) The preservation of fossil leaves in Cretaceous volcaniclastic rocks from Alexander Island, Antarctica. Geological Magazine 119:291–300
Johnson LAS, Briggs BG (1975) On the Proteaceae—the evolution and classification of a southern family. Botanical Journal of the Linnean Society 70:83–182
Jones DL, Cox A, Coney P, Beck M (1982) The growth of western North America. Scientific American 247:50–64
Kemp EM (1972a) Reworked palynomorphs from the West Ice Shelf area, East Antarctica, and their possible geological and palaeoclimatological significance. Marine Geology 13:145–157
Kemp EM (1972b) Recycled palynomorphs in continental shelf sediments from Antarctica. Antarctic Journal of the United States 7(5):190–191
Kemp EM (1976) Palynological observations in the Officer Basin, Western Australia. Bulletin of the Bureau of Mineral Resources, Geology and Geophysics, Australia 160:23–37
Kirchheimer F (1934) On pollen from the Upper Cretaceous dysodil of Banke, Namaqualand (South Africa). Transactions of the Royal Society of South Africa 21:41–50
Kröner A (1973) Comment on "Is the African Plate stationary?" Nature 243:29–30
Lidgard SH, Crane PR (1988) Quantitative analyses of the early angiosperm radiation. Nature 331:344–346
Lidgard SH, Crane PR (manuscript) Angiosperm diversication and Cretaceous floristic trends: a comparison of palynofloras and leaf macrofloras. Paleobiology (submitted)
Lukose NG (1974) Palynology of the subsurface sediments of Manhera Tibba structure, Jaisalmer, Western Rajasthan, India. The Palaeobotanist 21:285–297
Maheshwari HK (1974) Lower Cretaceous palynomorphs from the Bansa Formation, South Rewa Gondwana Basin, India. Palaeontographica B146:21–55
Maheshwari HK (1975) Palynology of the Athgarh Formation, near Cuttack, Orissa. The Palaeobotanist 22:23–28
Maheshwari HK (1986) *Thinnfeldia indica* Feistmantel and associated plant fossils from Tiruchirapalli District, Tamil Nadu. The Palaeobotanist 35:13–21
Maheshwari HK, Jana BN (1984) Cretaceous spore–pollen complexes from India. In Maheshwari HK (ed) Cretaceous of India, Indian Association of Palynostratigraphy, Lucknow
Mamgain VD, Sastry MVA, Subbaraman JV (1973) Report of ammonites from Gondwana plant beds at Terani, Tiruchirapalli District, Tamil Nadu. Journal of the Geological Society of India 14: 198–200
Martin ARH (1960) A Mesozoic microflora from South Africa. Nature 186:95
Martinez-Laborde JB (1988) Some comments on a recent classification of the Monimiaceae. Taxon 37:834–837
McDougall I, McElhinny MW (1970) The Rajmahal traps of India—K–Ar ages and palaeomagnetism. Earth and Planetary Sciences Letters 9:371–378
McLachlan IF, McMillan IK (1976) Review and stratigraphic significance of southern Cape Mesozoic palaeontology. Transactions of the Geological Society of South Africa 79:197–212

McLachlan IR, Pieterse A (1978) Preliminary palynological results: Site 361, Leg 40, Deep Sea Drilling Project. Initial Reports Deep Sea Drilling Project 40:857–881
McLachlan IR, McMillan IK, Brenner PN (1976) Micropalaeontological study of the Cretaceous beds at Mbotyi and Mngazana, Transkei, South Africa. Transactions of the Geological Society of South Africa 79:321–340
McQueen DR (1956) Leaves of Middle and Upper Cretaceous pteridophytes and cycads from New Zealand. Transactions of the Royal Society of New Zealand 83:673–685
Meyen SV (1987) Fundamentals of Palaeobotany, Chapman and Hall, London, 432 pp
Mildenhall DC (1980) New Zealand Late Cretaceous and Cenozoic plant biogeography: a contribution. Palaeogeography, Palaeoclimatology, Palaeoecology 31:197–233
Mitra ND, Rizvi SRA, Shah SC (1971) Advances in the study of the stratigraphy, sedimentation and structure of the Gondwana rocks of India. Records of the Geological Survey of India 101:144–161
Morgan R (1980) Palynostratigraphy of the Australian Early and Middle Cretaceous. Memoirs of the Geological Survey of New South Wales, Palaeontology 18:1–153
Müller H (1966) Palynological investigations of Cretaceous sediments north-eastern Brazil In: van Hinte JE (ed) Proceedings of the Second West African Micropaleontological Colloquum, Ibadan, Brill, Leiden 123–136
Muller J (1970) Palynological evidence on early differentiation of angiosperms. Biological Review 45:417–450
Muller J (1981) Fossil pollen records of extant angiosperms. Botanical Review 47:1–142
Muller J (1984) Significance of fossil pollen in angiosperm history. Annals of the Missouri Botanical Garden 71:419–443
Murthy NGK, Sastri VV (1962) Foraminifera from Sriperumbudur bed near Madras, India. Records of the Geological Survey of India 89:446–456
Nelson G, Platnick NI (1981) Systematics and Biogeography; Cladistics and Vicariance, Columbia University Press, New York
Nelson G, Rosen DE (eds) (1981) Vicariance Biogeography: A Critique, Columbia University Press, New York, 593 pp
Norvick MS, Burger D (1975) Palynology of the Cenomanian of Bathurst Island, Northern Territory, Australia. Bulletin of the Bureau of Mineral Resources, Geology and Geophysics, Australia 151:1–169
Nur A, Ben Avraham Z (1981) Lost Pacifica continent: a mobilistic speculation. In Nelson G, Rosen DE (eds) Vicariance Biogeography: A Critique, Columbia University Press, New York, pp 341–358
Oldham T, Morris J (1863) The fossil flora of the Rajmahal Series, the Rajmahal Hills, Bengal. Memoirs of the Geological Survey of India, Palaeontologia Indica, Series 2, 1:1–52
Platnick NI, Nelson G (1978) A method of analysis for historical biogeography. Systematic Zoology 27:1–16
Playford G, Cornelius KD (1967) Palynological and lithostratigraphic features of the Razorback Beds, Mount Morgan District, Queensland. Papers of the Department of Geology, University of Queensland 6:81–94
Playford G, Haig DW, Dettmann ME (1975) A mid-Cretaceous microfossil assemblage from the Great Artesian Basin, northwestern Queensland. Neues Jahrbuch für Geologie und Paläontologie, Abhandlungen 149:333–362
Plumstead EP (1963) Palaeobotany of Antarctica. In Adie RJ (ed) Antarctic Geology, North Holland, Amsterdam, pp 637–654
Raine JI (1984) Outline of a palynological zonation of Cretaceous to Paleogene terrestrial sediments in West Coast region South Island, New Zealand. Report of the New Zealand Geological Survey 109:1–81
Ramanujam CGK, Srisailam K (1974) Palynology of the carbonaceous shales from a borehole at Kattavakkam near Conjeevaram, Tamil Nadu, India. Pollen et Spores 16:67–102
Ramanujam CGK, Varma YNR (1977) Palynological evidence for the age of Sriperumbudur beds encountered in a borehole at Orikkai near Conjeevaram, Tamil Nadu. Journal of the Geological Society of India 18:429–435
Raven PH, Axelrod DI (1974) Angiosperm biogeography and past continental movements. Annals of the Missouri Botanic Garden 61:539–673
Riccardi AC (1987) Cretaceous paleogeography of southern South America. Palaeogeography, Palaeoclimatology, Palaeoecology 59:169–195
Robin G de Q (1975) Ice shelves and ice flow. Nature 253:168–172
Romero EJ (1973) Polen fósil de *Nothofagus* (*Nothofagidites*) del Cretácico y Paleoceno de Patagonia. Revista del Museo de la Plata (Nueva Serie), Paleontología 7:291–303
Romero EJ (1978) Paleoecologia y paleofitogeografia de las Tafofloras del Cenofitico de Argentina y areas vecinas. Ameghiniana 15:209–227
Romero EJ (1986a) Paleogene phytogeography and climatology of South America. Annals of the Missouri Botanical Garden 73:449–461
Romero EJ (1986b) Fossil evidence regarding the evolution of *Nothofagus* Blume. Annals of the Missouri Botanical Garden 73:276–283
Romero EJ, Archangelsky S (1986) Early Cretaceous angiosperm leaves from southern South America. Science 234:1580–1582
Romero EJ, Arguijo MH (1982) Analisis bioestratigráfico de las tafofloras del Cretácico Superior

de austro sudamérica. Cuencas sedimentarias del Jurásico y Cretácico de América del sur. 2:396–406
Romero EJ, Castro MT (1986) Material fungico y granos de polen de angiospermas de la Formación Río Turbio (Eoceno), Provincia de Santa Cruz, Republica Argentina. Ameghiniana 23: 101–118
Romero EJ, Hickey LJ (1976) A fossil leaf of Akaniaceae from Paleocene beds in Argentina. Bulletin of the Torrey Botanical Club 103:126–131
Rosen DE (1978) Vicariant patterns and historical explanation in biogeography. Systematic Zoology 27:159–188
Sah SCD, Jain, DP (1965) Jurassic spores and pollen grains from the Rajmahal Hills, Bihar, India, with a discussion on the age of the Rajmahal Intertrappean beds. The Palaeobotanist 13:264–290
Sahni B (1943) Indian silicified plants, II, *Enigmocarpon parijai*, a silicified fruit from the Deccan, with a review of the fossil history of the Lythraceae. Proceedings of the Indian Academy of Science 17B:59–96
Sahni B (1948) The Pentoxyleae: a new group of Jurassic gymnosperms from the Rajmahal Hills of India. Botanical Gazette 110:47–80
Sastri MVA, Mamgain VD (1971) The marine Mesozoic Formations of India. A review. Records of the Geological Survey of India 101:162–177
Schuster RM (1976) Plate tectonics and its bearing on the geographical origin and dispersal of angiosperms. In Beck CB (ed) Origin and Early Evolution of Angiosperms, Columbia University Press, New York, pp 48–138
Scott L (1971) Lower Cretaceous pollen and spores from the Algoa Basin (South Africa). Unpublished thesis, University of the Orange Free State, 80 p
Scott L (1972) Palynology of the Lower Cretaceous deposits (the Uitenhage Series) from the Algoa Basin. In van Zinderen Bakker EM (ed) Palaeoecology of Africa and of the Surrounding Islands and Antarctica 7:42–44
Scott L (1976) Palynology of Lower Cretaceous deposits from the Algoa Basin (Republic of South Africa). Pollen et Spores 18:563–609
Seward AC (1903) Fossil floras of Cape Colony. Annals of the South African Museum 4:1–122
Seward AC (1907) Notes on fossil plants from South Africa. Geological Magazine 54:481–487
Smith AG, Briden JC (1977) Mesozoic and Cainozoic Paleocontinental Maps, Cambridge University Press, Cambridge, 63 pp
Spath LF (1933) Revision of the Jurassic cephalopod fauna of Kach (Cutch), Part 6. Memoirs, Geological Survey of India, Palaeontologia Indica, n.s. 9:659–945
Stapleton RP, Beer EM (1977) Micropalaeontological age determination for the Brenton Beds. Geological Survey of South Africa, Bulletin 60:1–9
Stover LE, Evans PR (1973) Upper Cretaceous–Eocene spore–pollen zonation, offshore Gippsland Basin, Australia. Special Publication of the Geological Society of Australia 4:55–72
Stover LE, Partridge AD (1973) Tertiary and Late Cretaceous spores and pollen from the Gippsland Basin, southeastern Australia. Proceedings of the Royal Society of Victoria 85:237–286
Suryanarayana K (1954) Fossil plants from the Jurassic rocks of the Madras Coast, India. The Palaeobotanist 3:87–90
Suryanarayana K (1955) *Dadoxylon rajmahalense* Sahni from the Coastal Gondwanas of India. The Palaeobotanist 4:89–90
Tanai T (1986) Phytogeographic and phylogenetic history of the genus *Nothofagus* Bl. (Fagaceae) in the Southern Hemisphere. Journal of the Faculty of Science, Hokkaido University, Series IV, 21:505–582
Taylor TN, Archangelsky S (1985) The Cretaceous pteridosperms *Ruflorinia* and *Ktalenia* and implications on cupule and carpel evolution. American Journal of Botany 72:1842–1853
Thorne RF (1972) Major disjunctions in the geographic ranges of seed plants. Quarterly Review of Biology 47:365–411
Thorne RF (1973) Floristic relationships between tropical Africa and tropical America. In Meggers BJ, Ayensu ES, Duckworth WD (eds) Tropical Forest Ecosystems in Africa and South America: a Comparative Review, Smithsonian Institution Press, Washington DC, pp 27–47
Truswell EM (1982) Palynology of seafloor samples collected by the 1911–1914 Australasian Antarctic Expedition: implications for the geology of coastal East Antarctica. Journal of the Geological Society of Australia 29:343–356
Truswell EM (1983) Recycled Cretaceous and Tertiary pollen and spores in Antarctic marine sediments: a catalogue. Palaeontographica B186: 121–174
Truswell EM (1989) Cretaceous and Tertiary Vegetation of Antarctica: A Palynological Perspective. This volume, pp. 71–88
Truswell EM, Drewry DJ (1984) Distribution and provenance of recycled palynomorphs in surficial sediments of the Ross Sea, Antarctica. Marine Geology 59, 187–214
Vakhrameev VA, Dobruskina IA, Meyen SV, Zalinskaya ED (1978) Paläozoische und Mesozoische Floren Eurasiens und die Phytogeographie dieser Zeit, G Fisher Verlag, Jena, East Germany
Venkatachala BS (1977) Fossil floral assemblages in the East Coast Gondwanas—a critical review. Journal of the Geological Society of India 18:378–397

Venkatachala BS, Kar RK (1970) Palynology of the Mesozoic sediments of Kutch, Western India, 10, Palynological zonation of Katrol (Upper Jurassic) and Bhuj (Lower Cretaceous) sediments in Kutch, Gujarat. The Palaeobotanist 18:75–86
Venkatachala BS, Sharma KD (1974a) Palynology of the Cretaceous sediments from the subsurface of Vridhachalam area, Cauvery Basin. Geophytology 4:153–183
Venkatachala BS, Sharma KD (1974b) Palynology of the Cretaceous sediments from the subsurface of Pondicherry area, Cauvery Basin. New Botanist 1:170–200
Venkatachala BS, Sharma KD, Jain AK (1972) Palynological zonation of Jurassic–Lower Cretaceous sediments in the subsurface of Cauvery Basin. Seminars in Paleopalynology and Indian Stratigraphy, pp 172–187
Vishnu-Mittre (1957) Studies on the fossil flora of Nipania (Rajmahal Series), India—Pentoxyleae. The Palaeobotanist 6:31–46
Volkheimer W, Quatrocchio M (1975a) Palinología Estratigráfica del Titoniano (Formación Vaca Muerta) en el área de Caichigüe Cuenca Neuqúina. Parte A: especies terrestres. Ameghiniana 12:193–241
Volkheimer W, Quatrocchio M (1975b) Sobre el hallazgo de microfloras en el Jurásico superior del borde Austral de la Cuenca Neuqúina (República Argentina). Actas I Congreso Argentino Palinología Bioestratigráfica, Tucumán 1975, 1:589–615
Volkheimer W, Salas A (1975) Die ältesten Angiospermen-Palynoflora Argentiniens von der Typuslokalität der unterkretazischen Huitrin-Folge des Neuquén-Beckens. Mikrofloristische Assoziation und biostratigraphische Bedeutung. Neues Jahrbuch Geologie und Paläontologie, Monatshefte 7:424–436
Volkheimer W, Salas A (1976) Estudio palynológico de la Formación Huitrin Cretácico de la Cuenca Neuqúina, en su locadidad tip. Acta VI Congresso Geológico Argentina 1:433–456
Volkheimer W, Quatrocchio M, Salas A, Sepulveda EG (1975) Caracterización palinológica de formaciones del Jurásico Superior y Cretácico Inferior de la Cuenca Neuqúina (República Argentina). Abstr. VI Congresso Geológico, Argentina, Bahia Blanca 1975:57
Walker JW, Walker AG (1984) Ultrastructure of Lower Cretaceous angiosperm pollen and the origin and early evolution of flowering plants. Annals of the Missouri Botanical Garden 71:464–521
Walker JW, Brenner GJ, Walker AG (1983) Winteraceous pollen in the Lower Cretaceous of Israel: early evidence of a magnolialean angiosperm family. Science 220:1273–1275
Walkom AB (1918) Mesozoic Floras of Queensland, Part II, The Flora of the Maryborough (Marine) Series. Publication of the Geological Survey of Queensland 262:1–21
Walkom AB (1919) Mesozoic floras of Queensland, Parts 3 and 4, The floras of the Burrum and Styx River Series. Publication of the Geological Survey of Queensland 263:1–77
Weber R (1978) Some aspects of the Upper Cretaceous angiosperm flora of Coahuila, Mexico. Courier Forschungsinstitut Senckenberg 30:38–46
Zavada MS, Benson JM (1987) First fossil evidence for the primitive angiosperm family Lactoridaceae. American Journal of Botany 74:1590–1594
Zeba-Bano, Maheshwari HK, Bose MN (1979) Some plant remains from Pathargama, Rajmahal Hills, Bihar. The Palaeobotanist 26:144–156
Ziegler AM, Scotese CR, Barrett SF (1982) Mesozoic and Cainozoic paleogeographic maps. In Brosche P, Sundermann J (eds) Tidal Friction and the Earth's Rotation, Springer, New York, pp 240–252
Ziel W (1979) The Andes, a Geological Review. Gebrüder Borntraeger, Berlin–Stuttgart
Zinsmeister WJ (1987) Cretaceous paleogeography of Antarctica. Palaeogeography, Palaeoclimatology, Palaeoecology 59:197–206

Bibliography of Antarctic Paleobotany and Palynology

compiled by

Edith L. Taylor and Thomas N. Taylor

This bibliography was initially compiled in 1988 for the "Workshop on Antarctic Paleobotany and its Relationship to Reconstructions of Gondwana," which was held at The Ohio State University (OSU) from June 12–15, 1988. The version published here represents an updated version. We would like to thank Professor M.N. Bose for his suggestions of missing references, the Byrd Polar Research Center (OSU), the Office of Research and Graduate Studies (OSU), and the National Science Foundation (DPP-8713685) for their support of the workshop.

These references represent part of a larger bibliography of Gondwana paleobotany and palynology. Where possible, we have tried to include all papers that describe paleobotanical finds in Antarctica or that review those finds. All the paleobotanical entries that include "Antarctica" within the subject headings either mention fossil plants found on the continent or include descriptions of floras, etc. We have included a few additional papers on southern hemisphere floras and reconstructions of Gondwana, but no attempt has been made to be comprehensive in these areas. Because of the scattered nature of the literature on Antarctic research, there are no doubt references to fossil plants that were not included in this bibliography. As these come to your attention, please let us know so that the bibliography can be updated. Miscellaneous Publication No. 266 of the Byrd Polar Research Center (Ohio State University).

1. Adie, R.J. 1952. Representatives of the Gondwana system in Antarctica. In: Symp. sur les Series de Gondwana (19th Cong. Géol. Int., Algiers) (ed. C. Teichert, ed.), 393–399.

   Subjects: flora, Hope Bay, Antarctica, stratigraphy, Alexander Island, Jurassic, Cretaceous, Victoria Land, review, geology
2. Adie, R.J. 1958. Geological investigations in the Falkland Island Dependencies since 1940. Polar Record 9:3–17.

   Subjects: geology, South Shetland Islands, Antarctica, peninsula, Miocene, tuffs, flora, Nelson Island, Ezcurra Inlet, Dufayel Island, Tertiary
3. Adie, R.J. 1964. Geological history. In: Antarctic Research. A review of British scientific achievement in Antarctica (R. Priestley, R.J. Adie, and G. de Q. Robin, eds.), Butterworth & Co., London, 117–162.

   Subjects: Antarctica, history, Tertiary, floras, King George Island, Shetland Islands, peninsula, geology, stratigraphy, tectonics, Jurassic, Hope Bay, Graham Land, faunas, review
4. Adie, R.J. 1970. Review of Antarctic geology. In: Second Gondwana Symp. (South Africa, 1970), Proc. and Papers, Coun. for Sci. & Industr. Res., Pretoria, 15–22.

   Subjects: Antarctica, geology, stratigraphy, Transantarctic Mts., Victoria Land, Ellsworth Mts., Horlick Mts., Dronning Maud Land, peninsula
5. Adie, R.J. 1972. Recent advances in the geology of the Antarctic peninsula. In: Antarctic Geology and Geophysics (SCAR Symp. on Antarctic Geol. and Solid Earth Geophys., Oslo 1970) (R.J. Adie, ed.), Universitetetsforlaget, Oslo, 121–124.

   Subjects: Antarctica, peninsula, geology, review, flora, fauna, Triassic, Jurassic

6. Allen, A.D. 1962. Geological investigations in southern Victoria Land, Antarctica. Part 7--Formations of the Beacon Group in the Victoria Valley region. New Zealand J. Geol. Geophys. 5(2):278–291

*Subjects:*Antarctica, southern Victoria Land, geology, Beacon, Victoria Valley, Mt. Bastion coal measures, Permian, palynology, coal, stem fragments, leaves

7. Anderson, J.M. 1980. World Permo-Triassic correlations: Their biostratigraphic basis. In: Gondwana Five (5th Int. Gondwana Symp., Wellington, N.Z., 1980) (M.M. Cresswell and P. Vella, eds.), A.A. Balkema, Rotterdam, 3–10.

Subjects: stratigraphy, Permian, Triassic, Gondwana, faunas, palynology, floras

8. Andersson, J.G. 1906. On the geology of Graham Land. Bull. Geol. Inst. Upsala 7:19–71.

Subjects: Antarctica, geology, Graham Land, peninsula, Hope Bay, flora

9. Arber, E.A.N. 1905. Catalogue of the Fossil Plants of the *Glossopteris* Flora in the Department of Geology, British Museum (Natural History), British Museum, London, 255 pp., 8 pl.

Subjects: glossopterids, floras, Permian, Carboniferous, India, Gondwana, sphenophytes, ferns, lycopods, cordaites, cycadophytes, conifers, ginkgophytes

10. Arber, E.A.N. 1907. Report on the plant-remains from the Beacon Sandstone. In: Ferrar H.T., Report on the field-geology of the region explored during the "Discovery" Antarctic expedition, 1901–4. Brit. Natl. Antarctic Exp. Rept., 1901–1904, Natural Hist. 1:48.

Subjects: Antarctica, flora, Beacon, southern Victoria Land, Ferrar Glacier, fragments

11. Askin, R.A. 1979. Preliminary palynology investigation of Upper Paleozoic and Mesozoic rocks in the Antarctic Peninsula area. Antarctic Jour. of the U.S. 14(5):15.

Subjects: Antarctica, peninsula, palynology, South Shetland Islands, Jurassic, Paleozoic, Mesozoic

12. Askin, R.A. 1981. Jurassic-Cretaceous palynology of Byers Peninsula, Livingston Island, Antarctica. Antarctic Jour. of the U.S. 16(5):11–13.

Subjects: Antarctica, Livingston Island, peninsula, Jurassic, Cretaceous, palynology, stratigraphy, dinoflagellates, South Shetland Islands, Byers Peninsula

13. Askin, R.A. 1983. Tithonian (Uppermost Jurassic)–Barremian (Lower Cretaceous) spores, pollen, and microplankton from the South Shetland Islands, Antarctica. In: Antarctic Earth Sciences (4th Int. SCAR symp. on Antarctic Earth Sci. Adelaide, 1982) (R.L. Oliver, P.R. James and J.B. Jago, eds.), Cambridge Univ. Press, Cambridge, 295–297

Subjects: palynology, Cretaceous, Jurassic, Antarctica, South Shetland Islands, stratigraphy, peninsula, Byers Peninsula, Livingston Island, Snow Island, Argentina, Baqueŕo Formation

14. Askin, R.A. 1983. Campanian palynomorphs from James Ross and Vega Islands, Antarctic peninsula. Antarctic Jour. of the U.S. 18(5):63–64.

Subjects: James Ross Island, Vega Island, Antarctica, peninsula, palynology, Cretaceous

15. Askin, R.A. 1984. Palynological investigations of the James Ross Island basin and Robertson Island, Antarctic Peninsula. Antarctic Journal of the U.S. 19(5):6–7.

Subjects: palynology, Antarctica, Seymour Island, Cretaceous, Tertiary

16. Askin, R.A. 1985. Palynological studies in the James Ross Island basin, Antarctic Peninsula: A progress report. Antarctic Jour. of the U.S. 20(5):44–45.

Subjects: palynology, James Ross Island, Antarctica, peninsula, Cretaceous, Seymour Island, dinoflagellates, angiosperms, biostratigraphy

17. Askin, R.A. 1987. Palynology investigations of the James Ross Island basin, Antarctica. Antarctic Jour. of the U.S. 22(5): 13–14.

Subjects: palynology, Antarctica, peninsula, Cretaceous, Eocene, Paleocene, Tertiary, Seymour Island, biostratigraphy, dinoflagellates, acritarchs, ferns, podocarps, conifers

18. Askin, R.A. 1988. Campanian to Paleocene palynological succession of Seymour and adjacent islands, northeastern Antarctic Peninsula. In: Geology and Paleontology of Seymour Island, Antarctic Peninsula (R.M. Feldman and M.O. Woodburne, eds.). Geol. Soc. Mem. 169: 131–153.

Subjects: palynology, Cretaceous, Paleocene, Tertiary, Antarctica, peninsula, biostratigraphy, Seymour Island

19. Askin, R.A. 1988. The palynology record across the Cretaceous/Tertiary transition on Seymour Island, Antarctica. In: Geology and Paleontology of Seymour Island, Antarctic Peninsula (R.M. Feldman and M.O. Woodburne, eds.). Geol. Soc. Mem. 169: 155–162.

Subjects: palynology, Cretaceous, Tertiary, biostratigraphy, Seymour Island, Antarctica, peninsula

20. Askin, R.A., and D.H. Elliot. 1981. Recycled Permian and Triassic palynomorphs from Seymour Island, Antarctica and their geologic

implications. Palynology (Abstr. Proc. 13th Ann. Mtg. AASP) 5:231.

Subjects: palynology, Seymour Island, peninsula, Antarctica, recycled, Permian, Triassic, Cretaceous, Tertiary, dinoflagellates

21. Askin, R.A., and D.H. Elliot. 1982. Geologic implications of recycled Permian and Triassic palynomorphs in Tertiary rocks of Seymour Island, Antarctic Peninsula. Geology 10: 547–552.

Subjects: Antarctica, palynology, peninsula, Seymour Island, recycled, Permian, Triassic, Tertiary, reconstructions, Gondwanaland

22. Askin, R.A., and R.F. Fleming. 1982. Palynology investigations of Campanian to lower Oligocene sediments on Seymour Island, Antarctic Peninsula. Antarctic Jour. of the U.S. 17(5): 70–71.

Subjects: palynology, Seymour Island, Antarctica, peninsula, Oligocene, Cretaceous, Paleocene, Eocene, podocarps, conifers, angiosperms, *Nothofagus*

23. Askin, R.A., and V. Markgraf. 1986. Palynomorphs from the Sirius Formation, Dominion Range, Antarctica. Antarctic J. of the U.S. 21(5):34–35.

Subjects: palynology, Dominion Range, Antarctica, *Nothofagus,* recycled, Pliocene, Pleistocene

24. Askin, R.A., and J.M. Schopf. 1978. Palynologic studies in the Transantarctic Mountains. Antarctic Jour. of the U.S. 13(4):18–19.

Subjects: palynology, Antarctica, Transantarctic Mts., Nilsen Plateau, Beardmore, Shackleton Glacier, biostratigraphy, Permian, Triassic, correlation

25. Audley-Charles, M.G. 1983. Reconstruction of eastern Gondwanaland. Nature 306:48–50.

Subjects: Gondwanaland, reconstruction, floras, Triassic, Permian, Carboniferous

26. Axelrod, D.I. 1984. An interpretation of Cretaceous and Tertiary biota in polar regions. Palaeogeogr., Palaeoclimatol., Palaeoecol. 45:105–147.

Subjects: Cretaceous, Tertiary, floras, paleoclimate, Antarctic, Arctic

26a. Baldoni, A.M. 1986. Características generales de la megaflora, especialmente de la especie *Ptilophyllum antarcticum,* en el Jurásico superior-Cretácico inferior de Antártida y Patagonia, Argentina. Boletim IG-USP, Ser. Cient. 17:77–87

Subjects: Antarctica, peninsula, late Jurassic, early Cretaceous, Patagonia, Argentina, *Ptilophyllum,* paleogeography, cuticle, cycadophytes, Mesozoic, flora, South Shetlands, Mt. Flora

26b. Baldoni, A.M., Barreda, V. 1986. Estudio palinológico de las formaciones Lopez de Bertodano y Sobral, Isla Vicecomodoro Marambio, Antártida. Boletim IG-USP, Ser. Cient. 17:89–98

Subjects: palynology, Antarctica, peninsula, Seymour Island, Fagaceae, *Notofagidites,* podocarps, conifers, angiosperms, ferns, dinoflagellates, paleoclimate, Tertiary, Eocene

27. Ballance, P.F. 1977. The Beacon Supergroup in the Allan Hills, central Victoria Land, Antarctica. New Zealand J. Geol. Geophys. 20(6):1003–1016.

Subjects: Antarctica, Beacon, Victoria Land, Allan Hills, Permian, glossopterids, flora, Weller Coal Measures, Triassic, *Dicroidium,* Lashly Formation

28. Balme, B.E. 1980. Palynology and the Carboniferous–Permian boundary in Australia and other Gondwana continents. Palynology 4:43–55.

Subjects: palynology, Carboniferous, Permian, Australia, Gondwana, correlation, South America, India, Africa

29. Balme, B.E., and G. Playford. 1967. Late Permian plant microfossils from the Prince Charles Mountains, Antarctica. Revue de Micropaleontologie 10:179–192.

Subjects: Antarctica, Prince Charles Mts., palynology, Permian, stratigraphy

30. Banerji, J., and Y. Lemoigne. 1987. Significant additions to the Upper Triassic flora of Williams Point, Livingston Island, South Shetlands (Antarctica). Géobios 20(4):469–487.

Subjects: Antarctica, South Shetland Islands, Livingston Island, Triassic, flora, ferns, gymnosperms, conifers, pteridosperms, Caytoniales, Corystospermales

31. Banerji, J., Y. Lemoigne, and T. Torres. 1987. Significant additions to the Upper Triassic flora of Williams Point, Livingston Island, South Shetland Islands (Antarctica). Ser. Cient. INACH 36:33–58.

Subjects: Antarctica, South Shetland Islands, Livingston Island, Upper Triassic, flora, ferns, gymnosperms, conifers, sphenophytes, *Dicroidium,* wood, corystosperms

32. Barghoorn, E.S. 1961. A brief review of fossil plants of Antarctica and their geologic implications. Natl. Acad. Sci. Publ. 839:5–9.

Subjects: Antarctica, floras, paleoclimate, coal, review, Mt. Weaver, glossopterids, conifer, araucarians, collections

32a. Barker, P.F., Kennett, J.P., et al. (Leg 113 shipboard scientific party) 1987. Glacial history of Antarctica. Nature 328:115–116

Subjects: Antarctica, Weddell Sea, palynology, paleoclimate, Tertiary, glaciation, Cretaceous

33. Barrett, P.J. 1969. Stratigraphy and petrology of the mainly fluviatile Permian and Triassic Beacon rocks, Beardmore Glacier area, Antarctica. Ohio State Univ. Inst. Polar Studies Rept. 34:1–132.

    Subjects: stratigraphy, Permian, Triassic, Beacon, Beardmore, Antarctica

34. Barrett, P.J. 1970. Stratigraphy and paleogeography of the Beacon Supergroup in the Transantarctic Mountains, Antarctica. In: Second Gondwana Symposium (South Africa, 1970), Proc. and Papers, Council for Sci. & Ind. Res., Pretoria, 249–256.

    Subjects: stratigraphy, Antarctica, Transantarctic Mts., Beacon, paleography, geology, Devonian, Jurassic, Triassic, Permian, Victoria Group, floras, lycopods, correlation

35. Barrett, P.J. 1972. Stratigraphy and petrology of the mainly fluviatile Permian and Triassic part of the Beacon Supergroup, Beardmore Glacier Area. In: Antarctic Geology and Geophysics (SCAR Symp. on Antarctic Geol. and Solid Earth Geophys., Oslo 1970) (R.J. Adie, ed.), Universitetetsforlaget, Oslo, 365–372.

    Subjects: Permian, Triassic, Beardmore, Beacon, Antarctica, stratigraphy, petrology, Buckley, Fremouw, Triassic, Permian, glossopterids, leaves, wood, coal

36. Barrett, P.J. 1981. History of the Ross Sea region during the deposition of the Beacon Supergroup 400-180 million years ago. J. Roy. Soc. New Zealand 11(4):447–458.

    Subjects: Beacon, Ross Sea, Antarctica, paleogeography, Transantarctic Mts., palynology, stratigraphy, Permian, Triassic

37. Barrett, P.J. 1987. Oligocene sequence cored at CIROS-1, western McMurdo South. New Zealand Antarctic Record 7(3):1–7.

    Subjects: Oligocene, Antarctica, Ross Sea, McMurdo, diatoms, palynology, glaciation, stratigraphy

38. Barrett, P.J., and D.H. Elliot. 1973. Reconnaissance geologic map of the Buckley Island Quadrangle, Transantarctic Mountains, Antarctica. Antarctic Geologic Map (U.S.G.S.) A-3.

    Subjects: geology, Permian, Triassic, Beardmore, Antarctica

39. Barrett, P.J., D.H. Elliot, and J.F. Lindsay. 1986. The Beacon Supergroup (Devonian–Triassic) and Ferrar Group (Jurassic) in the Beardmore Glacier Area, Antarctica. In: Geology of the Central Transantarctic Mountains, Ant. Res. Ser., Amer. Geophys. Union, 36(14):339–428.

    Subjects: Antarctica, Devonian, Permian, Triassic, Jurassic, Beardmore, geology, stratigraphy, Beacon, glossopterids, *Dicroidium*, corystosperms, flora, fauna, review, Buckley, Fremouw, Falla

40. Barrett, P.J., G.W. Grindley, and P.N. Webb. 1972. The Beacon Supergroup of East Antarctica. In: Antarctic Geology and Geophysics (SCAR Symp. on Antarctic Geol. and Solid Earth Geophys., Oslo 1970) (R.J. Adie, ed.), Universitetetsforlaget, Oslo, 319–332.

    Subjects: Devonian, Permian, Triassic, Jurassic, Carboniferous, Antarctica, geology, stratigraphy, glossopterids, Transantarctic Mts., paleogeography

41. Barrett, P.J., and B.P. Kohn. 1975. Changing sediment transport directions from Devonian to Triassic in the Beacon Super-Group of South Victoria Land, Antarctica. In: Gondwana Geology (3rd Gondwana Symp.) (K.S.W. Campbell, ed.), Australian Natl. Univ. Press. Canberra, 15–35.

    Subjects: Beacon, Victoria Land, Antarctica, sedimentology, Triassic, Devonian, Permian, Transantarctic Mts.

42. Barrett, P.J., and R.A. Kyle. 1975. The early Permian glacial beds of South Victoria Land and the Darwin Mountains, Antarctica. In: Gondwana Geology (3rd Gondwana Symp.) (K.S.W. Campbell, ed.), Australian Natl. Univ. Press, Canberra, 333–346.

    Subjects: Antarctica, Permian, Victoria Land, Darwin Mts., palynology, *Gangamopteris*, glossopterids, glaciation, Mt. Fleming

43. Barron, J.A., and L.H. Burckle. 1987. Diatoms from the 1984 USGS antarctic cruise in the Ross Sea. In: Antarctic Continental Margin: Geology and Geophysics of the Western Ross Sea (A.K. Cooper and F.J. Davey, eds.), Circum-Pacific Council for Energy and Mineral Resources, Earth Science Series, Houston, Texas, 5B:225–230.

    Subjects: diatoms, palynology, Ross Sea, Antarctica, Holocene, Tertiary

44. Barton, C.M. 1964. Significance of the Tertiary fossil floras of King George Island, South Shetland Islands. In: Antarctic Geology (Proc. 1st Int. SCAR Symp. on Antarctic Geol., Cape Town 1963) (R.J. Adie, ed.), North-Holland Publ. Co., Amsterdam, 603–608

    Subjects: South Shetland Islands, King George Island, flora, Tertiary, Antarctica, peninsula, angiosperms, conifers, ferns, *Araucaria, Equisetum,* paleoclimate, *Nothofagus*

45. Birkenmajer, K., and E. Zastawniak. 1986. Plant remains of the Dufayel Island group (early Tertiary?), King George Island, South Shetland Islands (West Antarctica). Acta Palaeobotanica 26(1,2):33–54.

Subjects: flora, Tertiary, King George Island, South Shetland Islands, Antarctica, peninsula, angiosperms, Fagaceae, Cretaceous

46. Birnie, J.F., and D.E. Roberts. 1986. Evidence of Tertiary forest in the Falkland Islands (Islas Malvinas). Palaeogeog., Palaeoclimatol., Palaeoecol. 55:45–53.

    Subjects: wood, Tertiary, Falkland Islands, Antarctica, *Nothofagus,* palynology, paleoclimate, podocarps, conifers, in situ forest

47. Brady, H.T. 1977. *Thalassiosira torokina* n. sp. (diatom) and its significance in Late Cenozoic biostratigraphy. Antarctic Jour. of the U.S. 12(4):122–123.

    Subjects: diatoms, Antarctica, biostratigraphy, Pliocene, Dry Valleys

48. Brady, H.T. 1978. Miocene diatom flora from bottom cores at RISP site J-9. Antarctic Jour. of the U.S. 13(4):123–124.

    Subjects: Miocene, diatoms, palynology, Ross Sea, Antarctica, biostratigraphy, J-9

49. Brady, H.T. 1979. Diatom biostratigraphy in sediment cores from RISP site J-9. Antarctic Jour. of the U.S. 14:130.

    Subjects: diatoms, biostratigraphy, Ross Sea, Antarctica, palynology

50. Brady, H.T. 1983. Interpretation of sediment cores from the Ross Ice Shelf Site J-9, Antarctica. Nature 303:510–511.

    Subjects: Antarctica, Ross Sea, palynology, diatoms, silicoflagellates, biostratigraphy, Miocene, recycled

51. Brady, H., and H. Martin. 1979. Ross Sea region in the Middle Miocene: A glimpse into the past. Science 203:437–438.

    Subjects: Ross Sea, Antarctica, Miocene, palynology, diatoms, angiosperms, gymnosperms, ferns

52. Breed, W.J. 1971. Permian stromatolites from Coalsack Coal. Antarctic Jour. of the U.S. 6(5):189–190.

    Subjects: Permian, Beardmore, Antarctica, Coalsack, stromatolites, algae, Buckley

53. Briden, J.C. 1967. Recurrent continental drift of Gondwanaland. Nature 215: 1334–1339.

    Subjects: Gondwana, reconstruction, continental drift

54. Bunt, J. 1956. Living and fossil pollen from Macquarie Island. Nature 177(4530):337.

    Subjects: palynology, Macquarie Island, sub-Antarctic, lignite, Recent, Antarctic

55. Burckle, L.H., R.I. Gayley, M. Ram, and J.-R. Petit. 1988. Diatoms in Antarctic ice cores: Some implications for the glacial history of Antarctica. Geology 16:326–329.

    Subjects: diatoms, Antarctica, Pliocene, Sirius Group, marine, glaciology, recycled

56. Cande, S.C., and J.C. Mutter. 1982. A revised identification of the oldest sea-floor spreading anomalies between Australia and Antarctica. Earth and Planet. Sci. Letters 58:151–160.

    Subjects: paleomagnetism, plate tectonics, Australia, Antarctica, breakup, reconstruction, Cretaceous

57. Carlquist, S. 1987. Pliocene, *Nothofagus* wood from the Transantarctic Mountains. Aliso 11(4):571–583.

    Subjects: Pliocene, Fagaceae, *Nothofagus,* wood, Dominion Range, Antarctica, Transantarctic Mts., Pleistocene, flora

58. Case, J.A. 1988. Paleogene floras from Seymour Island, Antarctic Peninsula. In: Geology and Paleontology of Seymour Island, Antarctic Peninsula (R.M. Feldmann and M.O. Woodburne, eds.), Geol. Soc. Amer. Mem. 169:523–530.

    Subjects: flora, Tertiary, Paleocene, leaves, angiosperms, Seymour Island, Antarctica, peninsula, *Nothofagus,* araucarians, podocarps, conifers, gymnosperms, ferns, biogeography

59. Chatterjee, S. 1980. The paleoposition of Marie Byrd Land, West Antarctica. Antarctic Jour. of the U.S. 15(5):17–18.

    Subjects: reconstructions, Marie Byrd Land, West Antarctica, plate tectonics, Northern Victoria Land, Mesozoic

60. Christie, P. 1987. C:N ratios in two contrasting Antarctic peat profiles. Soil Biol. Biochem. 19(6):777–778

    Subjects: peat, chemistry, Antarctica, Recent, mosses

61. Ciesielski, P.F. 1983. The Neogene and Quaternary diatom biostratigraphy of subantarctic sediments. In: Initial Repts. of the Deep Sea Drilling Project, U.S. Govt. Printing Office, Washington, D.C., 71:635–655.

    Subjects: palynology, Miocene, Quaternary, diatoms, biostratigraphy, Antarctica, Pliocene, Pleistocene, Holocene

62. Colbert, E.H. 1980. The distribution of tetrapods and the breakup of Gondwana. In: Gondwana Five (5th Int. Gondwana Symp., Wellington, N.Z. 1980) (M.M. Cresswell and P. Vella, eds.), A.A. Balkema, Rotterdam, 277–282.

    Subjects: vertebrates, Triassic, Jurassic, Gondwana, reconstruction, biogeography, Antarctica, Africa, India, South America, Australia

63. Collinson, J.W., K.O. Stanley, and C.L. Vavra. 1980. Triassic fluvial depositional systems in the Fremouw Formation, Cumulus Hills, Antarctica. In: Gondwana Five (Proc. 5th Int. Gondwana Symp., Wellington, N.Z.

1980) (M.M. Cresswell and P. Vella, eds.), A.A. Balkema, Rotterdam, 141–148.

Subjects: sedimentology, Triassic, Antarctica, roots, Fremouw, Cumulus Hills

64. Cookson, I.C. 1947. Plant microfossils from the lignites of Kerguelen Archipelago. Brit., Austral., N.Z. Antarctic Res. Expd., Ser. A, 2(8):127–142.

Subjects: palynology, lignites, Kerguelen Island, angiosperms, gymnosperms, pteridophytes, fungi, Tertiary

65. Cooper, R.A., C.A. Landis, W.E. LeMasurier, and I.G. Speden. 1982. Geologic history and regional patterns in New Zealand and West Antarctica—their paleotectonic and paleogeographic significance. In: Antarctic Geoscience (C. Craddock, ed.), Univ. Wisconsin Press, Madison, 43–53.

Subjects: reconstructions, New Zealand, Antarctica, biogeography, plate tectonics, faunas, Triassic, Jurassic, Cretaceous, peninsula

66. Cortemiglia, G.C., P. Gastaldo, and R. Terranova. 1981. Studio di piante fossili trovate nella King George Island delle Isole Shetland del Sur (Antartide). Atti Soc. Ital. Sci. Nat. Museo Civ. Stor. Nat. Milano 122(1–2):37–61.

Subjects: King George Island, flora, peninsula, Antarctica, South Shetland Islands, Tertiary, Oligocene, Miocene, conifers, *Araucaria, Nothofagus,* angiosperms, wood

67. Cosgriff, J.W., and W.R. Hammer. 1979. New species of Dicynodontia from the Fremouw Formation. Antarctic Jour. of the U.S. 14(5):30–32.

Subjects: Fremouw, Antarctica, faunas, Triassic, Cumulus Hills, Transantarctic Mts.

68. Cosgriff, J.W., and W.R. Hammer. 1981. New skull of *Lystrosaurus curvatus* from the Fremouw Formation. Antarctic Jour. of the U.S. 16(5):52–53.

Subjects: Lystrosaurus, faunas, Triassic, Antarctica, Fremouw, Cumulus Hills

69. Cosgriff, J.W., and W.R. Hammer. 1983. The labyrinthodont amphibians of the earliest Triassic from Antarctica, Tasmania and South Africa. In: Antarctic Earth Science (4th Int. SCAR Symp. on Antarctic Earth Sci., Adelaide, 1982) (R.L. Oliver, P.R. James, and J.B. Jago, eds.), Cambridge Univ. Press, Cambridge, 590–592.

Subjects: Antarctica, Triassic, faunas

70. Cosgriff, J.W., and W.R. Hammer. 1984. New material of labyrinthodont amphibians from the Lower Triassic Fremouw Formation of Antarctica. J. Vertebrate Paleontol. 4(1): 47–56.

Subjects: faunas, Triassic, Antarctica, Fremouw

71. Cosgriff, J.W., W.R. Hammer, and W.J Ryan. 1982. The Pangaean reptile, *Lystrosaurus maccaigi,* in the Lower Triassic of Antarctica. J. Paleontol. 56(2):371–385.

Subjects: Antarctica, fauna, *Lystrosaurus,* Triassic

72. Cosgriff, J.W., W.R. Hammer, J.M. Zawiskie, and N.R. Kemp. 1978. New Triassic vertebrates from the Fremouw Formation of the Queen Maud Mountains. Antarctic Jour. of the U.S. 13(4):23–24.

Subjects: Triassic, faunas, Fremouw, Antarctica, Transantarctic Mts., Queen Maud Mts., Cumulus Hills

73. Couper, R.A. 1960. Southern Hemisphere Mesozoic and Tertiary Podocarpaceae and Fagaceae and their palaeogeographic significance. Proc. Roy. Soc. London, Series B 152: 491–500.

Subjects: podocarps, conifers, Tertiary, floras, Cretaceous, Australia, South America, Antarctica, Fagaceae, *Nothofagus,* palynology, review

74. Craddock, C. 1970. Antarctic geology and Gondwanaland. Antarctic Jour. of the U.S. 5(3):53–57.

Subjects: Antarctica, Gondwanaland, geology, reconstruction

75. Craddock, C. 1979. The evolution and fragmentation of Gondwanaland. In: Fourth International Gondwana Symposium: Papers (Calcutta, 1977) (B. Laskar and C.S. Raja Rao, eds.), Hindustan Publ. Co., Delhi, 2:711–719.

Subjects: Gondwana, reconstructions, plate tectonics, fragmentation

76. Craddock, C. 1982. Antarctica and Gondwanaland (review paper). In: Antarctic Geosciences (C. Craddock, ed.), Univ. Wisconsin Press, Madison, 3–13.

Subjects: Antarctica, Gondwana, reconstruction, plate tectonics, review

77. Craddock, C., T.W. Bastien, R.H. Rutford, and J. Anderson. 1965. *Glossopteris* discovered in West Antarctica. Science 148:634–637.

Subjects: glossopterids, Antarctica, Ellsworth, Permian, drift, paleobotany

78. Cranwell, L.M. 1959. Fossil pollen from Seymour Island, Antarctica. Nature 184:1782–1785.

Subjects: palynology, Antarctica, Seymour Island, peninsula, Cretaceous, Tertiary, conifers, *Araucaria, Nothofagus,* Fagaceae

79. Cranwell, L.M. 1964. Extra-Antarctic correlations in the dating of erratic deposits at

McMurdo Sound, Antarctica. Jour. Arizona Acad. Sci. 3(2):110–111.

Subjects: palynology, McMurdo Sound, Antarctica, Lower Tertiary, *Nothofagus,* podocarps, algae, Eocene

80. Cranwell, L.M. 1964. Hystrichospheres as an aid to Antarctic dating with special reference to the recovery of *Cordosphaeridium* in erratics at McMurdo Sound. Grana Palynol. 5(3):398–405.

Subjects: palynology, dinoflagellates, McMurdo Sound, Antarctica, Tertiary

81. Cranwell, L.M. 1964. Antarctica: Cradle or grave for its *Nothofagus*?. In: Ancient Pacific Floras: The Pollen Story (L.M. Cranwell, ed.), Univ. of Hawaii Press, Honolulu, 87–93.

Subjects: *Nothofagus,* Fagaceae, angiosperms, Antarctica, continental drift

82. Cranwell, L.M. 1966. Senonian dinoflagellates and microspores from Snow Hill and Seymour Island. J. Arizona Acad. Sci. 4:136.

Subjects: palynology, Antarctica, dinoflagellates, peninsula, Seymour Island, Cretaceous

83. Cranwell, L.M. 1969. Antarctic and circum-Antarctic palynological contributions. Antarctic Jour. of the U.S. 4:197–198.

Subjects: Antarctica, palynology, Snow Hill, McMurdo, Ross Sea, paleoclimate, Cretaceous, Eocene, recycled, Triassic, angiosperms, *Nothofagus*

84. Cranwell, L.M. 1969. Palynological intimations of some pre-Oligocene Antarctic climates. In: Palaeoecology of Africa and of the surrounding islands and Antarctica (E.M. Van Zinderen Bakker, ed.), Balkema, Cape Town, 5:1–19.

Subjects: palynology, Antarctica, paleoclimate, Tertiary, Cretaceous, West Antarctica, Seymour Island, peninsula, McMurdo Sound, review, paleoclimate

85. Cranwell, L.M., H.J. Harrington, and I.G. Speden. 1960. Lower Tertiary microfossils from McMurdo Sound, Antarctica. Nature 186:700–702.

Subjects: palynology, Tertiary, McMurdo, Antarctica, dinoflagellates, *Nothofagus*

86. Cridland, A.A. 1963. A *Glossopteris* flora from the Ohio Range, Antarctica. Amer. J. Bot. 50:186–195.

Subjects: glossopterids, flora, Ohio Range, Antarctica, Permian, paleobotany

87. Czajkowski, S., and O. Rösler. 1986. Plantas fósseis da Península Fildes, Ilha Rei Jorge (Shetlands do Sul): Morfografia das impressoes foliares [Fossil plants from the Fildes Peninsula, King George Island: Morphology of leaf impressions]. Anais Academia Brasileira de Ciências 58(1-Suppl.):99–110.

Subjects: flora, King George Island, South Shetlands, Antarctica, peninsula, early Tertiary, angiosperms, leaves, impressions, Seymour Island

87a. Daber, R., Weber, W. 1988. Neue Pflanzenfunde von der Antarktischen Halbinsel (New plant discoveries from the Antarctic Peninsula). Zeitschrift für geologische Wissenschaften. 16(6):560–564

Subjects: flora, Antarctica, peninsula, biostratigraphy, Guettard Range, southeastern Palmer Land

88. Dalziel, I.W.D. 1980. Comments and reply on "Mesozoic evolution of the Antarctic Peninsula and the southern Andes." Geology 8(6):260–261.

Subjects: Antarctica, peninsula, reconstruction, Mesozoic, tectonics, South America, breakup

89. Dalziel, I.W.D., and D.H. Elliot. 1982. West Antarctica: Problem child of Gondwanaland. Tectonics 1(1):3–19.

Subjects: tectonics, Antarctica, Gondwanaland, reconstruction, peninsula

90. Darrah, W.C. 1936. Antarctic fossil plants. Science 83:390–391.

Subjects: Antarctica, flora, coal, Mt. Weaver, Jurassic, *Taeniopteris, Sagenopteris, Araucarites,* Transantarctic Mts., Triassic

91. Darrah, W.C. 1941. Notas sobre la historia de la paleobotanica sudamericana. Lilloa (Rev. de Bot. del Inst. "Miguel Lillo") 6:213–239.

Subjects: South America, flora, Antarctica, Permian, glossopterids, Mesozoic

92. De Vore, M.L., and T.N. Taylor. 1988. Mesozoic seed plants: a pollen organ from the Triassic of Antarctica. Amer. J. Bot. 75(6, part 2):106.

Subjects: Triassic, peat, Fremouw Peak, corystosperms, *Pteruchus,* fructification, pollen, *Alisporites, Pteruchipollenites,* abstract

93. Del Valle, R.A., M.T. Diaz, and E.J. Romero. 1984. Informe preliminar sobre las sedimentitas de la Peninsula Barton, Isla 25 de Mayo, Islas Shetland del Sur, Antártida Argentina [Preliminary report on the sedimentites of Barton Peninsula, 25 de Mayo Island (King George Island), South Shetland Islands, Argentine Antarctica]. Contribu. Instituto Antártico Argentino 308:1–19.

Subjects: Antarctica, Shetland Islands, King George Island, Eocene, flora, Oligocene, geology, stratigraphy, leaves, angiosperms, biostratigraphy

94. Dettmann, M.E. 1987. Evidence for step-wise

migration of southern hemisphere Mesozoic floras. XIV Int. Botanical Congress Abstracts, Berlin, 24 July–1 August 1987, 287.

Subjects: floras, Gondwana, Cretaceous, podocarps, conifers

95. Dettmann, M.E., and M.R.A. Thomson. 1987. Cretaceous palynomorphs from the James Ross Island area, Antarctica—a pilot study. British Antarctic Surv. Bull. 77:13–59.

Subjects: Antarctica, palynology, Cretaceous, James Ross Island, peninsula, fungi, Dundee Island, peninsula, faunas, dinoflagellates, recycled, correlation, paleoclimate, *Nothofagus,* podocarps, araucarians, conifers, angiosperms, pteridophytes

96. Dibbern, M., and E.J. Romero. 1984. Revisión de los holotipos de las especies de Fagaceas estudiados por Dusén (Austrosudamerica, Eoceno–Paleoceno?). Actas III Cong. Argentino Paleontol. y Bioestrat., Corrientes, Argentina, 163–173.

Subjects: Fagaceae, angiosperms, Seymour Island, Antarctica, peninsula, South America, *Fagus, Nothofagus,* Patagonia, Tertiary, flora

97. Dibner, A.F. 1978. Palynocomplexes and age of the Amery Formation deposits, East Antarctica. Pollen et Spores 20:405–422.

Subjects: palynology, Antarctica, Prince Charles Mts., East Antarctica, Permian

98. Douglas, G.V. 1923. Geological results of the Shackleton–Rowett (*Quest*) Expedition (lecture) (with comments by W.T. Gordon on fossil wood). Quart. J. Geol. Soc. London 79:x–xi.

Subjects: St. Vincent, St. Paul's Islands, geology, *Araucarioxylon,* wood, Mesozoic, South Georgia, Antarctica, Bay of Isles, peninsula

99. Doumani, G.A., and W.E. Long. 1962. The ancient life of the Antarctic. Scientific Amer. 207(3):168–184.

Subjects: Antarctica, flora, fauna, glossopterids, continental drift, wood, geology, Paleozoic, Mesozoic, Tertiary

100. Doumani, G.A., and V.H. Minshew. 1965. General geology of the Mount Weaver area, Queen Maud Mountains, Antarctica. In: Geology and Paleontology of the Antarctic, Antarctic Res. Ser., Amer. Geophys. Union, 6: 127–139.

Subjects: Mt. Weaver, stratigraphy, geology, Antarctica, Queen Maud Mts., flora, glossopterids, Permian, wood, forests

101. Dusén, P. 1908. Über die Tertiäre Flora der Seymour-Insel. Wissenschaftliche Ergebnisse der Schwedischen Südpolar-Expedition, 1901–03, 3(3):1–27.

Subjects: Tertiary, flora, Seymour Island, paleobotany, Antarctica, angiosperms, leaves, conifers, ferns

102. Edwards, W.N. 1921. Fossil coniferous wood from Kerguelen Island. Ann. Bot. 35:609–617.

Subjects: Kerguelen Island, wood, conifers, *Cupressinoxylon, Dadoxylon,* Tertiary, araucarians, gymnosperms

103. Edwards, W.N. 1928. The occurrence of *Glossopteris* in the Beacon Sandstone of Ferrar Glacier, South Victoria Land. Geol. Mag. 65:323–327.

Subjects: glossopterids, Permian, Victoria Land, Antarctica, flora

104. Elliot, D.H. 1972. Aspects of Antarctic geology and drift reconstructions. In: Antarctic Geology and Geophysics (SCAR Symp. on Antarctic Geol. and Solid Earth Geophy., Oslo 1970) (R.J. Adie, ed.), Universitetetsforlaget, Oslo, 849–858.

Subjects: Antarctica, reconstructions, tectonics, glossopterids, Ellsworth Mts., Beacon, peninsula, West Antarctica, New Zealand, South America, Mesozoic, Paleozoic, Cretaceous

105. Elliot, D.H. 1975. Gondwana basins of Antarctica. In: Gondwana Geology (K.S.W. Campbell, ed.), Australian National Univ. Press, Canberra, 493–536.

Subjects: Antarctica, geology, stratigraphy

106. Elliot, D.H., and R.A. Askin. 1980. Geologic studies in the South Shetland Islands and at Hope Bay, Antarctic Peninsula: R/V Hero cruises 80-1 and 80-2. Antarctic Jour. of the U.S. 15(5):23–24.

Subjects: Antarctica, Hope Bay, South Shetland Islands, Jurassic, Cretaceous, flora, Byers Peninsula, peninsula, Snow Island, Livingston Island

107. Ferrar, H.T. 1907. Report on the field-geology of the region explored during the "Discovery" Antarctic Expedition, 1901–4. British National Antarctic Expedition Rept. 1901–1904, Nat. Hist. 1:1–100.

Subjects: Victoria Land, Antarctica, geology, plant debris

107a. Findlay, R.H., Jordan, H. 1984. The volcanic rocks of Mt. Black Prince and Lawrence Peaks, North Victoria Land, Antarctica. Geol. Jahrbuch 60B:143–151

Subjects: Antarctica, North Victoria Land, volcanics, Mt. Black Prince, Devonian, flora, *Protolepidodendropsis,* Paleozoic

108. Fleming, R.F., and R.A. Askin. 1982. Early Tertiary coal bed on Seymour Island, Antarctic Peninsula. Antarctic Jour. of the U.S. 17(5):67.

Subjects: palynology, coal, Tertiary, Seymour

Island, Antarctica, peninsula, podocarps, angiosperms, conifers

109. Florin, R. 1940. The Tertiary fossil conifers of south Chile and their phytogeographical significance, with a review of the fossil conifers of southern lands. Kungl. Svenska Vetenskapsakademiens Handlingar. Tredje Serien, 19(2):1–107.

Subjects: Tertiary, conifers, South America, Antarctica, biogeography, Hope Bay, Jurassic, Graham Land, flora, gymnosperms, Snow Hill Island, peninsula, Seymour Island, Cretaceous, Eocene, Kerguelen

110. Frakes, L.A., and J.C. Crowell. 1970. Geologic evidence for the place of Antarctica in Gondwanaland. Antarctic Jour. of the U.S. 5(3):67–69.

Subjects: Antarctica, Gondwana, reconstruction, geology, continental drift

111. Frakes, L.A., J.L. Matthews, and J.C. Crowell. 1971. Late Palaeozoic glaciation: Part III, Antarctica. Bull. Geol. Soc. Amer. 82:1581–1604.

Subjects: glaciation, Antarctica, Permian

112. Francis, J.E. 1986. Growth rings in Cretaceous and Tertiary wood from Antarctica and their palaeoclimatic implications. Palaeontology 29(4):665–684.

Subjects: Cretaceous, Tertiary, wood, Antarctica, paleoclimate, peninsula, flora

113. Fuenzalida, H. 1965. Serie sedimentaria volcánica con plantas en las Islas Snow y Livingston. Resumenes Sociedad Geologica Chile 10:3–4.

Subjects: flora, Antarctica, peninsula, Snow Island, Livingston Island

114. Fuenzalida, H., R. Araya, and F. Herve. 1972. Middle Jurassic flora from north-eastern Snow Island, South Shetland Islands. In: Antarctic Geology and Geophysics (SCAR Symp. on Antarctic Geol. & Solid Earth Geophys., Oslo 1970) (R.J. Adie, ed.), Universitetetsforlaget, Oslo, 93–97.

Subjects: flora, Jurassic, Snow Island, Shetland Islands, peninsula, Antarctica, cycadophytes

115. Gabites, H.I. 1985. Triassic paleoecology of the Lashly Formation, Transantarctic Mountains, Antarctica. Unpubl. M.S. Thesis, Victoria University of Wellington, New Zealand.

Subjects: geology, paleoecology, flora, Allan Hills, Lashly Fm., Triassic, Victoria Land, sedimentology, paleosols, Beacon Group

116. Gair, H.S., G. Norris, and J. Ricker. 1965. Early Mesozoic microfloras from Antarctica. New Zealand Jour. Geol. Geophys. 8:231–235.

Subjects: palynology, stratigraphy, Victoria, Triassic, Jurassic, Australia, Antarctica, Beacon, Ferrar

117. Gazdzicki, A. 1987. Paleontological studies on King George Island, West Antarctica, 1986. Polish Polar Research 8(1):85–92.

Subjects: King George Island, Antarctica, flora, fauna, Paleogene, glaciation, reconstruction, stratigraphy

118. Gee, C.T. 1984. Preliminary studies of a fossil flora from the Orville Coast, eastern Ellsworth Land, Antarctic Peninsula. Antarctic Jour. of the U.S. 19(5):36–37.

Subjects: flora, Antarctica, peninsula, Orville Coast, Jurassic, gymnosperms, conifers, pteridosperms, cycadeoids, wood

119. Gee, C.T. 1987. Revision of the Early Cretaceous flora from Hope Bay, Antarctica. Dissertation Abstr. Int. 48(5):1284-B.

Subjects: Antarctica, flora, Cretaceous, Hope Bay, peninsula, cycadophytes, pteridophyte, sphenophytes, seed ferns, conifers, South America, New Zealand, paleogeography, hepatophytes

120. Gee, C.T. 1987. Revision of the early Cretaceous flora from Hope Bay, Antarctica. XIV Int. Botanical Congress Abstracts, Berlin, 24 July–1 August 1987:333.

Subjects: flora, Hope Bay, Antarctica, peninsula, Cretaceous, abstract, conifers, gymnosperms, ferns, Jurassic

121. Gee, C.T., and B. Mohr. 1988. Early Cretaceous palynomorphs from the Weddell Sea, Antarctica (ODP Leg 113). Amer. J. Bot. 75(6, part 2):109.

Subjects: palynology, Early Cretaceous, Antarctica, Weddell Sea, biostratigraphy, dinoflagellates, ferns, podocarps, abstract

122. Gordon, W.T. 1930. A note on *Dadoxylon* (*Araucarioxylon*) from the Bay of Isles. In: Report on the Geological Collections Made During the Voyage of the "Quest" on the Shackleton–Rowett Expedition to the South Atlantic and Weddell Sea in 1921–1922 (W.C. Smith, ed.), British Museum (Natural History), London, 24–27.

Subjects: wood, *Dadoxylon, Araucarioxylon,* conifers, Bay of Isles, South Georgia, Antarctica, peninsula, Mesozoic

123. Gothan, W. 1908. Die fossilen Hölzer von der Seymour und Snow Hill Insel. Wissenschaftliche Ergebnisse der Schwedischen Südpolar-Expedition 1901–1903, 3(8):1–33.

Subjects: paleobotany, wood, Seymour, Snow Hill, Antarctica, podocarps, araucarians, angiosperms, Fagaceae, conifers, peninsula

124. Grindley, G.W. 1963. The geology of the Queen Alexandra Range, Beardmore Glacier,

Ross Dependency, Antarctica; with notes on the correlation of Gondwana sequences. New Zealand Jour. Geol. Geophys. 6:307–347.

Subjects: Buckley Formation, Beardmore, wood, *Dadoxylon,* Antarctica, Queen Alexandra Range, flora, glossopterids, Permian, Triassic, coal, correlation, sphenophytes, *Noeggerathiopsis,* Mesozoic, Paleozoic

125. Grindley, G.W., and F.J. Davey. 1982. The reconstruction of New Zealand, Australia, and Antarctica (review paper). In: Antarctic Geosciences (C. Craddock, ed.), Univ. Wisconsin Press, Madison, 15–29.

Subjects: reconstruction, New Zealand, Australia, Antarctica, Gondwana, tectonics

126. Grindley, G.W., V.R. McGregor, and R.I. Walcott. 1964. Outline of the geology of the Nimrod–Beardmore–Axel Heiberg Glaciers region, Ross Dependency. In: Antarctic Geology (Proc. 1st Int. SCAR Symp. on Antarctic Geol., Cape Town, 1963) (R.J. Adie, ed.), North-Holland Publ. Co., Amsterdam, 206–219.

Subjects: Antarctica, geology, Beardmore, Nimrod, Axel Heiberg Glaciers, Beacon, Buckley, glossopterids, wood, Dominion Coal Measures, Triassic, Permian, *Dicroidium,* corystosperms, Shackleton Glacier

127. Grindley, G.W., and D.C. Mildenhall. 1978. Discovery of fossils in Marie Byrd Land, Antarctica. New Zealand Geol. Soc. Newsl. 45:33–34.

Subjects: Antarctica, Marie Byrd Land, Devonian, flora

128. Grindley, G.W., and D.C. Mildenhall. 1980. Geologic background to a Devonian plant fossil discovery, Ruppert Coast, Marie Byrd Land, Antarctica. In: Gondwana Five, Proc. 5th Int. Gondwana Symp., Wellington, N.Z. (M.M. Cresswell and P. Vella, eds.), A.A. Balkema, Rotterdam, 23–30.

Subjects: Devonian, flora, Antarctica, Ruppert Coast, Marie Byrd Land, West Antarctica, stratigraphy

129. Grindley, G.W., D.C. Mildenhall, and J.M. Schopf. 1980. A mid–late Devonian flora from the Ruppert Coast, Marie Byrd Land, West Antarctica. J. Roy. Soc. New Zealand 10:271–285.

Subjects: Antarctica, flora, Devonian, Marie Byrd Land, paleobotany

130. Grindley, G.W., and G. Warren. 1964. Stratigraphic nomenclature and correlation in the western Ross Sea region. In: Antarctic Geology (Proc. 1st Int. Symp. on Antarctic Geol., Cape Town 1963) (R.J. Adie, ed.), North-Holland Publ. Co., Amsterdam, 314–333.

Subjects: Antarctica, geology, Victoria Land, stratigraphy, Transantarctic Mts., Permian, glossopterids, floras, Triassic, Jurassic, *Dicroidium,* review

131. Grunow, A.M., D.V. Kent, and I.W.D. Dalziel. 1987. Mesozoic evolution of West Antarctica and the Weddell Sea Basin: new paleomagnetic constraints. Earth and Planet. Sci. Letters 86:16–26.

Subjects: Antarctica, paleomagnetism, tectonics, reconstructions, West Antarctica, Weddell Sea, peninsula, Jurassic, Ellsworth Mts., Whitmore Mts., Cretaceous

132. Gunn, B.M., and G. Warren. 1962. Geology of Victoria Land between Mawson and Mulock Glaciers, Antarctica. New Zealand Geol. Surv. Bull. 71:1–157.

Subjects: Victoria Land, Antarctica, geology, Allan Hills, Permian, glossopterids, seeds, Shapeless Mt., Triassic, *Dicroidium,* corystosperms, cycadophytes, Horseshoe Mt., Mt. Fleming, Jurassic, Carapace Nunatak, Beacon Sandstone

133. Hall, S.A. 1975. Palynologic investigations of Quaternary sediment from Lake Vanda. Antarctic Jour. of the U.S. 10(4):173–174.

Subjects: Quaternary, palynology, Dry Valleys, Antarctica, Wright Valley, Lake Vanda, *Nothofagus*

134. Hall, S.A. 1977. Cretaceous and Tertiary dinoflagellates from Seymour Island, Antarctica. Nature 267:239–241.

Subjects: Antarctica, Seymour Island, Cretaceous, Tertiary, palynology, dinoflagellates, stratigraphy

135. Halle, T.G. 1911. On the geological structure and history of the Falkland Islands. Bull. Geol. Inst. Univ. Uppsala 11:115–129.

Subjects: Antarctica, Falkland Islands, geology, flora, Devonian, Permian, Carboniferous, glossopterids, wood, sphenophytes, conifers

136. Halle, T.G. 1912. The forest-bed of West Point Island. Bull. Geol. Inst. Univ. Uppsala 11:206–218.

Subjects: wood, Falkland Islands, Antarctica, conifers, Quaternary?, paleoclimate

137. Halle, T.G. 1913. Om de antarktiska trakternas Juraflora. Geologiska foreningens i Stockholm Forhandlinger 35, 2(289):105–106.

Subjects: Jurassic, flora, Antarctica

138. Halle, T.G. 1913. The Mesozoic flora of Graham Land. Wissenschaftliche Ergebnisse der Schwedischen Südpolar-Expedition, 1901–03, 3(14):3–124.

Subjects: flora, Graham Land, Antarctica, paleobotany, Mesozoic, conifers, cycadophytes, ferns, foliage

139. Hammer. W.R. 1986. Takrouna Formation

fossils of Northern Victoria Land. In: Geological Investigations in Northern Victoria Land (Ant. Res. Ser.) (E. Stump, ed.), Amer. Geophys. Union, 46:243–247.

Subjects: Victoria Land, Antarctica, glossopterids, trace fossils, Permian, fructification, flora

140. Hammer, W.R., and J.W. Cosgriff. 1981. *Myosaurus gracilis,* an anomodont reptile from the Lower Triassic of Antarctica and South Africa. J. Paleontol. 55(2):410–424.

Subjects: faunas, Triassic, Antarctica, South Africa

141. Hammer, W.R., and J.M. Zawiskie. 1982. Beacon fossils from northern Victoria Land. Antarctic Jour. of the U.S. 17(5):13–15.

Subjects: Victoria Land, Antarctica, Permian, flora, trace fossils, glossopterids, roots

142. Harrington, H.J. 1969. Fossiliferous rocks in moraines at Minna Bluff, McMurdo Sound. Antarctic Jour. of the U.S. 4(4):134–135.

Subjects: Antarctica, McMurdo Sound, Minna Bluff, Tertiary, palynology, erratics

143. Harrison, C.G.A., E.J. Barron, and W.W. Hay. 1979. Mesozoic evolution of the Antarctic peninsula and the southern Andes. Geology 7:374–378.

Subjects: Antarctica, peninsula, tectonics, Mesozoic, Gondwana, reconstruction, breakup

144. Hart, G.F. 1971. The Gondwana Permian palynofloras. Anais Academia Brasileira de Ciencias 43 (Suplemento):145–185.

Subjects: palynology, Permian, Gondwana, stratigraphy

145. Harwood, D.M. 1986. Diatoms. In: Antarctic Cenozoic history from the MSSTS-1 drillhole, McMurdo Sound (P.J. Barrett, ed.), Sci. Inf. Publ. Cent., Wellington, DSIR Bull. 237: 69–107.

Subjects: diatoms, McMurdo Sound, Antarctica, Miocene,Pliocene, Oligocene, Tertiary, microfossils, stratigraphy

146. Harwood, D.M. 1988. Upper Cretaceous and Lower Paleocene diatom and silicoflagellate biostratigraphy of Seymour Island, eastern Antarctic Peninsula. In: Geology and Paleontology of Seymour Island, Antarctic Peninsula (R.M. Feldmann and M.O. Woodburne, eds.), Geol. Soc. Amer. Mem. 169:55–129.

Subjects: Seymour Island, Antarctica, peninsula, diatoms, biostratigraphy, Cretaceous, Paleocene, Mesozoic, Tertiary, silicoflagellates

147. Helby, R.J., and C.T. McElroy. 1969. Microfloras from the Devonian and Triassic of the Beacon Group, Antarctica. New Zealand Jour. Geol. Geophys. 12:376–382.

Subjects: palynology, Devonian, Triassic, Victoria Land, Antarctica

148. Hernandez, P., and V. Azcarate. 1971. Estudio paleobotanico preliminar sobre restos de una tafoflora de la Peninsula Byers (Cerro Negro), Isla Livingston, Islas Shetland del Sur, Antártica. Instituto Antártico Chileno, Ser. Cientifica 2:15–50.

Subjects: flora, Byers Peninsula, Antarctica, Livingston Island, South Shetlands, Early Cretaceous, conifers, cycadophytes, Baqueró Fm., Argentina, pteridophytes, cycadeoids

148a. Hill, R.S. 1989. Fossil leaf with affinities to *Nothofagus gunnii.* In Report on CIROS-1 Drillhole, McMurdo Sound, Antarctica, DSIR, New Zealand, (in press)

Subjects: angiosperms, Tertiary, Fagaceae, McMurdo Sound, *Nothofagus,* leaf

149. Hjelle, A., and T. Winsnes. 1972. The sedimentary and volcanic sequence of Vestfjella, Dronning Maud Land. In: Antarctic Geology and Geophysics (Symp. on Ant. Geol. and Solid Earth Geophys., Oslo 1970) (R.J. Adie, ed.), Universitetetsforlaget, Oslo, 539–546.

Subjects: Antarctica, geology, Dronning Maud Land, East Antarctica, Vestfjella, Permian, flora, glossopterids, biostratigraphy

150. Hofmann, J., and W. Weber. 1983. A Gondwana reconstruction between Antarctica and South Africa. In: Antarctic Earth Sciences (4th Int. SCAR Symp. on Antarctic Earth Sci., Adelaide, 1982) (R.L. Oliver, P.R. James, and J.B. Jago, eds.), Cambridge Univ. Press, Cambridge, 584–589.

Subjects: Antarctica, reconstruction, Gondwana, South Africa, Paleozoic, Jurassic

151. Howell, D.G. 1980. Mesozoic accretion of exotic terranes along the New Zealand segment of Gondwanaland. Geology 8:487–491.

Subjects: New Zealand, Gondwanaland, Permian, Mesozoic, Cretaceous, Antarctica, Australia, reconstructions, terranes, tectonics

152. Jefferson, T.H. 1980. Angiosperm fossils in supposed Jurassic volcanogenic shales, Antarctica. Nature 285:157–158.

Subjects; Antarctica, Jurassic, flora, paleobotany, angiosperms

153. Jefferson, T.H. 1982. Fossil forests from the Lower Cretaceous of Alexander Island, Antarctica. Palaeontology 25:681–708.

Subjects: Cretaceous, wood, Alexander Island, peninsula, Antarctica, flora, paleoclimate, conifers, growth, gymnosperms

154. Jefferson, T.H. 1982. The preservation of fossil leaves in Cretaceous volcaniclastic rocks from Alexander Island, Antarctica. Geol. Mag. 119(3):291–300.

Subjects: Cretaceous, Alexander Island, flora,

Antarctica, peninsula, leaves, preservation, conifers, ferns

155. Jefferson, T.H. 1983. Palaeoclimatic significance of some Mesozoic Antarctic fossil floras. In: Antarctic Earth Sciences (4th Int. SCAR Symp. on Antarctic Earth Sci., Adelaide, 1982) (R.L. Oliver, P.R. James, and J.B. Jago, eds.), Cambridge Univ. Press, Cambridge, 593–599.

Subjects: paleoclimate, Antarctica, floras, Alexander Island, Cretaceous, wood, leaves, biogeography, growth, *Circoporoxylon*, podocarps

156. Jefferson, T.H. 1987. The preservation of conifer wood: examples from the Lower Cretaceous of Antarctica. Palaeontology 30(2): 233–249.

Subjects: wood, Cretaceous, Antarctica, conifers, Alexander Island, peninsula, preservation

157. Jefferson, T.H., and D.I.M. MacDonald. 1981. Fossil wood from South Georgia. British Antarctic Survey Bull. 54:57–64.

Subjects: wood, South Georgia, Antarctica, Cretaceous, preservation, conifers

158. Jefferson, T.H., M.A. Siders, and M.A. Haban. 1983. Jurassic trees engulfed by lavas of the Kirkpatrick Basalt Group, northern Victoria Land. Antarctic Jour. of the U.S. 18(5): 14–16.

Subjects: Jurassic, northern Victoria Land, Antarctica, trees, wood, paleoclimate, dendrochronology, Kirkpatrick basalt, conifers, growth, *Protocupressinoxylon*

159. Jefferson, T.H., and T.N. Taylor. 1983. Permian and Triassic woods from the Transantarctic Mountains: Paleoenvironmental indicators. Antarctic Jour. of the U.S. 18(5):55–57.

Subjects: wood, Permian, Triassic, Antarctica, paleoclimate, Beardmore area, Fremouw, Falla, Buckley, Fairchild, growth, *Araucarioxylon*

160. Kellogg, D.E., and T.B. Kellogg. 1984. Diatoms from the McMurdo Ice Shelf, Antarctica. Antarctic Jour. of the U.S. 19(5):76–77.

Subjects: diatoms, McMurdo Sound, Antarctica, Recent, ice shelf

161. Kellogg, D.E., and T.B. Kellogg. 1986. Diatom biostratigraphy of sediment cores from beneath the Ross Ice Shelf. Micropaleontology 32(1):74–94.

Subjects: diatoms, biostratigraphy, Antarctica, Ross Sea, recycled, Quaternary, Miocene, palynology

162. Kellogg, D.E., M. Stuiver, T.B. Kellogg, and G.H. Denton. 1980. Non-marine diatoms from late Wisconsin perched deltas in Taylor Valley, Antarctica. Palaeogeography, Palaeoclimatology, Palaeoecology 30:157–189.

Subjects: diatoms, Antarctica, southern Victoria Land, Dry Valleys, Quaternary, palynology

163. Kellogg, T.B., and D.E. Kellogg. 1981. Pleistocene sediments beneath the Ross Ice Shelf. Nature 293:130–133.

Subjects: Pleistocene, Antarctica, Ross Sea, Miocene, diatoms, palynology

164. Kellogg, T.B., and D.E. Kellogg. 1983. Reply to "Interpretation of sediment cores from the Ross Ice Shelf Site J-9, Antarctica" by H.T. Brady. Nature 303:511–513.

Subjects: palynology, Antarctica, Ross Sea, diatoms, Pliocene, Pleistocene, biostratigraphy

165. Kellogg, T.B., and R.S. Truesdale. 1979. Late Quaternary paleoecology and paleoclimatology of the Ross Sea: the diatom record. Marine Micropaleontology 4:137–158.

Subjects: Quaternary, paleoclimate, Ross Sea, Antarctica, diatoms, palynology, paleoecology

166. Kemp, E. 1969. Palynological examination of samples from the Beaver Lake area, Prince Charles Mountains, Antarctica. Bureau Mineral Resources, Geology and Geophysics, Australia Rec. 98:1–7.

Subjects: palynology, Prince Charles Mts., East Antarctica, Permian

167. Kemp, E.M. 1972. Recycled palynomorphs in continental shelf sediments from Antarctica. Antarctic Jour. of the U.S. 7(5):190–191.

Subjects: palynology, Antarctica, paleoclimate, Ross Sea, Permian, Triassic, Weddell Sea, Cretaceous, Tertiary, East Antarctica, Prydz Bay, Eocene

168. Kemp, E.M. 1972. Lower Devonian palynomorphs from the Horlick Formation, Ohio Range, Antarctica. Palaeontographica 139B: 105–124.

Subjects: palynology, Devonian, Antarctica, Ohio Range, Horlick Fm., chitinozoans, Emsian, Transantarctic Mts., monolete spores, biostratigraphy

169. Kemp, E.M. 1972. Reworked palynomorphs from the West Ice Shelf area, East Antarctica, and their possible geological and palaeoclimatological significance. Marine Geol. 13: 145–157.

Subjects: palynology, Antarctica, East Antarctica, paleoclimate, recycled, Permian, Cretaceous, Tertiary, Eocene, glaciation

170. Kemp, E.M. 1973. Permian flora from the Beaver Lake area, Prince Charles Mountains. 1. Palynological examination of samples. Austr., Bur. Miner. Resour., Geol. Geophys., Bull., Palaeont. Paper 126:7–12.

Subjects: palynology, Permian, Prince Charles Mts., Antarctica, East Antarctica

171. Kemp, E.M. 1975. The palynology of Late Paleozoic glacial deposits of Gondwanaland. In: Gondwana Geology (3rd Gondwana Symp., Canberra, 1973) (K.S.W. Campbell, ed.), Australian National Univ. Press, Canberra, 397–413.

Subjects: Gondwana, palynology, Permian, stratigraphy, Australia, biostratigraphy, Antarctica, Victoria Land, correlation, glaciation, Ohio Range, Buckeye Fm., Wisconsin Range

171a. Kemp, E.M. 1975. Palynology of Leg 28 drill sites, Deep Sea Drilling Project. In Hayes DE, Frakes, LA, et al. (eds), Initial Reports of the Deep Sea Drilling Project, U.S. Govt. Printing Office, Washington, D.C., 28:599–623.

Subjects: palynology, marine, glacial, dinoflagellates, Fagaceae, *Nothofagidites,* podocarps, conifers, angiosperms, Late Oligocene, paleoclimate

172. Kemp, E.M., B.E. Balme, R.J. Helby, R.A. Kyle, G. Playford, and P.L. Price. 1977. Carboniferous and Permian palynostratigraphy in Australia and Antarctica: a review. BMR Jour. of Austral. Geol. Geophys. 2:177–208.

Subjects: Permian, Carboniferous, palynology, stratigraphy, Australia, Antarctica, southern Victoria Land, central Transantarctic Mts., correlation

173. Kemp, E.M., and P.J. Barrett. 1975. Antarctic glaciation and early Tertiary vegetation. Nature 258:507–508.

Subjects: Antarctica, palynology, Ross Sea, Oligocene, glaciation, paleoclimate, *Nothofagus,* Fagaceae, Proteaceae, angiosperms

174. King, L., and T.W. Downard. 1963. Importance of Antarctica in the hypothesis of continental drift. In: Antarctic Geology (Proc. 1st Int. SCAR Symp. on Antarctic Geol., Cape Town, 1963) (R.J. Adie, ed.), North-Holland Publ. Co., Amsterdam, 727–735.

Subjects: Antarctica, reconstructions, Gondwana, continental drift

175. Korotkevich, E.S., and B.V. Timofeev. 1964. The age of the rocks of East Antarctica from spore analysis. In: Soviet Antarctic Expedition, Information Bulletin, Elsevier, Amsterdam, 2:63–69.

Subjects: palynology, Antarctica, East Antarctica

176. Kräusel, R. 1962. Antarctic fossil wood. In: Trans-Antarctic Expedition, 1955–1958, Sci. Rept. 9 (Geology), 133–154.

Subjects: Permian, Antarctica, wood, Allan Hills, *Dadoxylon, Taeniopitys,* southern Victoria Land

177. Kyle, R.A. 1974. *Plumsteadia ovata* n. sp., a glossopterid fructification from South Victoria Land, Antarctica (note). New Zealand Jour. Geol. Geophys. 17:719–721.

Subjects: glossopterids, fructification, Antarctica, southern Victoria Land, Permian, Mt. Feather

178. Kyle, R.A. 1976. Palaeobotanical studies of the Permian and Triassic Victoria Group (Beacon Supergroup) of South Victoria Land, Antarctica. Ph.D. Diss., Victoria Univ. of Wellington, 306 pp.

Subjects: Permian, Triassic, palynology, Victoria Land, Antarctica, flora, glossopterids, biostratigraphy, leaves

179. Kyle, R.A. 1977. Devonian palynomorphs from the basal Beacon Supergroup of South Victoria Land, Antarctica (note). New Zealand Jour. Geol. Geophys. 20:1147–1150.

Subjects: palynology, Early Devonian, southern Victoria Land, Antarctica, Table Mt.

180. Kyle, R.A. 1977. Palynostratigraphy of the Victoria Group of South Victoria Land, Antarctica. New Zealand Jour. Geol. Geophys. 20:1081–1102.

Subjects: Permian, Triassic, palynology, stratigraphy, Antarctica, Victoria Land, Nilsen Plateau, Ohio Range, Wisconsin Range, correlation, Australia

181. Kyle, R.A., and A. Fasola. 1978. Triassic palynology of the Beardmore Glacier area, Antarctica. Palinologia 1:313–319.

Subjects: palynology, Triassic, Antarctica, Beardmore, stratigraphy, Fremouw, Falla

182. Kyle, R.A., and J.M. Schopf. 1977. Palynomorph preservation in the Beacon Supergroup of the Transantarctic Mountains. Antarctic Jour. of the U.S. 12(4):121–122.

Subjects: Antarctica, palynology, Permian, Triassic, preservation, Queen Maud Mts., Nilsen Plateau, southern Victoria Land, Ohio Range

183. Kyle, R.A., and J.M. Schopf. 1982. Permian and Triassic palynostratigraphy of the Victoria Group, Transantarctic Mountains. In: Antarctic Geosciences (C. Craddock, ed.), Univ. Wisconsin Press, 649–659.

Subjects: Antarctica, stratigraphy, palynology, Triassic, Permian, Victoria Land, Beardmore, Ohio Range, Wisconsin Range, Transantarctic Mts., Beacon, glossopterids, *Dicroidium,* Fremouw, Falla

184. La Prade, K.E. 1972. Permian–Triassic Beacon Supergroup of the Shackleton Glacier Area, Queen Maud Range, Transantarctic

Mountains. In: Antarctic Geology and Geophysics (SCAR Symp. on Antarctic Geol. and Solid Earth Geophys., Oslo 1970) (R.J. Adie, ed.), Universitetetsforlaget, Oslo, 373–378.

Subjects: Antarctica, Beacon, Shackleton Glacier, Transantarctic Mts., Permian, Triassic, glossopterids, Buckley, geology, Triassic

185. Lacey, W.S. 1975. Some problems of "mixed" floras in the Permian of Gondwanaland. In: Gondwana Geology (3rd Gondwana Symp., Canberra, 1973) (K.S.W. Campbell, ed.), Australian Natl. Univ. Press, Canberra, 125–134.

Subjects: floras, Permian, Gondwana, "mixed," Africa, Argentina, New Guinea, glossopterids

186. Lacey, W.S., and R.C. Lucas. 1981. The Triassic flora of Livingston Island, South Shetland Islands. Brit. Antarctic Survey Bull. 53:157–173.

Subjects: Triassic, flora, Shetland Islands, Livingston Island, Antarctica, peninsula, bryophytes, ferns, cycadophytes, conifers, wood, palynology, anatomy

187. Lacey, W.S., and R.C. Lucas. 1981. A Lower Permian flora from the Theron Mountains, Coats Land. Brit. Antarctic Survey Bull. 53:153–156.

Subjects: flora, Permian, Theron Mts., Antarctica, Coats Land

188. Lambrecht, L.L., W.S. Lacey, and C.S. Smith. 1973. Observations on the Permian flora of the Law Glacier area, central Transantarctic Mountains. Bull. Soc. Belg. Géol., Paléont., Hydrol. 81(3–4):161–167.

Subjects: flora, Permian, Law Glacier, Transantarctic Mts., Antarctica, Coalsack Bluff, Mt. Sirius, Mt. Picciotto, Mt. Ropar, Beardmore area

189. Laudon, T.S., D.J. Lidke, T. Delevoryas, and C.T. Gee. 1985. Sedimentary rocks of the English Coast, eastern Ellsworth Land, Antarctica. Antarctic Jour. of the U.S. 20(5):38–40.

Subjects: Antarctica, English Coast, Ellsworth land, Permian, Triassic?, glossopterids, wood, conifers, *Elatocladus,* Jurassic

190. Laudon, T.S., D.J. Lidke, T. Delevoryas, and C.T. Gee. 1987. Sedimentary rocks of the English Coast, eastern Ellsworth Land, Antarctica. In: Gondwana Six: Structure, Tectonics, and Geophysics (G.D. McKenzie, ed.), Amer. Geophys. Union, Washington, D.C., 183–189.

Subjects: Antarctica, flora, English Coast, peninsula, Ellsworth Land, glossopterids, Permian, *Equisetum,* geology

191. Laudon, T.S., M.R.A. Thomson, P.L. Williams, K.L. Milliken, P.D. Rowley, and J.M. Boyles. 1983. The Jurassic Latady Formation, southern Antarctic peninsula. In: Antarctic Earth Sciences (4th Int. SCAR Symp. on Antarctic Earth Sci., Adelaide, 1982) (R.L. Oliver, P.R. James, and J.B. Jago, eds.), Cambridge Univ. Press, Cambridge, 308–314.

Subjects: Jurassic, Antarctica, peninsula, flora, fragments, Lassiter Coast, Behrendt Mts., wood, leaves, cycadophytes, conifers, gymnosperms, ginkgophytes

192. Lawver, L.A., J.G. Schlater, and L. Meinke. 1985. Mesozoic and Cenozoic reconstructions of the South Atlantic. Tectonophysics 114:233–254.

Subjects: reconstructions, Mesozoic, Cenozoic, plate tectonics, Antarctica, South America, South Africa

193. Lawver, L.A., and C.R. Scotese. 1987. A revised reconstruction of Gondwanaland. In: Gondwana Six: Structure, Tectonics, and Geophysics (G.D. McKenzie, ed.), Amer. Geophys. Union, Washington, D.C., 17–23.

Subjects: reconstruction, Gondwana, plate tectonics, Antarctica, peninsula

194. Lemoigne, Y. 1987. Confirmation de l'existence d'une flore triasique dans l'île Livingston des Shetland du sud (Ouest Antarctique). C.R. Acad. Sci. Paris, Sér. II 304(10):543–546.

Subjects: flora, Triassic, Antarctica, Livingston Island, South Shetland Islands, sphenophytes, corystosperms, *Dicroidium,* Caytoniales, conifers, gymnosperms, pteridosperms

195. Lemoigne, Y., and T. Torres. 1988. Paléoxylologie de l'Antarctide: *Sahnioxylon antarcticum* n.sp. et interprétation de la double zonation des cernes des bois secondaires de genre de structure (parataxon) *Sahnioxylon* Bose et Sah, 1954. C.R. Acad. Sci. Paris, Série II, 306:939–945.

Subjects: wood, Antarctica, *Sahnioxylon,* Upper Cretaceous, Livingston Island, South Shetlands, peninsula, anatomy, cycadophytes, bennettitales

196. Long, W.E. 1964. The stratigraphy of the Horlick Mountains. In: Antarctic Geology (Proc. 1st Int. SCAR Symp. on Antarctic Geol., Cape Town, 1963) (R.J. Adie, ed.), North-Holland Publ. Co., Amsterdam, 352–363.

Subjects: stratigraphy, Horlick Mts., Antarctica, Ohio Range, Wisconsin Range, Permian, flora, glossopterids

197. Long, W.E. 1965. Stratigraphy of the Ohio Range, Antarctica. In: Geology and Paleontology of the Antarctic, Antarctic Res. Ser., Amer. Geophys. Union 6:71–116.

Subjects: Ohio Range, Antarctica, stratigraphy, Permian, flora, glossopterids, wood, Mt. Glossopteris Formation

198. Lucas, R.C., and W.S. Lacey. 1981. A permineralized wood flora of probable Early Tertiary age from King George Island, South Shetland Islands. Brit. Antarctic Survey Bull. 53:147–151.

Subjects: Early Tertiary, wood, King George Island, South Shetland Islands, Antarctica, peninsula, gymnosperms, *Dadoxylon*

199. Lyra, C.S. 1986. Palinologia de sedimentos Terciários da Peninsula Fildes, Ilha Rei George (Ilhas Shetland do Sul, Antártica) e algumas consideraçoes paleoambientais [Tertiary sediment palynology at Fildes Peninsula, King George Island, South Shetland Islands, and some paleoenvironmental considerations]. Anais Academia Brasileira de Ciencias 58(1-Suppl.):137–147.

Subjects: early Tertiary, palynology, King George Island, South Shetlands, Antarctica, peninsula, angiosperms, pteridophytes, gymnosperms, conifers, podocarps, *Nothofagus,* coccoliths, Eocene, Oligocene, paleoclimate, podocarps

200. Maheshwari, H.K. 1972. Permian wood from Antarctica and revision of some Lower Gondwana wood taxa. Palaeontographica 138B: 1–43.

Subjects: wood, Permian, Antarctica, Gondwana, *Dadoxylon,* gymnosperms, *Araucarioxylon,* Horlick Mts., Ohio Range, Allan Hills, southern Victoria Land, Queen Maud Mts., *Protophyllocladoxylon, Damudoxylon, Megaporoxylon, Polysolenoxylon, ?Antarcticoxylon,* Mt. Schopf, Mt. Glossopteris, Mt. Nansen

201. Matz, D.B., P.R. Pinet, and M.O. Hayes. 1972. Stratigraphy and petrology of the Beacon Supergroup, southern Victoria Land. In: Antarctic Geology and Geophysics (SCAR Symp. on Antarctic Geol. and Solid Earth Geophys., Oslo 1970) (R.J. Adie, ed.), Universitetetsforlaget, Oslo, 353–358.

Subjects: Beacon, southern Victoria Land, Antarctica, stratigraphy, petrology, flora, Permian, Triassic, Mt. Fleming, wood, glossopterids, Robison Peak, plant debris, Mt. Bastion, sphenophytes, Allan Hills, *Dicroidium,* corystosperms, Weller Coal Measures

202. McElhinny, M.W., and B.J.J. Embleton. 1974. Australia palaeomagnetism and the Phanerozoic plate tectonics of eastern Gondwanaland. Tectonophysics 22:1–29.

Subjects: tectonics, Australia, Gondwanaland, reconstructions

202a. McElroy, C.T. 1969. Comparative lithostratigraphy of Gondwana sequences, eastern Australia and Antarctica. In Amos, AJ (ed) Gondwana Stratigraphy (1st IUGS Symp., Buenos Aires, 1967), UNESCO, Paris, 441–466

Subjects: Antarctica, Australia, geology, Paleozoic, Mesozoic, flora, glossopterids, stratigraphy, Permian, Triassic, sphenophytes, wood

203. McIntyre, D.J., and G.J. Wilson. 1966. Preliminary palynology of some Antarctic Tertiary erratics. New Zealand Jour. Bot. 4: 315–321.

204. McKelvey, B.C., P.N. Webb, M.P. Gorton, and B.P. Kohn. 1972. Stratigraphy of the Beacon Supergroup between the Olympus and Boomerang Ranges, Victoria Land. In: Antarctic Geology and Geophysics (SCAR Symp. on Antarctic Geol. and Solid Earth Geophys., Oslo 1970) (R.J. Adie, ed.), Universitetetsforlaget,Oslo, 345–352.

Subjects: Victoria Land, Antarctica, stratigraphy, Devonian, fauna, Permian, glossopterids, flora, Mt. Fleming, Beacon, Warren Range, *Haplostigma,* lycopods

204a. McPherson, J.G. 1979. Calcrete (caliche) palaeosols in fluvial redbeds of the Aztec Siltstone (Upper Devonian), southern Victoria Land, Antarctica. Sedimentary Geol. 22:267–285

Subjects: paleosols, Aztec Siltstone, Upper Devonian, south Victoria Land, Antarctica, paleoclimate, depositional environment, roots, palynology, Paleozoic

205. Meyer-Berthaud, B., and T.N. Taylor. 1989. The structure and affinities of woody stems from the Triassic of Antarctica. Amer. J. Bot. 76(6, part 2):170–171

Subjects: Triassic, Antarctica, peat, wood, anatomy, gymnosperms, abstract

206. Mildenhall, D.C. 1987. CIROS-1 drillhole, McMurdo Sound, Antarctica: Terrestrial palynology. New Zealand Activities in Antarctica 1986–1987, Prog. and Abstr. Antarctica Res. Ct., Victoria Univ. of Wellington, 10.

Subjects: palynology, Antarctica, McMurdo, recycled, Permian, Cretaceous, *Nothofagus,* angiosperms, lycopods, conifers, *Classopollis,* Oligocene

206a. Mildenhall, D.C. 1989. Terrestrial palynology. In Report on CIROS-1 Drillhole, McMurdo Sound, Antarctica, DSIR, New Zealand, (in press)

Subjects: palynology, Antarctica, Tertiary, McMurdo Sound

207. Millay, M.A. 1987. Triassic fern flora from Antarctica. Actas VII Simposio Argentino de Paleobot. y Palinol., Buenos Aires, 173–176.

Subjects: Antarctica, Triassic, Beardmore, ferns, peat, Fremouw

208. Millay, M.A., T.N. Taylor, and E.L. Taylor. 1987. Phi thickenings in fossil seed plants from Antarctica. IAWA Bull. 8:191–201.

Subjects: Fremouw, Triassic, Antarctica, Beardmore, peat, gymnosperms

209. Millay M.A., T.N. Taylor, and E.L. Taylor. 1987. Studies of Antarctic fossil plants: An association of ferns from the Triassic of Fremouw Peak. Antarctic Jour. of the U.S. 22(5):31–32.

Subjects: Antarctica, Fremouw, Beardmore, peat, ferns, Triassic

210. Miller, H. 1983. The position of Antarctica within Gondwana in the light of Palaeozoic orogenic development. In: Antarctic Earth Sciences (4th Int. SCAR Symp. on Antarctic Earth Sci., Adelaide, 1982) (R.L. Oliver, P.R. James, and J.B. Jago, eds.), Cambridge Univ. Press, Cambridge, 579–581.

Subjects: Antarctic, Gondwana, reconstructions, Paleozoic, Mesozoic, plate tectonics

211. Minshew, V.H. 1966. Stratigraphy of the Wisconsin Range, Horlick Mountains, Antarctica. Science 152(3722):637–638.

Subjects: Horlick Mts., Antarctica, Wisconsin Range, stratigraphy, Permian, flora, glossopterids

212. Mirsky, A., S.B. Treves, and P.E. Calkin. 1965. Stratigraphy and petrography, Mount Gran area, southern Victoria Land, Antarctica. In: Geology and Paleontology of the Antarctic (J.B. Hadley, ed.), Amer. Geophys. Union, Antarctic Res. Ser. 6:145–175.

Subjects: Antarctica, southern Victoria Land, stratigraphy, Mt. Gran, Alatna Valley, geology, Mt. Bastion Fm., glossopterids, Permian

213. Mitra, N.D., S.K. Bandyopadhyay, and U.K. Basu. 1979. Sedimentary framework of the Gondwana sequence of eastern India and its bearing on Indo-Antarctic fit. In: Fourth International Gondwana Symposium: Papers (Calcutta, 1977) (B. Laskar and C.S. Raja Rao, eds.), Hindustan Publ. Corp., Delhi, 1:37–41.

Subjects: Gondwana, India, Antarctica, plate tectonics, reconstruction, correlation

214. Mulligan, J.J., B.C. Parks, H.D. Hess, and J.M. Schopf. 1963. Mount Gran coal deposits, Victoria Land, Antarctica. U.S. Bureau Mines, Rept. Invest. 621B:1–66.

Subjects: Victoria Land, Antarctica, Mt. Gran, coal, geology, Stratigraphy, glossopterids, Permian

215. Mutter, J.C., K.A. Hegarty, S.C. Cande, and J.K. Weissel. 1985. Breakup between Australia and Antarctica: a brief review in the light of new data. Tectonophysics 114:255–279.

Subjects: Australia, Antarctica, tectonics, breakup, Gondwana, Cretaceous, Tertiary

216. Nathorst, A.G. 1904. Sur la flore fossile des régions antarctiques. C.R. Acad. Sci., Paris 138:1447–1450.

Subjects: flora, Antarctica, Jurassic, Tertiary, sphenophytes, Hope Bay, peninsula, Mt. Flora, ferns, cycadophytes, conifers, Seymour Island, wood, angiosperms, dicots, leaves

217. Nathorst, A.G. 1906. *Phyllotheca*-Reste aus den Falkland-Inseln. Bull. Geol. Inst. Uppsala 7:72–76.

Subjects: Falkland Islands, sphenophytes, flora, Permian, Carboniferous?, Antarctica

218. Nathorst, A.G. 1907. On the Upper Jurassic flora of Hope Bay. C.R. 10th Int. Geol. Congr., Mexico 2:1269–1270.

Subjects: Jurassic, flora, Antarctica, Hope Bay, peninsula, ferns, cycadophytes, sphenophytes, conifers

219. Nishida, M., N. Nishida, and T. Nasa. 1988. Anatomy and affinities of the petrified plants from the Tertiary of Chile. V. Bot. Mag 101:293–309.

Subjects: wood, angiosperms, gymnosperms, *Laurinoxylon,* Seymour Island, Tertiary, Cretaceous, Antarctica, peninsula, Chile

220. Nishida, M., N. Nishida, and M. Rancusi. 1988. Notes on the petrified plants from Chile (1). Jour. Jap. Bot. 63:39–48.

Subjects: Antarctica, Chile, wood, Tertiary, *Nothofagoxylon, Laurinoxylon,* Seymour Island, King George Island, Cretaceous, South Shetlands

221. Nordenskjöld, O. 1913. Antarktis. In: Handbüch der Regionalen Geologie (G. Steinmann and O. Wilckens, eds.), 7(6):1–29.

Subjects: Antarctica, geology, Jurassic, plants, flora

222. Norris, G. 1965. Triassic and Jurassic microspores and acritarchs from the Beacon and Ferrar Groups, Victoria Land, Antarctica. New Zealand Jour. Geol. Geophysics 8(2):236–277.

Subjects: palynology, Triassic, Jurassic, Beacon, Antarctica, Victoria Land, Rennick glacier, Priestley Glacier

223. Norton, I.O. 1982. Paleomotion between Africa, South America, and Antarctica, and implications for the Antarctic peninsula. In: Antarctic Geosciences (C. Craddock, ed.), Univ. Wisconsin Press, Madison, 99–106.

Subjects: reconstruction, Antarctica, South America, Africa, peninsula, plate tectonics

224. Oliver, R. 1979. The Antarctic coast between longitude 75° and 160°E in relation to the Gondwanaland jigsaw—a geological reconstruction. Jour. Geol. Soc. Australia 26:275–276.

Subjects: Gondwana, reconstruction, Antarctica

225. Orlando, H.A. 1963 (1964?). La flora fósil en las inmediaciones de la Peninsula Ardley, Isla 25 de Mayo, Islas Shetland del Sur. Contr. Inst. Antárt. Argentino 79:1–17.

Subjects: flora, King George Island, South Shetland Islands, Antarctica, peninsula, Tertiary, angiosperms, gymnosperms

226. Orlando, H.A. 1964. The fossil flora of the surroundings of Ardley Peninsula (Ardley Island), 25 de Mayo Island (King George Island), South Shetland Islands. In: Antarctic Geology (Proc. 1st Int. SCAR Symp. on Antarctic Geol., Cape Town, 1963) (R.J. Adie, ed.), North-Holland Publ. Co., Amsterdam, 629–636.

Subjects: flora, Tertiary, King George Island, Shetland Islands, peninsula, Miocene, angiosperms, gymnosperms, pteridophytes, leaves, Antarctica

227. Orlando, H.A. 1968. A new Triassic flora from Livingston Island, South Shetland Islands. Brit. Antarctic Survey Bull. 16:1–13.

Subjects: Triassic, flora, Livingston Island, Shetland Islands, Antarctica, peninsula

228. Osborn, J.M., and T.N. Taylor. 1989. Structurally preserved sphenophytes from the Triassic of Antarctica: vegetative remains of *Spaciinodum*. Amer. J. Bot. (in press).

Subjects: sphenophytes, Antarctica, Triassic, anatomy, peat, Fremouw Peak, *Spaciinodum*, stems

229. Osborn, J.M., T.N. Taylor, and J.F. White. 1989. *Palaeofibulus*, gen. nov., a clamp-bearing fungus from the Triassic of Antarctica. Mycologia (in press).

Subjects: fungi, Triassic, Antarctica, peat, Fremouw, anatomy, basidiomycetes, evolution, *Palaeofibulus*

230. Patriat, P., J. Segoufin, J. Goslin, and P. Beuzart. 1985. Relative position of Africa and Antarctica in the Upper Cretaceous: evidence for non-stationary behaviour of fracture zones. Earth and Planetary Sci. Letters 75:204–214.

Subjects: Africa, Antarctica, tectonics, Cretaceous, reconstruction

231. Pavlov, V.V. 1958. Result'taty palinologicheskogo-analiza obraztsov iz otlozhenio o sadochnoval canicheskoy serii bikon (Antarkida, Zemlya Kordya Georga V Mys Blaff). Nauchno-Issledovatel'skiy Institut Geologii Arktik Spornik Staley po Paleontologii i Biostratigrafi. 12:77–79.

Subjects: palynology, George V Coast, East Antarctica, Wilkes Land, Triassic

232. Perovich, N.E., and E.L. Taylor. 1989. Structurally preserved fossil plants from Antarctica. IV. Triassic ovules. Amer. J. Bot. 76:992–999.

Subjects: Triassic, seeds, peat, Fremouw, *Ignotospermum*, anatomy, Antarctica

233. Pickard, J., and R.D. Seppelt. 1984. Holocene occurrence of the moss *Bryum algens* Card. in the Vestfold Hills, Antarctica. Jour. Bryol. 13:209–217.

Subjects: Holocene, mosses, *Bryum*, Vestfold Hills, Antarctica, East Antarctica

234. Pierart, P. 1980. Stratigraphical and geographical distribution of Gondwana megaspores. In: Gondwana Five (Proc. 5th Int. Gondwana Symp., Wellington, N.Z. 1980) (M.M. Creswell and P. Vella, eds.), A.A. Balkema, Rotterdam, 19–22.

Subjects: palynology, Gondwana, megaspores, glossopterids, Permian, Carboniferous, correlation, biostratigraphy, Africa, South America, India, Australia

235. Pigg, K.B. 1987. Structurally preserved Gondwana plants: *Glossopteris* and *Dicroidium*. XIV Int. Botanical Congress Abstracts, Berlin, 24 July–1 August 1987, 287.

Subjects: abstract, glossopterids, corystosperms, Antarctica, flora, Skaar Ridge, Permian, Triassic, Fremouw Peak, peat, leaves

236. Pigg, K.B., and T.N. Taylor. 1985. Anatomically preserved *Glossopteris* from the Beardmore Glacier area of Antarctica. Antarctic Jour. of the U.S. 20(5):8–10.

Subjects: glossopterids, leaves, Beardmore, Antarctica, Permian, Skaar Ridge, peat

237. Pigg, K.B., and T.N. Taylor. 1987. Anatomically preserved *Glossopteris* from Antarctica. Actas VII Simposio Argentino de Paleobot. y Palinol., Buenos Aires, 177–180.

Subjects: Antarctica, Permian, glossopterids, Skaar, peat, Beardmore, leaves

238. Pigg, K.B., and T.N. Taylor. 1987. Anatomically preserved *Dicroidium* from the Transantarctic Mountains. Antarctic Jour. of the U.S. 22(5):28–29.

Subjects: Antarctica, Triassic, foliage, corystosperms, *Dicroidium*, anatomy, Fremouw Peak, peat, foliage, Beardmore Glacier area

239. Pirrie, D., and J.B. Riding. 1988. Sedimentology, palynology and structure of Humps Island, northern Antarctic peninsula. British Antarct. Surv. Bull. 80:1–19.

Subjects: Antarctica, peninsula, Humps Island, palynology, sedimentology, Late Cretaceous, Lopez de Bertodano Formation, dinoflagellates

240. Plumstead, E.P. 1962. Fossil floras of Antarctica. In: Trans-Antarctic Expedition, 1955–1958, Sci. Rept. 9 (Geology), 1–132.

Subjects: floras, Antarctica, Permian, Triassic, Devonian, southern Victoria Land, Theron Mts., Shackleton Range, Whichaway Nunataks, glossopterids, sphenophytes, fructifications, *Dicroidium,* cycadophytes, conifers, Jurassic, Allan Nunatak, Carapace Nunatak

241. Plumstead, E.P. 1964. Palaeobotany of Antarctica. In: Antarctic Geology (Proc. 1st Int. SCAR Symp. on Antarctic Geol., Cape Town, 1963) (R.J. Adie, ed.), North-Holland Publ. Co., Amsterdam, 637–654.

Subjects: Antarctica, floras, Permian, Triassic, Devonian, glossopterids (= review of Plumstead, 1962 = same subjects)

242. Plumstead, E.P. 1965. Glimpses into the history and prehistory of Antarctica. Antarktiese Bull. No. 9 (May):1–5.

Subjects: Antarctica, floras, history, glossopterids, Permian, Devonian, review, wood, Triassic, conifers

243. Plumstead, E.P. 1970. Recent progress and the future of palaeobotanical correlation in Gondwanaland. In: Second Gondwana Symp. (South Africa, 1970), Proc. and Papers, Coun. for Sci. & Indus. Res., Pretoria, 139–148.

Subjects: floras, Gondwanaland, biogeography, correlation

244. Plumstead, E.P. 1973. The Late Palaeozoic *Glossopteris* flora. In: Atlas of Paleobiogeography (A. Hallam, ed.), Elsevier Publ. Co., Amsterdam, 187–205.

Subjects: flora, Permian, glossopterids, Gondwana, biogeography, lycopods, ferns, sphenophytes, cycadophytes, conifers, Antarctica

245. Plumstead, E.P. 1975. A new assemblage of plant fossils from Milorgfjella, Dronning Maud Land. Brit. Antarctic Survey Sci. Rept. 83: 1–30.

Subjects: flora, Dronning Maud Land, Antarctica, wood, sphenophytes, lycopods, conifers, glossopterids, cordaites, gingkophytes, Permian, Carboniferous

246. Rao, A.R. 1953. Some observations on the Rajmahal flora. The Palaeobotanist 2:25–28.

Subjects: flora, Jurassic, India, Hope Bay, Graham Land, Antarctica, peninsula, biogeography, podocarps

247. Rees, P.M. 1987. A diverse Middle Jurassic–Lower Cretaceous flora from the northern Antarctic peninsula. XIV Int. Botanical Congress Abstracts, Berlin 24 July–1 August 1987, 287.

Subjects: Jurassic, Cretaceous, flora, Antarctica, peninsula, conifers, ferns, cycadophytes, Hope Bay, Botany Bay, abstract

247a. Ricker, J. 1964. Outline of the geology between Mawson and Priestley Glaciers, Victoria Land. In Adie RJ (ed) Antarctic Geology, North Holland Publ. Co., Amsterdam, pp. 265–275

Subjects: Antarctica, geology, Mawson, Priestley Glacier, northern Victoria Land, Timber Peak, Triassic, palynology, wood, *Antarcticoxylon* (described in Seward 1914), Beacon Group

248. Rigby, J.F. 1968. Some recent developments in the study of the Permian *Glossopteris* flora in Australia and Antarctica. Ciencia e Cultura 20(2):158.

Subjects: glossopterids, sphenopsids, Australia, Antarctica, flora

249. Rigby, J.F. 1969. Permian sphenopsids from Antarctica. U.S. Geol. Survey Professional Paper 613-F:1–12.

Subjects: Permian, sphenopsids, Antarctica, flora

250. Rigby, J.F. 1972. The Gondwana palaeobotanical province at the end of the Palaeozoic. 24th Int. Geol. Congr. Sect. 7, Paleontol.:324–330.

Subjects: Gondwana, floras, Permian, Carboniferous, glossopterids, reconstructions, lycopsids, sphenopsids, pteridosperms, conifers, biogeography, ferns, Antarctica

251. Rigby, J.F. 1984. The origin of the *Glossopteris* flora—some thoughts based on macrophyte remains. In: Evolutionary Botany and Biostratigraphy (A.K. Ghosh Commem. Vol.), 19–28.

Subjects: glossopterids, floras, Permian, Gondwana, leaves, origins

252. Rigby, J.F. 1985. Some Triassic (Middle Gondwana) floras from South Victoria Land, Antarctica. In: Hornibrook Symp. Extended Abstr. (Christchurch, 1985) (R. Cooper, ed.), Dept. Scientif. Industr. Res., New Zealand, 78–79.

Subjects: floras, Antarctica, Victoria Land, Triassic, Shapeless Mt., Horseshoe Mt., *Dicroidium*

253. Rigby, J.F., and J.M. Schopf. 1969. Stratigraphic implications of Antarctic paleobotanical studies. In: Gondwana Stratigraphy (1st IUGS Gondwana Symp., Buenos Aires, 1967) (A.J. Amos, ed.), UNESCO, Paris, 2 (Earth Sciences): 91–106.

Subjects: stratigraphy, floras, Antarctica, Permian, Triassic

254. Romero, E.J. 1986. Fossil evidence regarding the evolution of *Nothofagus* Blume. Ann. Missouri Bot. Gard. 73:276–283.

Subjects: angiosperms, biogeography, floras, Fagaceae, *Nothofagus,* Gondwana, Cretaceous, Tertiary

255. Romero, E.J. 1986. Paleogene phytogeography and climatology of South America. Ann. Missouri Bot. Gard. 73:449–461.

Subjects: South America, biogeography, Paleogene, Tertiary, floras, angiosperms, paleoclimate, Antarctica

256. Ronan, T.E., P.N. Webb, J.H. Lipps, and T.E. DeLaca. 1978. Miocene glaciomarine sediments from site J-9, Ross Ice Shelf, Antarctica. Antarctic Jour. of the U.S. 13(4):121–123.

Subjects: diatoms, Miocene, Antarctica, Ross Sea, J-9, biostratigraphy

257. Rösler, O., and D. Mussa. 1985. Nota preliminar sobre novas coorrençias de troncos fósseis da península Fildes, Arquipelago Shetland do Sud, Antártica. Paleobotanica Latinamericana 7(1):9.

Subjects: Antarctica, wood, South Shetlands, Livingston Island, Fildes peninsula

258. Ross, C.A., and J.R.P. Ross. 1985. Carboniferous and Early Permian biogeography. Geology 13:27–30.

Subjects: Carboniferous, Permian, reconstructions, Gondwana, faunas

259. Scharnberger, C.K., and L. Scharon. 1972. Palaeomagnetism and plate tectonics of Antarctica. In: Antarctic Geology and Geophysics (SCAR Symp. on Antarctic Geol. and Solid Earth Geophys., Oslo 1970) (R.J. Adie, ed.), Universitetetsforlaget, Oslo, 843–847.

Subjects: Antarctica, plate tectonics, paleomagnetism, reconstructions, Marie Byrd Land, Cretaceous

260. Schopf, J.M. 1962. A preliminary report on plant remains and coal of the sedimentary section in the central range of the Horlick Mountains, Antarctica. Ohio St. Univ. Inst. Polar Studies Rept. 2:1–61.

Subjects: coal, Horlick Mts., Antarctica, Permian, flora, *Antarcticoxylon,* wood, Mt. Glossopteris

261. Schopf, J.M. 1965. Anatomy of the axis in *Vertebraria*. In: Geology and Paleontology of the Antarctic (J.B. Hadley, ed.), Amer. Geophys. Union, Antarctic Res. Ser. 6:217–228.

Subjects: Antarctica, *Vertebraria,* glossopterids, Permian, anatomy, Ohio Range, wood

262. Schopf, J.M. 1966. Antarctic paleobotany and palynology. Antarctic Jour. of the U.S. 1(4):135.

Subjects: Antarctica, palynology, glossopterids, wood, floras, correlation, Pensacola Mts., Ellsworth Mts., Permian

263. Schopf, J.M. 1967. Antarctic fossil plant collecting during the 1966–1967 season. Antarctic Jour. of the U.S. 2:114–116.

Subjects: Antarctica, flora, Permian, glossopterids, fertile, fructification, Ohio Range, Sentinel Range, Polarstar Fm.

264. Schopf, J.M. 1968. Studies in Antarctic paleobotany. Antarctic Jour. of the U.S. 3(5):176–177.

Subjects: Antarctica, Devonian, floras, Permian, glossopterids, *Buriadia,* Triassic, wood, Transantarctic Mts., Ellsworth Mts., Horlick Mts., Theron Mts., Victoria Land, Pensacola Mts., Whichaway Nunataks, Ohio Range, *Haplostigma*

265. Schopf, J.M. 1969. Ellsworth Mountains: position in West Antarctica due to sea-floor spreading. Science 164:63–66.

Subjects: Ellsworth Mts., Antarctica, plate tectonics.

266. Schopf, J.M. 1970. Gondwana paleobotany. Antarctic Jour. of the U.S. 5(3):62–66.

Subjects: Antarctica, Permian, flora, glossopterids, sphenopsids, reconstructions, continental drift, Gondwana

267. Schopf, J.M. 1970. Petrified peat from a Permian coal bed in Antarctica. Science 169: 274–277.

Subjects: Permian, Antarctica, flora, peat, Beardmore, Skaar Ridge, *Vertebraria,* glossopterids, leaves, fungi, palynology

268. Schopf, J.M. 1970. Antarctic collections of plant fossils, 1969–1970. Antarctic Jour. of the U.S. 5:89.

Subjects: Permian, Antarctica, Beardmore, peat, flora, Skaar Ridge, Victoria Land, Allan Hills, Carapace Nunatak, Coalsack Bluff, Transantarctic Mts., glossopterids

269. Schopf, J.M. 1971. Notes on plant tissue preservation and mineralization in a Permian deposit of peat from Antarctica. Amer. J. Sci. 271:522–543.

Subjects: Permian, peat, Antarctica, Beardmore, flora, preservation, Skaar Ridge

270. Schopf, J.M. 1973. Coal, climate and global tectonics. In: Implications of Continental Drift to the Earth Sciences (D.H. Tarling and S.K. Runcorn, eds.), Academic Press, N.Y., 609–622.

Subjects: coal, paleoclimate, reconstruction, plate tectonics, wood, Antarctica, peninsula, rings, dendrochronology, gymnosperms, peat

271. Schopf, J.M. 1973. The contrasting plant assemblages from Permian and Triassic deposits in southern continents. In: The Permian and Triassic Systems and Their Mutual Boundary (A. Logan and L.V. Hills, eds.) 379–397.

Subjects: floras, Permian, Triassic, glossopterids, corystosperms, *Dicroidium,* Gondwana, Antarctica, wood

272. Schopf, J.M, 1973. Plant material from the Miers Bluff Formation of the South Shetland Islands. Ohio St. Univ. Inst. Polar Studies Rept. 45:1–45.

Subjects: flora, South Shetland Islands, Livingston Island, Antarctica, peninsula, fragments, seeds?, coal, Paleozoic?, Mesozoic?

273. Schopf, J.M. 1976. Morphologic interpretation of fertile structures in glossopterid gymnosperms. Rev.Palaeobot. Palynol. 21:25–64.

Subjects: glossopterids, Antarctica, Permian, fertile, paleobotany, morphology, fructifications, peat, Gondwana

274. Schopf, J.M. 1977. Coal forming elements in permineralized peat from Mt. Augusta (Queen Alexandra Range). Antarctic Jour. of the U.S. 12(4):110–112.

Subjects: Permian, coal, peat, Beardmore, Antarctica, preservation, Skaar Ridge, Queen Alexandra Range, central Transantarctic Mts., Mt. Augusta

275. Schopf, J.M. 1978. An unusual osmundaceous specimen from Antarctica. Can. J. Bot. 56:3083–3095.

Subjects: Osmundaceae, Antarctica, Triassic, Fremouw Peak, Beardmore, flora, peat, *Osmundacaulis,* ferns, Mesozoic

276. Schopf, J.M. 1982. Forms and facies of *Vertebraria* in relation to Gondwana coal. In: Geology of the Central Transantarctic Mountains (M.D. Turner and J.E. Splettstoesser, eds.), Antarctic Res. Ser., Amer. Geophys. Union, Washington, D.C., 36(3):37–62.

Subjects: glossopterids, *Vertebraria,* Antarctica, coal, Permian, flora, paleoclimate, preservation, Graphite Peak, Queen Maud Mts.

277. Schopf, J.M., and R.A. Askin. 1980. Permian and Triassic floral biostratigraphic zones of southern land masses. In: Biostratigraphy of Fossil Plants (D.L. Dilcher and T.N. Taylor, eds.), Dowden, Hutchinson & Ross, Inc., Stroudsburg, Pa., 119–152.

Subjects: Permian, Triassic, palynology, stratigraphy, floras, Gondwana

278. Schopf, J.M., and W.E. Long. 1966. Coal metamorphism and igneous associations in Antarctica. In: Coal Sciences (R.F. Gould, ed.), Amer. Chem. Soc., 156–195.

Subjects: coal, Permian, Antarctica, Transantarctic Mts., petrology

279. Schuchert, C. 1932. Permian floral provinces and their interrelations. Amer. J. Sci. 24:405–413.

Subjects: Permian, floras, glossopterids, Gondwana

280. Seward, A.C. 1913. An extinct Antarctic flora. New Phytol. 12:188–190.

Subjects: Antarctica, flora, Jurassic, sphenophytes, ferns, Graham Land, peninsula, Hope Bay, cycadophytes, conifers, Mesozoic

281. Seward, A.C. 1914. Antarctic fossil plants. Brit. Antarct. Terra Nova Exped. (Geol.) 1:1–49.

Subjects: Antarctica, flora, Permian, Beardmore, glossopterids, Priestley Glacier, *Antarcticoxylon,* Victoria Land, Buckley Island, wood, palynology, *Pitysporites, Vertebraria,* anatomy, biogeography

282. Seward, A.C. 1933. An Antarctic pollen grain, fact or fancy? New Phytol. 32:311–313.

Subjects: Antarctica, palynology, Transantarctic Mts., *Pityosporites,* Priestley Glacier, *Antarcticoxylon, Rhexoxylon,* Triassic, corystosperms, Victoria Land

283. Seward, A.C., and V. Conway. 1934. A phytogeographical problem: fossil plants from the Kerguelen Archipelago. Ann. Bot. 48:715–741.

Subjects: biogeography, Kerguelen Island, *Araucaria,* conifers, Tertiary, diatoms, mosses, ferns, angiosperms, leaves, dicots, gymnosperms, *Araucarites*

284. Seward, A.C., and J. Walton. 1923. On a collection of fossil plants from the Falkland Islands. Quarterly Jour. Geol. Soc. London 79:313–333.

Subjects: flora, Falkland Islands, Permian, Middle Devonian, lycopods, sphenophytes, glossopterids, wood, *Dadoxylon, Phyllotheca, Neocalamites, Callixylon, Rhexoxylon,* Triassic, Antarctica, peninsula

285. Sharman, G., and E.T. Newton. 1894. Note on some fossils from Seymour Island, in the Antarctic regions, obtained by Dr. Donald. Trans. Roy. Soc. Edinburgh 39:707–709.

Subjects: Antarctica, Seymour Island, conifers, wood, fauna, Tertiary

286. Sharman, G., and E.T. Newton. 1898. Notes on some additional fossils collected at Seymour Island, Graham's Land, by Dr. Donald and Captain Larsen. Proc. Roy. Soc. Edinburgh 22:58–61.

Subjects: wood, conifers, Seymour Island, Antarctica, peninsula, Tertiary, fauna, invertebrates

286a.Skidmore, M.J. 1972. The geology of South Georgia: III. Prince Olav Harbour and Stromness Bay areas. British Antarctic. Surv. Sci. Rept. 73:1–50

Subjects: Antarctica, peninsula, South Georgia, geology, stratigraphy, petrology, wood, trace fossils, invertebrates, algae

287. Skinner, D.N.B., and J. Ricker. 1968. The geology of the region between the Mawson and Priestley Glaciers, North Victoria Land, Antarctica. Part II—Upper Paleozoic to Quaternary geology. New Zealand Jl. Geol. Geophys. 11:1041–1075.

Subjects: wood, Timber Peak, northern Victoria Land, Antarctica, gymnosperms, Permian, *Vertebraria,* Mawson, Preistley Glaciers, geology, coal, Triassic, flora, Beacon Group, Paleozoic, Mesozoic

287a. Skottsberg, C. 1949. Influence of the Antarctic continent on the vegetation of southern lands. Proc. Pacific Sci. Congr. 5:92–99

Subjects: review, wood, Mesozoic, Tertiary, exploration, Antarctica, flora, Seymour Island, Cretaceous, Jurassic, conifers, araucarians, biogeography, angiosperms

288. Smith, A.G., and A. Hallam. 1970. The fit of the southern continents. Nature 225(5278): 139–144

Subjects: reconstructions, Gondwana, continental drift

289. Smoot, E.L., and T.N. Taylor. 1986. Evidence of simple polyembryony in Permian seeds from Antarctica. Amer. J. Bot. 73:1077–1079.

Subjects: peat, Permian, seeds, Skaar, Antarctica, Beardmore

290. Smoot, E.L., and T.N. Taylor. 1986. Structurally preserved fossil plants from Antarctica. II. A Permian moss from the Transantarctic Mountains. Amer. J. Bot. 73:1683–1691.

Subjects: Antarctica, Beardmore, Skaar Ridge, Permian, moss, peat

291. Smoot, E.L., T.N. Taylor, and J.W. Collinson. 1987. Lower Triassic plants from Antarctica: Diversity and paleoecology. Actas VII Simposio Argentino Paleobotanica y Palinologia, Buenos Aires, 193–197.

Subjects: Antarctica, Triassic, peat, Fremouw, deposition, Beardmore, flora.

292. Smoot, E.L., T.N. Taylor, and T. Delevoryas. 1985. Structurally preserved fossil plants from Antarctica. I. *Antarcticycas,* gen. n., a Triassic cycad stem from the Beardmore Glacier area. Amer. J. Bot. 71:410–423.

Subjects: cycads, Triassic, Beardmore, Antarctica, Fremouw, peat, *Antarcticyas*

293. Stubblefield, S.P., and T.N. Taylor. 1985. Fossil fungi in antarctic wood. Antarctic Jour. of the U.S. 20(5):7–8.

Subjects: fungi, wood, Permian, Triassic, Antarctica, Beardmore, Fremouw, Skaar Ridge, peat, *Vertebraria,* glossopterids, *Araucarioxylon,* gymnosperms

294. Stubblefield, S.P., and T.N. Taylor. 1986. Wood decay in silicified gymnosperms from Antarctica. Bot. Gaz. 147:116–125.

Subjects: fungi, Antarctica, wood, Permian, Triassic, peat, Beardmore, Fremouw, Skaar, gymnosperms

295. Stubblefield, S.P., T.N. Taylor, and R.L. Seymour. 1987. A possible endogonaceous fungus from the Triassic of Antarctica. Mycologia 79:905–906.

Subjects: fungi, Antarctica, Endogonaceae, Triassic, peat, Fremouw Peak, chlamydospores, *Sclerocystis,* Mesozoic

296. Stubblefield, S.P., T.N. Taylor, and J.M. Trappe, 1987. Fossil mycorrhizae: a case for symbiosis. Science 237:59–60.

Subjects: Antarctica, fungi, Triassic, peat, Fremouw, Beardmore, mycorrhizae

297. Stubblefield, S.P., T.N. Taylor, and J.M. Trappe, 1987. Vesicular–arbuscular mycorrhizae from the Triassic of Antarctica. Amer. J. Bot. 74(12):1904–1911.

Subjects: Triassic, peat, fungi, Antarctica, Fremouw, mycorrhizae

298. Stuchlik, L. 1981. Tertiary pollen spectra from the Ezcurra Inlet Group of Admiralty Bay, King George Island (South Shetland Islands, Antarctica). Studia Geol. Polonica 72:109–132.

Subjects: palynology, Tertiary, South Shetland, Islands, Antarctica, peninsula, King George Island, *Nothofagus,* pteridophytes, angiosperms, Eocene, Oligocene

299. Stump, E., A.J.R. White, and S.G. Borg. 1986. Reconstruction of Australia and Antarctica: evidence from granites and recent mapping. Earth and Planetary Sci. Letters 79:348–360.

Subjects: Antarctica, Australia, reconstruction, stratigraphy, Cretaceous, Tasmania, Victoria Land

300. Tarling, D.H. 1981. Models for the fragmentation of Gondwana. In: Gondwana Five (5th Int. Gondwana Symp., Wellington, N.Z. 1980) (M.M. Cresswell and P. Vella eds.), A.A. Balkema, Rotterdam, 261–266.

Subjects: Gondwana, reconstruction, fragmentation, Antarctica

301. Tasch, P. 1969. Antarctic paleobiology: New fossil data and their significance. Antarctic Jour. of the U.S. 4(5):198–199.

Subjects: Antarctica, Sentinel Mts., Polarstar Fm., algae?, insect, trace fossil, paleobiogeography, reconstructions

302. Tasch, P. 1977. Ancient Antarctic freshwater ecosystems. In: Adaptations Within Antarctic

Ecosystems (Proc. 3rd SCAR Symp. on Antarctic Biol.) (G.A. Llano, ed.), 1077–1130.

Subjects: Antarctica, fauna, Paleozoic, Mesozoic, paleoenvironment, paleoclimate, flora, Devonian, Lashly Mts., Portal Mt., Jurassic, Transantarctic Mts., Permian, Ohio Range, coal swamp, lacustrine, fish

303. Tasch, P. 1977. Intercontinental correlation by conchostracans and palynomorphs from Antarctica, western Australia, India, and Africa. Antarctic Jour. of the U.S. 12(4):121.

Subjects: palynology, Antarctica, Australia, India, Africa, correlation, Jurassic, Carapace Nunatak, Storm Peak

304. Tasch, P. 1978. Permian palynomorphs (Coalsack Bluff, Mt. Sirius, Mt. Picciotto) and other studies. Antarctic Jour. of the U.S. 13(4):19–20.

Subjects: Permian, Antarctica, Coalsack Bluff, palynology, Mt. Sirius, Mt. Picciotto, Triassic, biostratigraphy, correlation

305. Tasch, P., and E.L. Gafford. 1984. Central Transantarctic Mountains nonmarine deposits. In: Geol. of the Central Transantarctic Mountains, Amer. Geophys. Union (Ant. Res. Ser.), Washington, D.C., 36(5):75–96.

Subjects: geology, Transantarctic Mts., Antarctica, Jurassic, flora, algae, wood, plant debris

306. Tasch, P., and J.M. Lammons. 1978. Palynology of some lacustrine beds of the Antarctic Jurassic. Palinología Num. Extra. 1:455–460.

Subjects: palynology, Antarctica, Jurassic, biostratigraphy, Transantarctic Mts., correlation, Australia, Coalsack Bluff, Storm Peak, Carapace Nunatak, Beardmore area, southern Victoria Land

307. Taylor, B.J. 1971. Thallophyte borings in phosophatic fossils from the Lower Cretaceous of south-east Alexander Island, Antarctica. Palaeontology 14:294–302.

Subjects: Antarctica, fungi, Lower Cretaceous, Alexander Island, borings

308. Taylor, B.J., M.R.A. Thomson, and L.E. Willey. 1979. The geology of the Ablation Point–Keystone Cliffs area, Alexander Island. Brit. Antarctic Surv. Scientif. Rept. 82:1–65.

Subjects: geology, Jurassic, Cretaceous, Mesozoic, Antarctica, peninsula, Alexander Island, fauna, floras, cycadophytes, cycadeoids, Ablation Point, Keystone Cliffs

309. Taylor, E.L. 1987. *Glossopteris* reproductive organs: An analysis of structure and morphology. XIV Int. Botanical Congress Abstracts, Berlin, 24 July–1 August 198, 286.

Subjects: glossopterids, fructifications, Antarctica, peat, Permian, Skaar Ridge, Beardmore area, flora, abstract

310. Taylor, E.L., and E.R. McCallister. 1989. Tree-ring structure and implications for paleoclimate in the Triassic of Antarctica. Amer. J. bot. 76(6, part 2) p. 175.

Subjects: Triassic, Antarctica, peat, Fremouw Peak, rings, wood, paleoclimate, anatomy, abstract, Mesozoic

311. Taylor, E.L., and T.N. Taylor. 1988. Late Triassic flora from Mt. Falla, Queen Alexandra Range. Antarctic Jour. of the U.S. 23(5) (in press).

Subjects: Triassic, flora, Mt. Falla, Beardmore, corystosperms, *Dicroidium*, gymnosperms, Antarctica, Mesozoic

312. Taylor, E.L., T.N. Taylor, and J.W. Collinson. 1989. Depositional setting and paleobotany of Permian and Triassic permineralized peat from the central Transantarctic Mountains, Antarctica. Int. Jour. Coal Geol. 657–679.

Subject: Permian, Triassic, sedimentology, geology, peat, Beardmore, Antarctica, depositional environment, Skaar Ridge, Fremouw Peak, Buckley Fm., Paleozoic, Mesozoic

313. Taylor, E.L., T.N. Taylor, J.W. Collinson, and D.H. Elliot. 1986. Structurally preserved Permian plants from Skaar Ridge, Beardmore Glacier region. Antarctic Jour. of the U.S. 21(5):27–28.

Subjects: Antarctica, Beardmore, Permian, Skaar, peat, anatomy

314. Taylor, T.N., and E.L. Smoot. 1983. Structurally preserved plants from the Beardmore Glacier area. Antarctic Jour. of the U.S. 18(5):57–58.

Subjects: Antarctica, peat, Beardmore, Skaar, Fremouw, Triassic, Permian

315. Taylor, T.N., and E.L. Smoot. 1984. Fossil plants from the Beardmore Glacier area. Antarctic Jour. of the U.S. 19(5):12.

Subjects: Antarctica, Beardmore, peat, Permian, Triassic, Fremouw, Skaar Ridge

316. Taylor, T.N., and E.L. Smoot. 1985. A new Triassic cycad from the Beardmore Glacier area of Antarctica. Antarctic Jour. of the U.S. 20(5):5–7.

Subjects: cycads, Triassic, Beardmore, Antarctica, Fremouw, peat

317. Taylor, T.N., and E.L. Smoot. 1985. Plant fossils from the Ellsworth Mountains. Antarctic Jour. of the U.S. 20(5):48–49.

Subjects: Ellsworth Mts., Antarctica, Permian, flora, glossopterids

318. Taylor, T.N., and E.L. Smoot. 1989. Permian plants from the Ellsworth Mountains, West Antarctica. In: Geology and Paleontology of the Ellsworth Mountains (C. Craddock, J.

Splettstoesser and G. Webers, eds.), Geol. Soc. Amer. Mem. (in press).

Subjects: Antarctica, Ellsworth, Permian, flora, glossopterids, sphenopsids

319. Taylor, T.N., and S.P. Stubblefield. 1987. A fossil mycoflora from Antarctica. Actas VII Simposio Argentino de Paleobotanica y Palinologia, Buenos Aires, 187–191.

Subjects: Antarctica, Beardmore, fungi, wood, Permian, Triassic, Skaar Ridge, Fremouw

320. Taylor, T.N., and E.L. Taylor. 1987. Structurally preserved fossil plants from Antarctica. III. Permian seeds. Amer. J. Bot. 74:904–913.

Subjects: Antarctica, Beardmore, peat, Permian, Skaar Ridge, seeds, Buckley Fm., *Plectilospermum*

321. Taylor, T.N., and E.L. Taylor. 1987. An unusual gymnospermous reproductive organ of Triassic age. Antarctic Jour. of the U.S. 22(5):29–30.

Subjects: Triassic, Fremouw, peat, Antarctica, Beardmore, gymnosperms, seed ferns

322. Taylor, T.N., E.L. Taylor, and J.W. Collinson. 1986. Paleoenvironment of Lower Triassic plants from the Fremouw Formation. Antarctic Jour. of the U.S. 21(5):26–27.

Subjects: Antarctica, Beardmore, Triassic, Fremouw, peat, deposition

323. Taylor, T.N., and J.F. White. 1989. Fossil fungi (Endogonaceae) from the Triassic of Antarctica. Amer. J. Bot.: 76:389–396.

Subjects: Antarctica, Triassic, fungi, Endogonaceae, Fremouw Peak, peat, Mesozoic

323a. Tessensohn, F., Madler, K. 1987. Triassic plant fossils from North Victoria Land, Antarctica. Geol. Jahrbuch 66:187–201

Subjects: Antarctica, Late Triassic, Vulcan Hills, Campbell Glacier, north Victoria Land, Permian, Mt. Lugering, floras, *Glossopteris, Dicroidium,* palynology, wood, corystosperms, *Linguifolium,* Paleozoic, Mesozoic, glossopterids

324. Thomson, M.R.A. 1982. Mesozoic paleogeography of West Antarctica. In: Antarctic Geoscience (C. Craddock, ed.), Univ. Wisconsin Press, Madison, 331–337.

Subjects: West Antarctica, peninsula, Triassic, Cretaceous, Jurassic, geology, stratigraphy, reconstructions, faunas, biogeography

325. Thomson, M.R.A. 1983. Antarctica. In: The Phanerozoic Geology of the World II. The Mesozoic, B (M. Moullade and A.E.M. Nairn, eds.), Elsevier, Amsterdam, 391–422.

Subjects: Antarctica, Triassic, Jurassic, Cretaceous, stratigraphy, flora, palynology, correlation, Victoria Land, Transantarctic Mts., Beardmore, peninsula, review, reconstructions, Mesozoic

326. Thomson, M.R.A., and R.W. Burn. 1977. Angiosperm fossils from latitude 70°S. Nature 269:139–141.

Subjects: angiosperms, flora, Alexander Island, Antarctica, peninsula, *Nothofagus,* leaves, Tertiary

327. Thomson, M.R.A., R.J. Pankhurst, and P.D. Clarkson. 1983. The Antarctic peninsula—a late Mesozoic–Cenozoic arc (review). In: Antarctic Earth Science (4th Int. SCAR Symp. on Antarctic Earth Sci., Adelaide, 1982) (R.L. Oliver, P.R. James, and J.B. Jago, eds.), Cambridge Univ. Press, Cambridge, 289–294.

Subjects: Antarctica, peninsula, tectonics, reconstruction, Cretaceous, Cenozoic

328. Thomson, M.R.A., and T.H. Tranter. 1986. Early Jurassic fossils from central Alexander Island and their geological setting. British Antarctic Surv. Bull. 70:23–39.

Subjects: Jurassic, Antarctica, Alexander Island, geology, fauna, flora, conifers, gymnosperms, peninsula, ferns?, anatomy, permineralizations, *Brachyphyllum*

329. Tokarski, A.K. 1986. Polskie badania geologiczne na Wyspie Krola Jerzego (Antarktyka Zachodnia) w sezonie 1985–1986 [Polish geological investigations on King George Island in summer 1985–1986]. Przeglad geologiczny 11(403):617–621.

Subjects: King George Island, Antarctica, peninsula, flora, Eocene, Oligocene, leaves, angiosperms

330. Tokarski, A.K., W. Danowski, and E. Zastawniak. 1987. On the age of fossil flora from Barton Peninsula, King George Island, West Antarctica. Polish Polar Research 8(3):293–302.

Subjects: flora, Antarctica, Barton Peninsula, King George Island, South Shetlands, Paleocene, *Nothofagus,* leaves, Tertiary, angiosperms

331. Torres, T. 1984. *Nothofagoxylon antarcticus* n. sp. madera fósil del Terciário de la isla Rey Jorge, islas Shetland del Sur, Antártica. Serie Cientifica INACH 31:39–52.

Subjects: wood, *Nothofagus,* King George Island, Shetland Islands, Antarctica, peninsula, Tertiary, *Nothofagoxylon,* angiosperms, conifers, podocarps, Fagaceae

332. Torres, T. 1985. Plantas fósiles en la Antártica. Bot. Antárt. Chileno 5(2):17–31.

Subjects: floras, Antarctica, peninsula, King George Island, South Shetlands, Paleozoic, Mesozoic, Cenozoic

333. Torres, T., and Y. Lemoigne. 1988. Maderas fósiles terciárias de la Formación Caleta Arctowski, Isla Rey Jorge, Antártica. Ser. Cient. INACH 37:69–107.

Subjects: wood, Tertiary, King George Island, Antarctica, peninsula, South Shetlands, Admiralty Bay, *Araucarioxylon, Nothofagoxylon, Phyllocladoxylon,* anatomy, conifers, araucarians, gymnosperms, Paleogene, Fagaceae, biogeography

334. Torres, T., M.A. Hansen, F.L. Troian, H.C. Fensterseifer, and A. Linn. 1984. Flora fósil de alrededores de Punta Suffield, Isla Rey Jorge, Shetland del Sur. Bol. Antárt. Chileno 4(2):1–7.

Subjects: flora, King George Island, South Shetland Islands, Antarctica, peninsula

335. Torres, T., A. Roman, C. Rivera, and A. Deza. 1984. Anatomía, mineralogía y termoluminiscencia de madera fósil del Terciário de la isla Rey Jorge, Islas Shetland del Sur. Memoria III Cong. Latinoamericano de Paleontología, Mexico 2:566–574.

Subjects: King George Island, Antarctica, peninsula, South Shetland Islands, wood, Tertiary, *Cupressinoxylon,* conifers

336. Torres, T., E. Valenzuela, and I. Gonzalez. 1982. Paleoxilología de peninsula Byers, Isla Livingston, Antártica. Actas III Cong. Geol. Chileno 2:A321–A342.

Subjects: wood, Livingston Island, Antarctica, peninsula, South Shetlands, Tertiary, Byers Peninsula

337. Townrow, J.A. 1967. On a conifer from the Jurassic of East Antarctica. Roy. Soc. Tasmania, Papers and Proc. 101:137–148.

Subjects: Jurassic, conifer, Antarctica, Victoria Land, Carapace Nunatak, fructification, *Masculostrobus,* podocarps, *Nothodacrium*

338. Townrow, J.A. 1967. Fossil plants from the Allan and Carapace Nunataks and from the Upper Mill and Shackleton Glaciers, Antarctica. New Zealand Jour. Geol. Geophys. 10:456–473.

Subjects: Antarctica, Triassic, Jurassic, flora, Victoria Land, Allan Hills, Carapace Nunatak, Permian, Mt. Bumstead, Shackleton Glacier, glossopterids, corystosperms, *Dicroidium,* cycadophytes, podocarps, conifers

339. Troncoso, A. 1986. Nuevas organo-especies en la tafoflora terciária inferior de peninsula Fildes, isla Rey Jorge, Antárctica. Ser. Cient. INACH 34:23–46.

Subjects: flora, Tertiary, King George Island, Antarctica, peninsula, South Shetlands

340. Truswell, E.M. 1980. Permo-Carboniferous palynology of Gondwana: Progress and problems in the decade to 1980 (abstract). In: Gondwana Five (Proc. 5th Int. Gondwana Symp., Wellington, N.Z. 1980) (M.M. Creswell and P. Vella, eds.), A.A. Balkema, Rotterdam, 15–17.

Subjects: Permian, Carboniferous, palynology, Gondwana, correlation, biostratigraphy, Antarctica, India, Africa, South America

341. Truswell, E.M. 1980. Permo-Carboniferous palynology of Gondwanaland: progress and problems in the decade to 1980. BMR Jour. of Australian Geol. & Geophys. 5:95–111.

Subjects: palynology, Permian, Carboniferous, Gondwana, stratigraphy, Antarctica, Australia, India, Africa, South America, floras, glossopterids, biogeography

342. Truswell, E.M. 1982. Palynology of seafloor samples collected by the 1911–14 Australasian Antarctic Expedition: Implications for the geology of coastal East Antarctica. J. Geol. Soc. Australia 29:343–356.

Subjects: palynology, Antarctica, East Antarctica, Tertiary, Permian, Jurassic, Cretaceous, reconstruction, Australia

343. Truswell, E.M. 1982. Antarctica: the vegetation of the past and its climatic implications. Australian Meteorological Mag. 30:169–173.

Subjects: Antarctica, paleoclimate, floras, Permian, Triassic, Tertiary, review

344. Truswell, E.M. 1983. Geological implications of recycled palynomorphs in continental shelf sediments around Antarctica. In: Antarctic Earth Sciences (4th Int. SCAR Symp. on Antarctic Earth Sci., Adelaide, 1982) (R.L. Oliver, P.R. James, and J.B. Jago, eds.), Cambridge Univ. Press, Cambridge, 394–399.

Subjects: palynology, Antarctica, recycled, Permian, Jurassic, Cretaceous, Tertiary, Ross Sea, Wilkes Shelf, dinoflagellates

345. Truswell, E.M. 1983. Recycled Cretaceous and Tertiary pollen and spores in Antarctic marine sediments: A catalogue. Palaeontographica 186B:121–174, Pl. 1–6.

Subjects: palynology, Cretaceous, Tertiary, Antarctica, Ross Sea, Weddell Sea, East Antarctica, recycled

346. Truswell, E.M. 1986. Palynology. In: Antarctic Cenozoic history from the MSSTS-1 drillhole, McMurdo Sound (P.J. Barrett, ed.), Sci. Inf. Publ. Cent., Wellington, DSIR Bull. 237:131–134.

Subjects: Antarctica, McMurdo Sound, palynology, Tertiary, Permian, recycled, dinoflagellates, Oligocene, Miocene, *Nothofagus,* Jurassic, Cretaceous, Eocene

347. Truswell, E.M. 1987. The palynology of core samples from the *S.P. Lee* Wilkes Land cruise. In: The Antarctic Continental Mar-

gin: Geology and Geophysics of Offshore Wilkes Land (S.L. Eittreim and M.A. Hampton, eds.), Circum-Pacific Council for Energy and Mineral Resources, Earth Science Series, Houston, Texas, 5A:215–218.

Subjects: palynology, Antarctica, Cretaceous, Tertiary, recycling, Wilkes Land

348. Truswell, E.M., and J.B. Anderson. 1984. Recycled palynomorphs and the age of sedimentary sequences in the eastern Weddell Sea. Antarctic Jour. of the U.S. 19(5):90–92.

Subjects: palynology, Weddell Sea, Antarctica, Jurassic, Cretaceous, Tertiary, recycled

349. Truswell, E.M., and D.J. Drewry. 1984. Distribution and provenance of recycled palynomorphs in surficial sediments of the Ross Sea, Antarctica. Marine Geol. 59:187–214.

Subjects: Ross Sea, Antarctica, palynology, Cretaceous, Tertiary, recycled

350. Van der Voo, R. 1988. Paleozoic paleogeography of North America, Gondwana, and intervening displaced terranes: Comparisons of paleomagnetism with paleoclimatology and biogeographical patterns. Geol. Soc. Amer. Bull. 100:311–324.

Subjects: Gondwana, reconstructions, paleomagnetism, paleoclimate, Silurian, Devonian, paleobiogeography

351. Vavra, C.L., K.O. Stanley, and J.W. Collinson. 1980. Provenance and alteration of Triassic Fremouw Formation, Central Transantarctic Mountains. In: Gondwana Five, Proc. 5th Int. Gondwana Symp., Wellington, N.Z. (M.M. Cresswell and P. Vella, ed.), A.A. Balkema, Rotterdam, 149–153.

Subjects: Triassic, Antarctica, Beardmore, depositional environment

352. Veevers, J.J. 1986. Breakup of Australia and Antarctica estimated as mid-Cretaceous (95 + 5 Ma) from magnetic and seismic data at the continental margin. Earth and Planet. Sci. Letters 77:91–99.

Subjects: Australia, Antarctica, plate tectonics, reconstructions, Cretaceous

353. Volkheimer, W. 1969. Palaeoclimatic evolution in Argentina and relations with other regions of Gondwana. In: Gondwana Stratigraphy (1st IUGS Gondwana Symp., Buenos Aires, 1967) (A.J. Amos, ed.), UNESCO, Paris, 551–587.

Subjects: paleoclimate, Gondwana, Argentina, Permian, Triassic, Jurassic, Cretaceous, Tertiary

354. Wace, N.M. 1965. Vascular plants. In: Biogeography and Ecology in Antarctica (J. Van Mieghem and P. Van Oye, eds.), Dr. W. Junk, The Hague, Monographiae Biologicae, Vol. XV:201–266.

Subjects: floras, Antarctica, Paleozoic, Mesozoic, Tertiary, conifers, palynology, review, wood, peninsula, Transantarctic Mts., peninsula, angiosperms, gymnosperms

355. Walton, J. 1923. On *Rhexoxylon*, Bancroft—a Triassic genus of plants exhibiting a liane-type of vascular organisation. Phil. Trans. Roy. Soc. London 212B:79–109.

Subjects: corystosperms, Triassic, anatomy, Priestley Glacier, Victoria Land, Antarctica, *Rhexoxylon*, Beacon sandstone

356. Webb, P.N., and D.M. Harwood. 1987. The terrestrial flora of the Sirius Formation: its significance in interpreting late Cenozoic glacial history. Antarctic J. of the U.S. 22(4):7–11.

Subjects: flora, Pliocene, Pleistocene, glaciation, wood, Antarctica, Sirius Fm., Dominion Range, palynology, *Nothofagus*, palynology, paleoclimate

357. Webb, P.N., B.C. McKelvey, D.M. Harwood, M.C.G. Mabin, and J.H. Mercer. 1987. Sirius Formation of the Beardmore Glacier region. Antarctic Jour. of the U.S. 22(1/2):8–13.

Subjects: stratigraphy, geology, Antarctica, Sirius Fm., Beardmore, stems, roots, recycled, palynology, diatoms, foraminifera

358. Webb, P.N., T.E. Ronan, J.H. Lipps, and T.E. DeLaca. 1979. Miocene glaciomarine sediments from beneath the southern Ross Ice Shelf, Antarctica. Science 203:435–437.

Subjects: Antarctica, Ross Sea, Miocene, sedimentology, diatoms, foraminifera, palynology

359. Webers, G.F. 1970. Paleontological investigations in the Ellsworth Mountains, West Antarctica. Antarctic Jour. of the U.S. 5(5):162–163.

Subjects: Antarctica, Ellsworth Mts., Permian, flora, fauna, Polarstar Fm., glossopterids

360. Webers, G.F., and K.B. Sporli. 1983. Palaeontological and stratigraphic investigations in the Ellsworth Mountains, West Antarctica. In: Antarctic Earth Sciences (4th Int. SCAR Symp. on Antarctic Earth Sci., Adelaide, 1982) (R.L. Oliver, P.R. James, and J.B. Jago, eds.), Cambridge Univ. Press, Cambridge, 261–264.

Subjects: fauna, flora, Antarctica, Ellsworth Mts., West, Permian, Cambrian, glossopterids, Polarstar Fm.

361. White, J.F., and T.N. Taylor 1988. Triassic fungus from Antarctica with possible ascomycetous affinities. Amer. J. Bot. 75:1495–1500.

Subjects: fungi, Triassic, ascomycetes, Antarctica, Fremouw Peak, peat, *Endochaetophora,* Mesozoic

362. White, J.F., and T.N. Taylor. 1989. A trichomycete-like fossil from the Triassic of Antarctica. Mycologia (in press).

Subjects: fungi, Triassic, Antarctica, Fremouw Peak, peat, trichomycetes, Mesozoic

363. White, M.E. 1973. Permian flora from the Beaver Lake area, Prince Charles Mountains, Antarctica. 2. Plant fossils. Australian Bur. Mineral Resources, Geol. and Geophys. Bull. 126, Palaeont. Papers 1969:13–18.

Subjects: Antarctica, flora, Prince Charles Mts., East Antarctica, glossopterids, Permian, wood, *Vertebraria*

364. Wilson, G.J. 1967. Some new species of Lower Tertiary dinoflagellates from McMurdo Sound, Antarctica. New Zealand Jour. Bot. 5:57–87.

Subjects: palynology, Tertiary, dinoflagellates, McMurdo, Antarctica

365. Wilson, G.J. 1968. On the occurrence of fossil microspores, pollen grains and microplankton in bottom sediments of the Ross Sea, Antarctica. New Zealand Jour. Marine and Freshwater Res. 2:381–389.

Subjects: palynology, Ross Sea, Antarctica

366. Wise, S.W., P.F. Ciesielski, D.T. MacKenzie, F.H. Wind, K.E. Busen, A.M. Gombos, B.U. Haq, G.P. Lohmann, R.C. Tjalsma, W.K. Harris, R.W. Hedlund, C.N. Beju, D.L. Jones, G. Plafker, and W.V. Sliter. 1982. Paleontologic and paleoenvironmental synthesis of the southwest Atlantic Ocean basin based on Jurassic to Holocene faunas and floras from the Falkland Plateau. In: Antarctic Geoscience (C. Craddock, ed.), Univ. Wisconsin Press, Madison, 155–163.

Subjects: DSDP, Holocene, Jurassic, Cretaceous, Mesozoic, Tertiary, palynology, diatoms, glaciation, Falkland Plateau

367. Wolfe, J.A. 1985. Probabilities of high-latitude glaciers during the Tertiary. 98th Ann. Mtg. Geol. Soc. Amer. 17(7):753.

Subjects: Tertiary, Arctic, Antarctica, paleoclimate, floras, leaves

368. Wrenn, J.H. 1981. Preliminary palynology of the RISP Site J-9, Ross Sea. Antarctic Jour. of the U.S. 16:72–74.

Subjects: palynology, Antarctica, Ross Sea, Tertiary

368a. Young, D.J., Ryburn, R.J. 1968. The geology of Buckley and Darwin Nunataks, Beardmore Glacier, Ross Dependency, Antarctica. New Zealand Jour. Geol. Geophys. 11:922–939

Subjects: Antarctica, geology, Beardmore, Buckley Coal Measures, Darwin Nunataks, glossopterids, Permian, biostratigraphy, *Noeggerathiopsis,* spehnophytes, wood

369. Zastawniak, E. 1981. Tertiary leaf flora from the Point Hennequin Group of King George Island (South Shetland Islands, Antarctica). Preliminary report. Studia Geol. Polonica 72:97–108.

Subjects: flora, South Shetland Islands, King George Island, Antarctica, Tertiary, *Nothofagus,* conifers, podocarps, angiosperms

370. Zastawniak, E., R. Wrona, A. Gazdzicki, and K. Birkenmajer. 1985. Plant remains from the top part of the Point Hennequin Group (Upper Oligocene), King George Island (South Shetland Islands, Antarctica). Studia Geol. Polonica 81:143–164.

Subjects: flora, Antarctica, King George Island, South Shetland Island, peninsula, Oligocene, Tertiary, *Nothofagus,* podocarps, conifers, angiosperms

371. Zawiskie, J.M., and J.M. Collinson. 1983. Trace fossils of the Permian–Triassic Takrouna Formation, northern Victoria Land, Antarctica. In: Antarctic Earth Science (4th Int. SCAR Symp. on Antarctic Earth Sci., Adelaide, 1982) (R.L. Oliver, P.R. James, and J.B. Jago, eds.), Cambridge Univ. Press, Cambridge, 215–220.

Subjects: Antarctica, Permian, Triassic, flora, roots, Victoria Land, trace fossils

372. Ziegler, A.M., R.K. Bambach, J.T. Parrish, S.F. Barrett, E.H. Gierlowski, W.C. Parker, A. Raymond, and J.J. Sepkoski, Jr. 1981. Paleozoic biogeography and climatology. In: Paleobotany, Paleoecology and Evolution (K.J. Niklas, ed.), Praeger 2:231–266.

Subjects: biogeography, paleoclimate, reconstructions, Gondwana, Permian, Carboniferous, Devonian

373. Zinsmeister, W.J. 1987. Cretaceous paleogeography of Antarctica. Palaeogeogr., Palaeoclimatol., Palaeoecol. 59:197–206.

Subjects: Cretaceous, paleogeography, Antarctica, peninsula, Gondwana, breakup, reconstructions, faunas, Tertiary

374. Zinsmeister, W.J. 1988. Early geological exploration of Seymour Island, Antarctica. In: Geology and Paleontology of Seymour Island, Antarctic Peninsula (R.M. Feldmann and M.O. Woodburne, eds.), Geol. Soc. Amer. Mem. 169:1–16.

Subjects: Seymour Island, Antarctica, peninsula, flora, leaves, Tertiary, Jurassic, Hope Bay, angiosperms, gymnosperms, cycadophytes, ferns, Snow Hill Island, wood

Index